# Research on Metallofullerenes

# Research on Metallofullerenes


Guest Editors

**Filip Uhlík**

**Takeshi Akasaka**

**Zdeněk Slanina**

**Xing Lu**



MDPI

Basel • Beijing • Wuhan • Barcelona • Belgrade • Novi Sad • Cluj • Manchester

*Guest Editors*
Filip Uhlík
Department of Physical Chemistry
Charles University
Prague
Czech Republic

Takeshi Akasaka
TARA Center
University of Tsukuba
Ibaraki
Japan

Zdeněk Slanina
Department of Chemistry and Biochemistry
University of Arizona
Tucson
USA

Xing Lu
School of Materials Science and Engineering
Huazhong University of Science and Technology
Wuhan
China

*Editorial Office*
MDPI AG
Grosspeteranlage 5
4052 Basel, Switzerland

This is a reprint of the Special Issue, published open access by the journal *Inorganics* (ISSN 2304-6740), freely accessible at: https://www.mdpi.com/journal/inorganics/special_issues/4X5Y71H3FX.

For citation purposes, cite each article independently as indicated on the article page online and as indicated below:

| Lastname, A.A.; Lastname, B.B. Article Title. *Journal Name* **Year**, *Volume Number*, Page Range. |
|---|

**ISBN 978-3-7258-6778-3 (Hbk)**
**ISBN 978-3-7258-6779-0 (PDF)**
**https://doi.org/10.3390/books978-3-7258-6779-0**

# Contents

# About the Editors

**Filip Uhlík**

Filip Uhlík was born in Brno, Czech Republic. He is a Full Professor at Charles University in Prague, Czech Republic. His research interests include the study of fullerenes and carbon materials, atmospheric complexes, Ge/Si quantum dots, and polymer nanostructures using theoretical methods of quantum chemistry and statistical thermodynamics. He has co-authored over 200 publications.

**Takeshi Akasaka**

Takeshi Akasaka was born in Kyoto, Japan. He received his Ph.D. Degree from the University of Tsukuba, Japan, in 1979. After working as a Postdoctoral Fellow at Brookhaven National Laboratory, he returned to the University of Tsukuba in 1981. In 1996, he began work at Niigata University as a Professor. From 2001 to 2013, he was a Professor of the TARA Center and Department of Chemistry, University of Tsukuba. Following his retirement from the University of Tsukuba in 2013, he was appointed as a senior researcher at the Foundation for Advancement of International Science (FAIS) and a Professor Emeritus of the University of Tsukuba. He received the Chemical Society of Japan Award for Creative Work (2000); the Commendation of Science and Technology (Research Category) from the Ministry of Education, Culture, Sports, Science and Technology of Japan in 2011; and the Chemical Society of Japan Award in 2013. His research interests include the chemistry of nanocarbons.

**Zdeněk Slanina**

Zdeněk Slanina was born in what was formerly Czechoslovakia. He received his M.A. degree from Charles University in Prague and Ph.D. from the (former) Czechoslovak Academy of Sciences, where he was also employed between 1971 and 1991, up to the principal senior-scientist level. Since the arrival of the era of fullerene science, he has been active in computational research, also teaching at various fullerene research centers like Hokkaido University, Sapporo; the Max-Planck Institute for Chemistry, Mainz; the University of Arizona, Tucson; the Institute for Molecular Science, Okazaki; the University of Tsukuba; and Huazhong University of Science and Technology, Wuhan.

**Xing Lu**

Xing Lu is a Full Professor at Hainan University and an Adjunct Professor at Huazhong University of Science and Technology. He received his Ph.D. from Peking University in 2004 and then worked as a COE Postdoctoral researcher at Nagoya University. From 2006 to 2013, he was a Senior Scientist at the University of Tsukuba. He was the recipient of the Ambassador Award from the Chinese Embassy in Japan (2009), the Osawa Award from the Fullerenes and Nanotubes Research Society of Japan (2011), and the National Science Fund for Distinguished Young Scholars (2019). His research interests lie in the rational design and facile generation of novel hybrid carbon materials with applications in energy storage and conversion. He has published over 220 peer-reviewed papers in such journals as *Nat. Syn., Nat. Commun., J. Am. Chem. Soc.,* and *Angew. Chem. Int. Ed.*

# Preface

The 40th anniversary is approaching of the first experimental observation of fullerenes in a laser-vaporized graphite cluster beam mass spectrum, which was made by H. W. Kroto, J. R. Heath, S. C. O'Brien, R. F. Curl, and R. E. Smalley in 1985. This will be the anniversary not only for fullerenes and buckminsterfullerene in particular, but also for the first endohedral metallofullerene observed, La@$C_{60}$, just a few days later. There has been tremendous progress in the field since then, documented by tens of thousands of publications. The production of fullerenes in macroscopic quantities, i.e., hundreds of observed and isolated new species, has influenced their characterization both experimentally and theoretically, determining their various applications as MRI and X-ray contrast agents, radiotracers, photovoltaic cells, nanoelectronics, superconductors, and more.

The field of metallofullerene research is leading, with hundreds of new publications each year; therefore, this Reprint on Metallofullerenes in Inorganics, discussing the chemical and physical properties of metallofulleres studied both experimentally and theoretically, is very timely.

**Filip Uhlík, Takeshi Akasaka, Zdeněk Slanina, and Xing Lu**
*Guest Editors*

inorganics

MDPI

*Article*

# $La_2C_2@D_5(450)$-$C_{100}$: Calculated High Energy Gain in Encapsulation

**Zdeněk Slanina [1,2,*], Filip Uhlík [3], Takeshi Akasaka [2], Xing Lu [2] and Ludwik Adamowicz [1]**

1 Department of Chemistry and Biochemistry, University of Arizona, Tucson, AZ 85721-0041, USA
2 State Key Laboratory of Materials Processing and Die & Mould Technology, School of Material Science and Engineering, Huazhong University of Science and Technology, Wuhan 430074, China
3 Department of Physical and Macromolecular Chemistry, Faculty of Science, Charles University, Albertov 6, 128 43 Prague, Czech Republic
* Correspondence: zdeneks@email.arizona.edu

**Abstract:** The structure and energetics of the clusterfullerene $La_2C_2@D_5(450)$-$C_{100}$ are calculated at the B3LYP/6-31G*~SDD level (including counterpoise correction for the basis set superposition error), and the observed features are confirmed. Its stability is explained by substantial energy gain connected with the encapsulation, viz. 140 kcal/mol per atom of the encapsulate, actually higher than previously found for comparable systems.

**Keywords:** metallofullerenes; clusterfullerenes; $La_2C_2@D_5(450)$-$C_{100}$; theory-experiment comparisons; quantum-chemical calculations

## 1. Introduction

In fullerene and metallofullerene research, calculations have from the beginning helped experimental search and understanding of new nanocarbons [1–4], more recently even with species containing around one hundred carbons [5–7]. In the group of such large metallofullerenes [8–18], another giant system has recently been characterized [14] by X-ray crystal diffraction, namely $La_2C_2@D_5(450)$-$C_{100}$. In this study, the giant clusterfullerene is calculated using the density functional theory (DFT) in order to further enhance knowledge of its structure and encapsulation energetics.

## 2. Calculations

The $D_5(450)$-$C_{100}$ fullerene is one of the 450 isolated-pentagon-rule (IPR) cages of $C_{100}$ [19] (if enantiomers are considered as equivalent, 862 IPR isomers if enantiomers are regarded as distinct). Interestingly, the total number of all $C_{100}$ isomeric cages (i.e., IPR and non-IPR cages) is 285,913 if enantiomers are considered equivalent [4]. The $D_5(450)$-$C_{100}$ fullerene exhibits $D_5$ topological symmetry which is however reduced owing to the Jahn-Teller distortion. At our computational level the topological symmetry is in geometry optimization lowered to the $C_2$ point group of symmetry (Figure 1, one can speak on a near $D_5$-symmetry). Interestingly enough, there are other $C_{100}$ IPR cages with a higher topological symmetry like $D_{5d}$ or $T$.

The $La_2C_2@D_5(450)$-$C_{100}$ geometrical structure was optimized with the Becke's three parameter DFT functional with the non-local Lee-Yang-Parr correlation functional [20–22], the approach is conventionally labeled as B3LYP. The DFT calculations were at first performed in a combined atomic basis set: the standard 3-21G basis for C atoms while for La atoms the SDD basis set with the SDD effective core potential [23] (B3LYP/3-21G~SDD). The geometry optimization started from the X-ray structure [14] and was further refined with the B3LYP/6-31G*~SDD treatment using an extended 6-31G*~SDD basis set, followed by the GF harmonic vibrational analysis. The vibrational analysis allows for a check of the nature of the stationary points found, i.e., if local energy minima were really found (and not activated complexes or even higher types of hypersurface stationary points). Moreover, the analysis can also be used for simulations of the IR and Raman vibrational spectra. For numerical integrations with the DFT functional the so called ultrafine grid was used. Moreover, the tight SCF (self consistent field) convergency criterion was applied. Finally, the wavefunction stability was tested throughout in order to eliminate unstable SCF solutions (without a real physical significance).

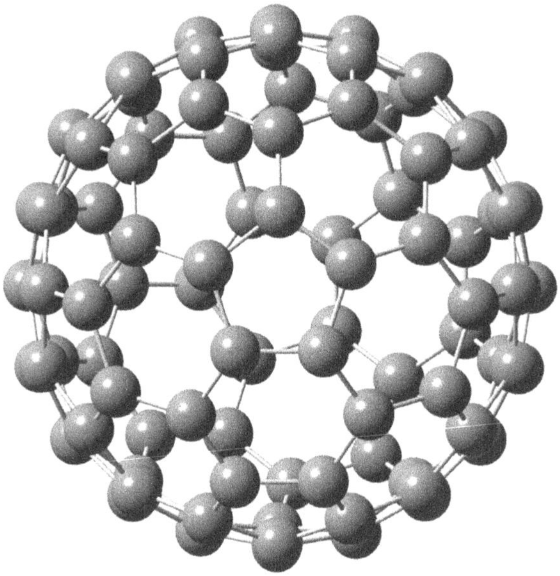

**Figure 1.** The B3LYP/6-31G* optimized structure of $D_5$(450)-$C_{100}$.

The so-called basis set superposition error (BSSE) was considered accordingly, namely by application of the approximative Boys-Bernardi counterpoise method (also known as CP2, or CPn, depending on the number of the reaction components in the treated association process) [24], since recently considered [25] also with metallofullerenes. The Boys-Bernardi approach to the BSSE correction term is an approximative treatment that ensures that every reaction component of a studied chemical process is formally described by the same number of the atomic basis set functions (so that the calculated energies are directly comparable). The required basis-set consistence is realized by rather formal, artificial atoms (thus also called ghost) possessing no electrons. The BSSE problem essentially originates in the fact that the considered basis sets are always finite. In other words, the BSSE correction would disappear in a situation of a basis set with infinite number of atomic basis functions (of course, such an infinite scheme cannot practically be handled). The BSSE correction term represents an important addition also to the energetics of any encapsulation process. Without the BSSE correction the energy gain created by encapsulation would be exaggerated (so that endohedrals would artificially be over-stabilized). Thus, the BSSE consideration is needed for reliable predictions of energetics, indeed.

All the reported calculations were performed with the program package Gaussian 09 [26]. The calculations were carried out with computers operating in parallel regime, mostly with 8–24 processors (with computational frequency up 3 GHz each and with the available operational memory up to 60 GB).

## 3. Results and Discussion

The B3LYP/6-31G*~SDD optimized structure of $La_2C_2$@$D_5$(450)-$C_{100}$ shown in Figure 2 basically agrees with the X-ray results [14]. The calculated C-C bond length in the $La_2C_2$

encapsulate is 1.26 Å while the C-La bond 2.44 Å, the La-C shortest contact with the cage 2.64 Å. The B3LYP/3-21G SDD computed Wiberg bond index for the C-C bond is 2.58 (i.e., between double and triple bond), for C-La bond 0.69 (B3LYP/6-31G*~SDD values C-C 2.59, C-La 0.63). It should be realized that the calculated values are obtained for the gas-phase free (isolated) endohedral. Moreover, the calculated terms correspond to the equilibrium geometries with motionless nuclei. The X-ray values are however obtained for the crystal-state endohedral.

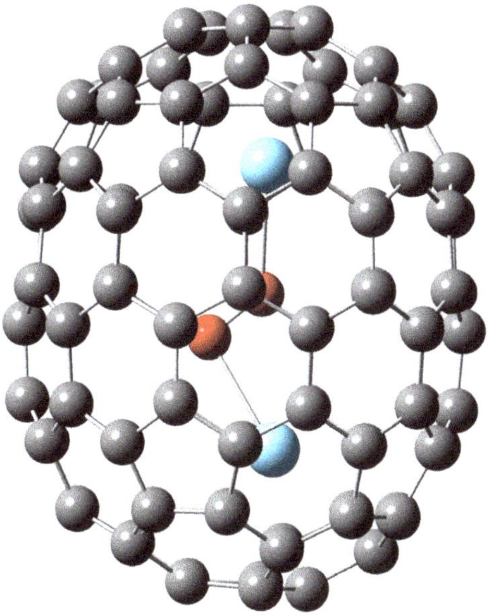

**Figure 2.** The B3LYP/6-31G*~SDD optimized structure of $La_2C_2@D_5(450)\text{-}C_{100}$ (carbons of $La_2C_2$ are in red).

The Mulliken atomic charges on atoms are (owing to their historical introduction) to be constructed at a simpler level, i.e., B3LYP/3-21G~SDD. The Mulliken charges on atoms calculated with the 3-21G basis set are known [27] to produce for metallofullerenes a good agreement with the available observed charges [28]. Moreover, there are also general methodological reasons [29,30] why some larger basis sets should not be applied in the calculations of the Mulliken atomic charges on atoms—as they can sometimes produce rather unphysical values. The Mulliken charges on the La atoms are 2.41 and 2.47 (of the electron elementary charge) while on the carbons of the $La_2C_2$ encapsulate the charges are −0.32 and −0.34. Hence, the total charge on the encapsulate is 4.22. In overall, the charge transfer from the encapsulate to the cage amounts to 4.22 electrons. Thus, energetics of tetra-anions of empty cages could serve for stability estimations with encapsulations of $La_2C_2$ by other $C_{100}$ cages.

Although the ground state of $La_2C_2@D_5(450)\text{-}C_{100}$ is singlet electronic state, the first electronic triplet state is rather low lying. At the B3LYP/6-31G*~SDD level the triplet state is located only about 1.88 kcal/mol above the ground-state singlet.

The energy change connected with the (formal thermodynamic) gas-phase encapsulation process:

$$2La(g) + 2C(g) + D_5(450) - C_{100}(g) = La_2C_2@D_5(450) - C_{100}(g) \quad (1)$$

is a useful characteristics known as encapsulation energy. Calculated reaction thermodynamic changes are mostly treated as the reaction potential-energy change $\Delta E_{pot,enc}$.

However, a more adequate description refers to the enthalpy change at absolute zero temperature $\Delta H^o_{0,enc}$, i.e., the potential energy change corrected for the vibrational zero-point energies upon encapsulation $\Delta ZPE_{enc}$:

$$\Delta H^o_{0,enc} = \Delta E_{pot,enc} + \Delta ZPE_{enc}. \tag{2}$$

The encapsulation potential-energy change corrected $\Delta E_{pot,enc}$ is to be refined by inclusion of the basis set superposition error [24,26] (BSSE/CP5) and so-called steric correction [25] that reflects the cage distortion upon encapsulation [31]. Table 1 presents the energy terms for reaction (1) considered as reduced or relative values, i.e., per one atom of the $La_2C_2$ encapsulate (thus in the units of kcal/mol/atom). Such reduced values are convenient for comparisons with encapsulations of species with different numbers of atoms. For example, the reduced $\Delta H^o_{0,enc}$ term of $-138.4$ kcal/mol/atom from Table 1 represents the encapsulation-energy gain larger by 21.1 kcal/mol/atom than found [18] for $Y_2C_2$@$C_1$(1660)-$C_{108}$ at the same computational level. Similarly, the reduced $\Delta E_{pot,enc}$ term of $-140.2$ kcal/mol/atom in Table 1 gives by 3.7 kcal/mol/atom better energy gain than calculated [32] at the same level for $Ti_3C_3$@$I_h$-$C_{80}$. In overall, $La_2C_2$@$D_5$(450)-$C_{100}$ is a new interesting member of the family of metallofullerenes with relatively large energy gains upon encapsulation.

**Table 1.** The B3LYP/6-31G*~SDD computed [a] encapsulation energetics for $La_2C_2$@$D_5$(450)-$C_{100}$.

| Energy Term | (kcal/mol/atom) |
|---|---|
| $\Delta E_{pot,enc}$ without BSSE | −148.7 |
| $\Delta E_{pot,enc}$ with BSSE | −143.2 |
| $\Delta E_{pot,enc}$ with BSSE & steric | −140.2 |
| $\Delta H^o_{0,enc}$ | −138.4 |

[a] See process (1), the BSSE correction evaluated with the CP5 approach.

The encapsulation enthalpy $\Delta H^o_{0,enc}$ (enhanced by the related entropic contributions) allows for statistical thermodynamic evaluations of the equilibrium concentrations [33,34], in our case by means of the encapsulation equilibrium constant for process (1):

$$K_{p,enc}(1) = \frac{p_{La_2C_2@D_5(450)-C_{100}}}{p^2_{La} p^2_C p_{D_5(450)-C_{100}}} \tag{3}$$

defined by the partial pressures $p$ of the individual reaction components in reaction (1). The encapsulation equilibrium constant from Equation (3) can indeed be expressed [34, 35] by a compact formula using the molecular (or atomic) partition functions $q^o_i$ (the partition functions have to correspond to the selected standard state, e.g., ideal gas phase at 1 atm = 101,325 Pa pressure) and the standard encapsulation enthalpy change at the absolute zero temperature $\Delta H^o_{0,enc}$:

$$K_{p,enc}(1) = \frac{\frac{q^o_{La_2C_2@D_5(450)-C_{100}}}{N_A}}{[\frac{q^o_{La}}{N_A}]^2 [\frac{q^o_C}{N_A}]^2 \frac{q^o_{D_5(450)-C_{100}}}{N_A}} exp[-\frac{\Delta H^o_{0,enc}}{RT}], \tag{4}$$

where symbol $N_A$ stands for the Avogadro number. The molecular partition functions $q^o_i$ primarily consist of the rotational-vibrational partition functions based on the molecular parameters from the GF harmonic vibrational analysis.

In principle, the encapsulation energy can also be defined [34] by a more straightforward encapsulation stoichiometry, namely:

$$La_2C_2(g) + D_5(450) - C_{100}(g) = La_2C_2@D_5(450) - C_{100}(g) \tag{5}$$

when the basis set superposition error [24,26] is evaluated by the simpler BSSE/CP2 approach which at the B3LYP/6-31G*~SDD level yields the reduced energy gain of −58.7 kcal/mol/atom (−62.5 kcal/mol/atom without the BSSE correction). Obviously, the energy gain must be now smaller than with the BSSE/CP5 approach as in the latter case the formation of the bonds of the encapsulate is included.

However, if one is interested in the thermodynamic-equilibrium populations of non-isomeric endohedrals [33], approach according to Equation (1) should be preferred. Incidentally, in addition to our starting system $La_2C_2@D_5(450)$-$C_{100}$ another pertinent species is known [10], viz. $La_2@D_5(450)$-$C_{100}$. The relative populations of the two non-isomeric endohedrals can be described using the equilibrium constant $K_{p,enc}(1)$ defined by Equation (3) and its equivalent $K_{p,enc}(6)$ for parent process in Equation (6):

$$2La(g) + D_5(450) - C_{100}(g) = La_2@D_5(450) - C_{100}(g) \tag{6}$$

The population ratio of the two endohedrals is given by formula [33]:

$$\frac{p_{La_2C_2@D_5(450)-C_{100}}}{p_{La_2@D_5(450)-C_{100}}} = p_C^2 \frac{K_{p,enc}(1)}{K_{p,enc}(6)}. \tag{7}$$

Obviously, the relative equilibrium populations of such non-isomeric nanocarbons represent distinctly more complex problem compared to the situation of just isomeric species [34–37]. The equilibrium constant $K_{p,enc}(6)$ can again be evaluated by the partition functions and calculated molecular parameters as in Equation (4). The partial pressure of carbon $p_C$ can be approximated by the carbon saturated pressure [34]. However, the actual carbon pressure $p_C$ depends on particular experimental setup and thus varies from case to case. This aspect is one of the reasons why the population ratio of the two endohedrals can be different in different experiments, too. The interesting relative stability problem for non-isomeric nanocarbons should further be calculated in future.

The encapsulation of $La_2C_2$ into the $D_5(450)$-$C_{100}$ cage creates six new vibrational modes, actually with rather low frequencies—at the B3LYP/6-31G*~SDD level between 24 and 73 $cm^{-1}$. The low frequencies basically belong to motions of the whole $La_2C_2$ encapsulate inside the cage, they could roughly be viewed as relatively free rotations. Such low frequency motions would particularly be important for the endohedral thermodynamics [35,36]. The new vibrational frequencies created by the encapsulation also contribute to the reaction change $\Delta ZPE_{enc}$ of the vibrational zero-point energy for process (1). On the other hand, the highest computed vibrational frequency has a value of 1926 $cm^{-1}$ and belongs to the bond stretching of the C-C bond in $La_2C_2$ (though the vibrational mode exhibits rather negligible IR intensity as well as Raman activity). Let us note that the gas-phase $C_2$ molecule has [38] in the singlet ground state observed vibrational frequency of 1855 $cm^{-1}$ (the $C_2$ lowest triplet state lies higher only by 2.05 kcal/mol). For another comparison, the highest vibrational frequency of the empty $D_5(450)$-$C_{100}$ cage at the B3LYP/6-31G* level has value of 1615 $cm^{-1}$. Figure 3 presents the B3LYP/6-31G*~SDD calculated harmonic IR and Raman spectra of $La_2C_2@D_5(450)$-$C_{100}$. The spectra should be useful in identification of the species.

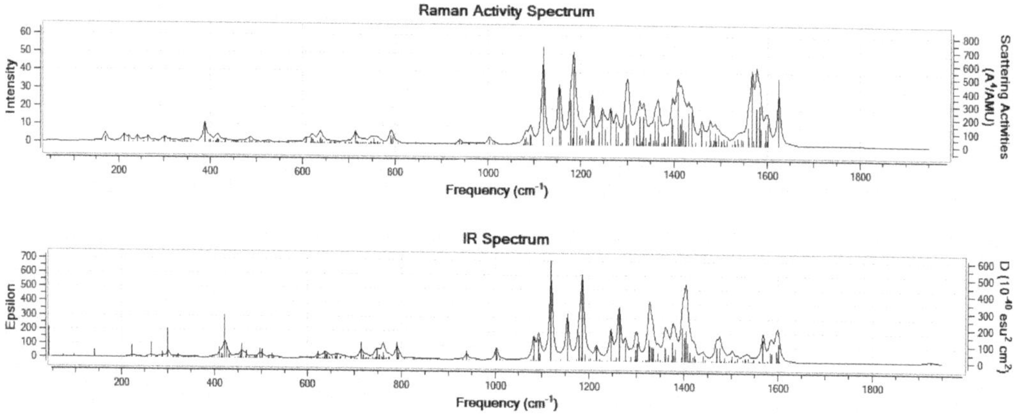


**Figure 3.** The B3LYP/6-31G*∼SDD calculated IR (**bottom**) and Raman (**top**) spectra of $La_2C_2@D_5(450)\text{-}C_{100}$.

## 4. Conclusions

The structure and energetics of the metallofullerene (or clusterfullerene) $La_2C_2@D_5(450)\text{-}C_{100}$ are evaluated using the DFT approach at the B3LYP/6-31G*∼SDD level with inclusion of the correction for the basis set superposition error calculated by the counterpoise method. The calculated encapsulation potential-energy change $\Delta E_{pot,enc}$ amounts to a sizeable value of −140.2 kcal/mol/atom (i.e., per atom of the encapsulate), or at the $\Delta H^o_{0,enc}$ level to −138.4 kcal/mol/atom. The found energy gains connected with encapsulation are larger than previously reported for other clusterfullerene systems like $Y_2C_2@C_1(1660)\text{-}C_{108}$ or $Ti_3C_3@I_h\text{-}C_{80}$. Possible descriptions of encapsulation energetics are discussed and a thermodynamically consistent evaluation of the relative equilibrium populations of non-isomeric metallofullerenes is suggested with an aim to compare the $La_2C_2@D_5(450)\text{-}C_{100}$ equilibrium population with that of a related observed species $La_2@D_5(450)\text{-}C_{100}$. The approach can be of a use also for studies of other interesting complex nanocarbon systems [39–45].

**Author Contributions:** Conceptualization, Z.S., F.U., T.A., X.L. and L.A.; methodology, Z.S., F.U. and L.A.; hardware and software, Z.S., F.U. and L.A.; models validation, Z.S., F.U., T.A., X.L. and L.A.; analysis and interpretation, Z.S., F.U., T.A., X.L. and L.A.; writing—original draft preparation, Z.S. and F.U. All authors have read and agreed to the published version of the manuscript.

**Funding:** The reported research has been supported by the National Natural Science Foundation of China (21925104 and 92261204), the Hubei Provincial Natural Science Foundation of China (No. 2021CFA020), and the International Cooperation Key Project of Science and Technology Department of Shaanxi; and by the Charles University Centre of Advanced Materials/CUCAM (CZ.02.1.01/0.0/0.0/15 003/0000417), the MetaCentrum ((LM2010005) and CERIT-SC (CZ.1.05/3.2.00/08.0144) computing facilities. A very initial phase of the research line was supported by the Alexander von Humboldt-Stiftung and the Max-Planck-Institut für Chemie (Otto-Hahn-Institut), too.

**Data Availability Statement:** The data presented in this study are available in article.

**Conflicts of Interest:** The authors declare no conflict of interest.

## References

1. Bochvar, D.A.; Galpern, E.G. Hypothetical systems: Carbododecahedron, s-Icosahedron, and carbo-s-Icosahedron. *Dokl. Akad. Nauk SSSR* **1973**, *209*, 610–612.
2. Pitzer, K.S.; Clementi, E. Large molecules in carbon vapor. *J. Am. Chem. Soc.* **1959**, *81*, 4477–4485. [CrossRef]
3. Slanina, Z. Remark on the present applicability of quantum chemistry to the calculations of equilibrium and rate constants of chemical reactions. *Radiochem. Radioanal. Lett.* **1975**, *22*, 291–298.
4. Slanina, Z.; Lee, S.-L.; Yu, C.-H. Computations in treating fullerenes and carbon aggregates. *Rev. Comput. Chem.* **1996**, *8*, 1–62.

5. Zhao, X.; Slanina, Z.; Goto, H. Theoretical studies on the relative stabilities of $C_{96}$ IPR fullerenes. *J. Phys. Chem. A* **2004**, *108*, 4479–4484. [CrossRef]
6. Zhao, X.; Slanina, Z. $C_{98}$ IPR isomers: Gibbs-energy based relative stabilities. *J. Mol. Struct. (Theochem)* **2003**, *636*, 195–201. [CrossRef]
7. Zhao, X.; Goto, H.; Slanina, Z. $C_{100}$ IPR fullerenes: Temperature-dependent relative stabilities based on the Gibbs function. *Chem. Phys.* **2004**, *306*, 93–104. [CrossRef]
8. Diener, M.D.; Alford, J.M. Isolation and properties of small-bandgap fullerenes. *Nature* **1998**, *393*, 668–671. [CrossRef]
9. Mercado, B.Q.; Jiang, A.; Yang, H.; Wang, Z.; Jin, H.; Liu, Z.; Olmstead, M.M.; Balch, A.L. Isolation and structural characterization of the molecular nanocapsule $Sm_2@D_{3d}(822)$-$C_{104}$. *Angew. Chem. Int. Ed.* **2009**, *48*, 9114–9116. [CrossRef]
10. Beavers, C.M.; Jin, H.; Yang, H.; Wang, Z.; Wang, X.; Ge, H.; Liu, Z.; Mercado, B.Q.; Olmstead, M.M.; Balch, A.L. $La_2C_{90}$ to $La_2C_{138}$: Isolation and crystallographic characterization of $La_2@D_5(450)$-$C_{100}$. *J. Am. Chem. Soc.* **2011**, *133*, 15338–15341. [CrossRef] [PubMed]
11. Rodríguez-Fortea, A.; Balch, A.L.; Poblet, J.M. Endohedral metallofullerenes: A unique host-guest association. *Chem. Soc. Rev.* **2011**, *40*, 3551–3563. [CrossRef] [PubMed]
12. Popov, A.A.; Yang, S.; Dunsch, L. Endohedral fullerenes. *Chem. Rev.* **2013**, *113*, 5989–6113. [CrossRef] [PubMed]
13. Yang, S.; Wang, S.; Troyanov, S.I. The most stable isomers of giant fullerenes $C_{102}$ and $C_{104}$ captured as Chlorides, $C_{102}(603)Cl_{18/20}$ and $C_{104}(234)Cl_{16/18/20/22}$. *Chem. Eur. J.* **2014**, *20*, 6875–6878. [CrossRef] [PubMed]
14. Cai, W.; Bao, L.; Zhao, S.; Xie, Y.; Akasaka, T.; Lu, X. Anomalous compression of $D_5(450)$-$C_{100}$ by encapsulating $La_2C_2$ Cluster Instead of $La_2$. *J. Am. Chem. Soc.* **2015**, *137*, 10292–10296. [CrossRef] [PubMed]
15. Cai, W.; Li, F.-F.; Bao, L.; Xie, Y.; Lu, X. Isolation and crystallographic characterization of $La_2C_2@C_s(574)$-$C_{102}$ and $La_2C_2@C_2(816)$-$C_{104}$: Evidence for the top-down formation mechanism of fullerenes. *J. Am. Chem. Soc.* **2016**, *138*, 6670–6675. [CrossRef] [PubMed]
16. Wang, S.; Yang, S.; Kemnitz, E.; Troyanov, S.I. New giant fullerenes identified as chloro derivatives: Isolated-pentagon-rule $C_{108}(1771)Cl_{12}$ and $C_{106}(1155)Cl_{24}$ as well as nonclassical $C_{104}Cl_{24}$. *Inorg. Chem.* **2016**, *55*, 5741–5743. [CrossRef] [PubMed]
17. Pan, C.; Bao, L.; Yu, X.; Fang, H.; Xie, Y.; Akasaka, T.; Lu, X. Facile Access to $Y_2C_{2n}$ ($2n$ = 92–130) and crystallographic characterization of $Y_2C_2@C_1(1660)$-$C_{108}$: A giant nanocapsule with a linear carbide cluster. *ACS Nano* **2018**, *12*, 2065–2069. [CrossRef] [PubMed]
18. Slanina, Z.; Uhlík, F.; Pan, C.; Akasaka, T.; Lu, X.; Adamowicz, L. Computed stabilization for a giant fullerene endohedral: $Y_2C_2@C_1(1660)$-$C_{108}$. *Chem. Phys. Lett.* **2018**, *710*, 147–149. [CrossRef]
19. Shao, N.; Gao, Y.; Yoo, S.; An, W.; Zeng, X.C. Search for lowest-energy fullerenes: $C_{98}$ to $C_{110}$. *J. Phys. Chem. A* **2006**, *110*, 7672–7676. [CrossRef] [PubMed]
20. Becke, A.D. Density-functional exchange-energy approximation with correct asymptotic behavior. *Phys. Rev. A* **1988**, *38*, 3098–3100. [CrossRef] [PubMed]
21. Lee, C.; Yang, W.; Parr, R.G. Development of the Colle-Salvetti correlation-energy formula into a functional of the electron density. *Phys. Rev. B* **1988**, *37*, 785–789. [CrossRef] [PubMed]
22. Becke, A.D. Density-functional thermochemistry. III. The role of exact exchange. *J. Chem. Phys.* **1993**, *98*, 5648–5652. [CrossRef]
23. Cao, X.Y.; Dolg, M. Segmented contraction scheme for small-core lanthanide pseudopotential basis sets. *J. Mol. Struct. (Theochem)* **2002**, *581*, 139–147. [CrossRef]
24. Boys, S.F.; Bernardi, F. The calculation of small molecular interactions by the difference of separate total energies. Some procedures with reduced errors. *Mol. Phys.* **1970**, *19*, 553–566. [CrossRef]
25. Slanina, Z.; Uhlík, F.; Lee, S.-L.; Adamowicz, L.; Akasaka, T.; Nagase, S. Computed stabilities in metallofullerene series: $Al@C_{82}$, $Sc@C_{82}$, $Y@C_{82}$, and $La@C_{82}$. *Int. J. Quant. Chem.* **2011**, *111*, 2712–2718. [CrossRef]
26. Frisch, M.J.; Trucks, G.W.; Schlegel, H.B.; Scuseria, G.E.; Robb, M.A.; Cheeseman, J.R.; Scalmani, G.; Barone, V.; Mennucci, B.; Petersson, G.A.; et al. *Gaussian 09, Rev. D.01*; Gaussian Inc.: Wallingford, CT, USA, 2013.
27. Slanina, Z.; Uhlík, F.; Nagase, S.; Akasaka, T.; Adamowicz, L.; Lu, X. $Eu@C_{72}$: Computed Comparable Populations of Two Non-IPR Isomers. *Molecules* **2017**, *22*, 1053. [CrossRef] [PubMed]
28. Okubo, S.; Kato, T.; Inakuma, M.; Shinohara, H. Separation and characterization of ESR-active lanthanum endohedral fullerenes. *New Diam. Front. Carbon Technol.* **2001**, *11*, 285–294.
29. Hehre, W.J. *A Guide to Molecular Mechanics and Quantum Chemical Calculations*; Wavefunction: Irvine, CA, USA, 2003; p. 435.
30. Jensen, F. *Introduction to Computational Chemistry*; Wiley: Chichester, UK, 2017; p. 319.
31. Slanina, Z.; Uhlík, F.; Lu, X.; Akasaka, T.; Lemke, K.H.; Seward, T.M.; Nagase, S.; Adamowicz, L. Calculations of the water-dimer encapsulations into $C_{84}$. *Fullerenes Nanotub. Carbon Nanostruct.* **2016**, *24*, 1–7. [CrossRef]
32. Yu, P.; Shen, W.; Bao, L.; Pan, C.; Slanina, Z.; Lu, X. Trapping an unprecedented $Ti_3C_3$ unit inside the icosahedral $C_{80}$ fullerene: A crystallographic survey. *Chem. Sci.* **2019**, *10*, 10925–10930. [CrossRef] [PubMed]
33. Slanina, Z.; Uhlík, F.; Lee, S.-L.; Wang, B.-C.; Adamowicz, L.; Suzuki, M.; Haranaka, M.; Feng, L.; Lu, X.; Nagase, S.; et al. Towards relative populations of non-isomeric metallofullerenes: $La@C_{76}(T_d)$ vs. $La_2@C_{76}(C_s,17490)$. *Fullerenes Nanotub. Carbon Nanostruct.* **2014**, *22*, 299–306. [CrossRef]
34. Slanina, Z.; Uhlík, F.; Adamowicz, L. Theoretical predictions of fullerene stabilities. In *Handbook of Fullerene Science and Technology*; Lu, X., Akasaka, T., Slanina, Z., Eds.; Springer: Singapore, 2022; pp. 111–179.

35. Slanina, Z. Breakdown of the conventional formula for the partition function of free internal rotation. *J. Phys. Chem.* **1982**, *86*, 4782–4786. [CrossRef]
36. Slanina, Z.; Zhao, X.; Kurita, N.; Gotoh, H.; Uhlík, F.; Rudziński, J.M.; Lee, K.H.; Adamowicz, L. Computing the relative gas-phase populations of $C_{60}$ and $C_{70}$: Beyond the traditional $\Delta H^{o}_{f,298}$ scale. *J. Mol. Graph. Mod.* **2001**, *19*, 216–221. [CrossRef] [PubMed]
37. Slanina, Z.; Uhlík, F.; Hu, S.; Akasaka, T.; Lu, X.; Adamowicz, L. Calculated isomeric populations of Er@$C_{82}$. *Fullerenes Nanotub. Carbon Nanostruct.* **2024**, *32*. [CrossRef]
38. Chase, M.W., Jr.; Davies, C.A.; Downery, J.R., Jr.; Frurip, D.J.; McDonald, R.A.; Syverud, A.N. JANAF thermochemical tables, Third edition, Vols. 1,2. *J. Phys. Chem. Ref. Data* **1985**, *14* (Suppl. 1), 646.
39. Gueorguiev, G.K.; Stafström, S.; Hultman L. Nano-wire formation by self-assembly of silicon-metal cage-like molecules. *Chem. Phys. Lett.* **2008**, *458*, 170–174. [CrossRef]
40. Basiuk, V.A.; Basiuk, E.V. Noncovalent complexes of $I_h$-$C_{80}$ fullerene with phthalocyanines. *Fullerenes Nanotub. Carbon Nanostruct.* **2018**, *26*, 69–75. [CrossRef]
41. Basiuk, V.A.; Tahuilan-Anguiano, D.E. Complexation of free-base and 3d transition metal(II) phthalocyanines with endohedral fullerene $Sc_3N$@$C_{80}$. *Chem. Phys. Lett.* **2019**, *722*, 146–152. [CrossRef]
42. Wang, S.; Chang, Q.; Zhang, G.; Li, F.; Wang, X.; Yang, S.; Troyanov, S.I. Structural studies of giant empty and endohedral fullerenes. *Front. Chem.* **2020**, *8*, 607712. [CrossRef] [PubMed]
43. Li, M.; Zhao, R.; Dang, J.; Zhao, X. Theoretical study on the stabilities, electronic structures, and reaction and formation mechanisms of fullerenes and endohedral metallofullerenes. *Coor. Chem. Rev.* **2022**, *471*, 214762. [CrossRef]
44. Li, Y.B.; Biswas, R.; Kopcha, W.P.; Dubroca, T.; Abella, L.; Sun, Y.; Crichton, R.A.; Rathnam, C.; Yang, L.T.; Yeh, Y.-W.; et al. Structurally defined water-soluble metallofullerene derivatives towards biomedical applications. *Angew. Chem. Int. Ed. Engl.* **2023**, *62*, e202211704. [CrossRef] [PubMed]
45. Menon, A.; Kaur, R.; Guldi, D.M. Merging carbon nanostructures with porphyrins. In *Handbook of Fullerene Science and Technology*; Lu, X., Akasaka, T., Slanina, Z., Eds.; Springer: Singapore, 2022; pp. 219–264.

inorganics

Article

# On the Structural and Vibrational Properties of Solid Endohedral Metallofullerene Li@$C_{60}$

**Martina Vrankić [1,†], Takeshi Nakagawa [2,†], Melita Menelaou [3,†], Yasuhiro Takabayashi [4,†], Naoya Yoshikane [5], Keisuke Matsui [6], Ken Kokubo [7,‡], Kenichi Kato [8], Saori Kawaguchi-Imada [9], Hirokazu Kadobayashi [9], John Arvanitidis [10], Yoshiki Kubota [5] and Kosmas Prassides [5,11,*]**

1 Division of Materials Physics, Ruđer Bošković Institute, 10000 Zagreb, Croatia
2 Center for High-Pressure Science & Technology Advanced Research, Beijing 100094, China
3 Department of Mechanical Engineering and Materials Science and Engineering, Cyprus University of Technology, Limassol 3036, Cyprus
4 Department of Physical Science and Engineering, Nagoya Institute of Technology, Nagoya 466-8555, Japan
5 Department of Physics, Graduate School of Science, Osaka Metropolitan University, Osaka 599-8531, Japan
6 Department of Materials Science, Graduate School of Engineering, Osaka Metropolitan University, Osaka 599-8531, Japan
7 Division of Applied Chemistry, Graduate School of Engineering, Osaka University, Osaka 565-0871, Japan
8 RIKEN SPring-8 Center, 1-1-1 Kouto, Sayo-gun, Sayo-cho 679-5148, Japan
9 Japan Synchrotron Radiation Research Institute (JASRI), 1-1-1 Kouto, Sayo-gun, Sayo-cho 679-5198, Japan
10 Physics Department, Aristotle University of Thessaloniki, 54124 Thessaloniki, Greece
11 Faculty of Engineering, Kyoto University of Advanced Science, Kameoka 621-8555, Japan
* Correspondence: k.prassides@omu.ac.jp
† These authors contributed equally to this work.
‡ Present address: Nano Carbon Device Research Center, National Institute of Advanced Industrial Science and Technology (AIST), Tsukuba 305-8565, Japan.

**Abstract:** The endohedral lithium fulleride, $Li^+@C_{60}{}^{\bullet-}$, is a potential precursor for new families of molecular superconducting and electronic materials beyond those accessible to date from $C_{60}$ itself. Solid Li@$C_{60}$ comprises $(Li@C_{60})_2$ dimers, isostructural and isoelectronic with the $(C_{59}N)_2$ units found in solid azafullerene. Here, we investigate the structural and vibrational properties of Li@$C_{60}$ samples synthesized by electrolytic reduction routes. The resulting materials are of high quality, with crystallinity far superior to that of their antecedents isolated by chemical reduction. They permit facile, unambiguous identification of both the reduced state of the fulleride units and the interball C-C bonds responsible for dimerization. However, severe orientational disorder conceals any crystal symmetry lowering due to the presence of dimers. Diffraction reveals the adoption of a hexagonal crystal structure (space group $P6_3/mmc$) at both low temperatures and high pressures, typically associated with close-packing of spherical monomer units. Such a situation is reminiscent of the structural behavior of the high-pressure Phase I of solid dihydrogen, $H_2$.

**Keywords:** endofullerenes; dimerization; vibrational spectroscopy; synchrotron X-ray diffraction; high pressure

## 1. Introduction

Following the serendipitous discovery of the fullerene family of molecular carbon nearly 40 years ago, research on the chemical, physical, and materials properties of its members and those of their derivatives has proceeded unabated. In brief, there are four distinct ways of manipulating the molecular and crystalline forms of the fullerenes. Firstly, direct chemical functionalization of the $\pi$-electron molecular framework leads to a rich organic chemistry, affording numerous derivatives of both $C_{60}$ and its higher fullerene antecedents that have applications in diverse fields from molecular electronics to solar cells [1].

Secondly, reductive intercalation of metal ions into the voids of the crystalline structures of fullerenes—most prominently, that of $C_{60}$—leads to sensitively modifying the electronic structure and has resulted in a plethora of fulleride salts with $M_xC_{60}$ stoichiometry and $x$ ranging from a low value of 1 (e.g., in $RbC_{60}$) to a high value of 12 (e.g., in $Li_{12}C_{60}$). In particular, exohedrally modified fulleride materials with $A_3C_{60}$ (A = alkali metal) stoichiometry have provided examples of high-$T_c$ molecular superconductors with transition temperatures as high as 38 K and with the zero-resistance performance surviving to record high magnetic fields (exceeding 90 T). Of significance is also the unconventional, highly correlated nature of superconductivity, which is proximate to an antiferromagnetic Mott insulating state out of which it emerges upon lattice contraction via the application of chemical and/or physical pressure [2–5].

Thirdly, the electronic structure can be tuned by on-ball substitution of heteroatoms such as B or N. The most prominent derivative has been the azafullerene radical, $C_{59}N^{\bullet}$, which dimerizes to afford the non-magnetic $(C_{59}N)_2$ moiety [6]. Further manipulation of the properties of such heterofullerenes can be achieved by exohedral metal intercalation, resulting in salts such as $K_6C_{59}N$, in which electron doping is achieved synergistically both by on- and in-between-the-balls donors [7].

Finally, a large family of fullerene derivatives can be accessed by endohedral encapsulation of atomic or molecular units to afford $A@C_n$ moieties. For a long while, this field of fullerene research was dominated by metal encapsulation of higher fullerenes [8,9], with bulk availability of endohedral $C_{60}$ units remaining elusive. This situation has changed dramatically following two significant discoveries: (a) the development of the "molecular surgery" synthetic technique [10], which has led to numerous endohedral $C_{60}$ fullerenes incorporating noble gas atoms (e.g., He, Ne, Ar, Kr) or small molecules (e.g., $H_2$, HF, CO, $H_2O$, $CH_4$) [11,12] and (b) the successful synthesis of fullerene salts, $(Li^+@C_{60})X^-$ ($X^-$ = $PF_6^-$, $SbCl_6^-$), which comprise $Li^+@C_{60}$ endohedral molecular ions and $X^-$ spacer units, which serve a charge-balancing role [13–15]. Remarkably, both chemical and electrochemical reduction of the $Li^+@C_{60}$ cation lead to the formation of the neutral radical anion, $Li^+@C_{60}^{\bullet-}$ [16,17], which, in analogy with the on-ball-substituted azafullerene $C_{59}N^{\bullet}$ radical, dimerizes to afford $(Li@C_{60})_2$, isostructural and isoelectronic with the $(C_{59}N)_2$ molecule [18,19].

Structural and electronic characterization of the $(Li@C_{60})_2$ solid obtained by chemical reduction confirmed its dimeric structure in the solid state [17]. However, the as-obtained material lacked good crystallinity, resulting in moderate quality diffraction and spectroscopic data. Attempts to sublime the material led to the escape of the metal ion from the endofullerene space, precluding any improvement in crystallinity, in contrast to its azafullerene antecedent [19]. Here, we report on the structural and vibrational properties of the $(Li@C_{60})_2$ solid obtained by electrochemical reduction—this is a highly crystalline material, which permits detailed study of the structural and electronic properties. While vibrational spectroscopy readily identifies the vibrational signatures of interball C-C bond formation consistent with dimerization, the material is shown to adopt a highly symmetric hexagonal-close-packed crystal structure—typically associated with monomer sphere packing—as a result of disorder both at ambient pressure down to 3.4 K and at ambient temperature up to 13.1 GPa.

## 2. Results and Discussion

### 2.1. Vibrational Properties

The Raman spectrum of $(Li@C_{60})_2$ obtained by electrochemical reduction of the $(Li^+@C_{60})(PF_6^-)$ precursor, along with those of monomeric $C_{60}$, $(Li@C_{60})(PF_6)$, and dimeric $(C_{59}N)_2$ solids, is illustrated in Figure 1—the corresponding molecular units are generated by VESTA [20]. The frequencies of the observed Raman peaks are summarized in Table S1. In the case of the monomeric $C_{60}$ and $(Li@C_{60})(PF_6)$ fullerene materials, the 785 nm laser line was used for excitation to prevent photoinduced polymerization upon illumination with visible or ultraviolet light [21]. On the other hand, owing to the stability of the dimeric

$(Li@C_{60})_2$ and $(C_{59}N)_2$ fullerenes against photoinduced polymerization, the 633 nm line of a He-Ne laser was used for their excitation to take advantage of the improved throughput of the Raman spectrometer in the corresponding spectral region. The Raman spectra of $C_{60}$ solid and $(Li@C_{60})(PF_6)$ are nearly identical to each other with small frequency shifts ($\leq 3$ cm$^{-1}$) between the corresponding peaks and are compatible with the monomeric nature of the $C_{60}$ units. Both spectra are dominated by the ten Raman-active intramolecular $C_{60}$ modes, attributed to the two totally symmetric $A_g(1)$ and $A_g(2)$ modes and the eight fivefold-degenerate $H_g(1)$-$H_g(8)$ of the $C_{60}$ molecule in icosahedral $I_h$ symmetry [22]. The weak crystal field in solid $C_{60}$ causes only the splitting of the radial $H_g(1)$ mode into two components (at 266 and 272 cm$^{-1}$), while the appearance of a weak peak at 1458 cm$^{-1}$ can be attributed to the presence of traces of polymerized $C_{60}$ ($A_g(2)$ mode of polymeric chains [23]).

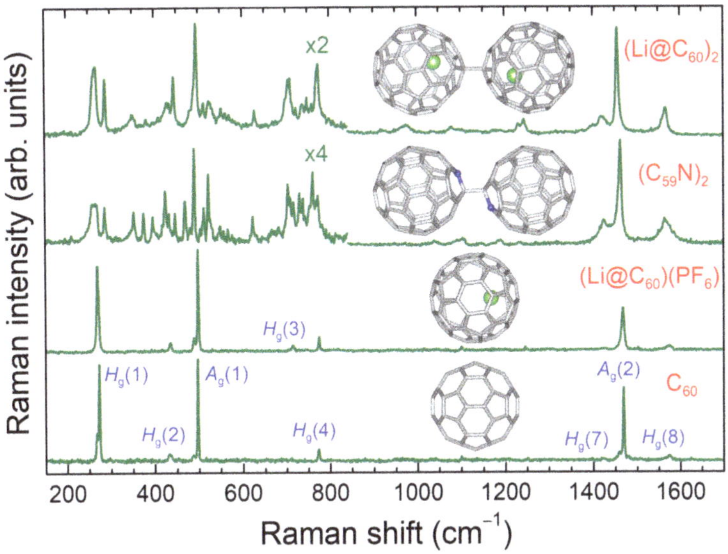


**Figure 1.** The Raman spectrum of the $(Li@C_{60})_2$ sample prepared by electrochemical reduction ($\lambda_{exc}$ = 633 nm) is displayed together with those of $C_{60}$ ($\lambda_{exc}$ = 785 nm), monomeric $(Li@C_{60})(PF_6)$ ($\lambda_{exc}$ = 785 nm), and dimeric $(C_{59}N)_2$ ($\lambda_{exc}$ = 633 nm) solids. The low-frequency spectral region in the dimeric molecular solids has been suitably enhanced to improve visibility. The mode assignment refers to the irreducible representations of the $C_{60}$ molecule in $I_h$ symmetry. Each spectrum is accompanied by the structure of the corresponding molecular unit drawn by VESTA [20].

The Raman spectrum of dimeric $(C_{59}N)_2$ solid is far richer in structure compared to the monomeric systems, as the cage distortion and the concomitant lowering of the molecular symmetry upon dimerization cause the appearance of additional Raman peaks. These originate from the splitting of the degenerate $H_g$ modes and the activation and splitting of inactive vibrational modes of the $C_{60}$ molecule in $I_h$ symmetry [24,25]. The plethora of Raman split components appearing in the spectrum of $(C_{59}N)_2$ is in fair agreement with those earlier reported in the literature and assigned to intramolecular $C_{60}$ modes in $I_h$ symmetry [24]. The same assignment of the observed Raman peaks in the present work was also adopted in Table S1. As inferred from Figure 1 and Table S1, the similarity of the Raman spectrum profile and the frequencies of the constituent Raman peaks of $(Li@C_{60})_2$ obtained by electrochemical reduction with those of dimeric $(C_{59}N)_2$ is evident, establishing its dimeric nature unambiguously. Moreover, comparison of the overall Raman spectrum profile of the present $(Li@C_{60})_2$ sample with that of the neutral $Li@C_{60}$ obtained by chemical reduction of the $Li^{+}@C_{60}$ cation by decamethylferrocene (Figure S1) clearly illustrates the stronger intensities and narrower lineshapes of the split Raman components in $(Li@C_{60})_2$ obtained by electrochemical reduction—this is in addition to the similarities regarding the appearance of the Raman peaks associated with the dimerization of the $C_{60}$ cages. The present results are compatible with the superior crystallinity of the present material, in accordance with its powder XRD data (*vide infra*), contrary to the situation encountered in $(Li@C_{60})_2$ obtained by chemical reduction, where a small fraction of monomers is also present [17].

The frequency of the $A_g(2)$ pentagonal pinch mode of the $C_{60}$ molecule deserves further attention. In particular, the $A_g(2)$ frequency in $C_{60}$-based materials is extremely sensitive to both doping and intermolecular covalent bonding [23–28]. In monomeric alkali fullerides, it downshifts quasi-linearly by 6–7 $cm^{-1}$ per electron transferred from the metal atoms to the $C_{60}$ units [26–28], due to the elongation of the intramolecular bonds induced by the charge transfer process and the resulting softening of the force constants [29,30]. Moreover, in neutral polymeric $C_{60}$-based fullerene materials—including dimerized systems—the $A_g(2)$ mode also softens due to the intermolecular covalent bond formation and the concomitant decrease in the mean intramolecular bond strength. The observed softening depends strongly on the number of the $sp^3$-like coordinated carbon atoms per molecular cage [31]. Accordingly, and as evident in Table S1 (numbers in bold), there is a softening of the $A_g(2)$ peak in neutral $(C_{59}N)_2$ compared to that in $C_{60}$ solid by ~6 $cm^{-1}$, which is attributed to the formation of the intermolecular covalent bond due to dimerization [25]. In the case of $(Li@C_{60})_2$, the softening of this peak is even larger, ~13 and ~10 $cm^{-1}$ relative to monomeric $C_{60}$ and $(Li@C_{60})(PF_6)$, respectively. This can be understood in terms of the combined effect of covalent intermolecular bonding and uptake of one electron by the $C_{60}$ cage in $(Li@C_{60})_2$—here, electrons are more delocalized over the radical carbanion cage than in the case of $(C_{59}N)_2$, where the electron pairs are localized on the nitrogen atoms [17].

The dimeric nature of $(Li@C_{60})_2$ obtained by electrochemical reduction is further supported by its IR transmission spectrum illustrated in Figure 2 in comparison with that of monomeric $C_{60}$ solid. More specifically, the IR spectrum of $C_{60}$ solid is dominated by the four IR active modes of the $C_{60}$ molecule in $I_h$ symmetry $\{F_{1u}(1)\text{-}F_{1u}(4)\}$, appearing at 525, 576, 1182, and 1428 $cm^{-1}$, in excellent agreement with the literature [32]. The additional weaker spectral features are due to forbidden fundamentals, activated by the distortion of the molecular $I_h$ symmetry in the $C_{60}$ crystal. As in the case of the corresponding Raman spectrum, in the IR spectrum of $(Li@C_{60})_2$, several split components appear due to intercage bonding and symmetry lowering, similar to the case of $(C_{59}N)_2$ [33]. Despite similarities of the IR spectrum with that of monomeric $C_{59}HN$, the appearance of characteristic features in the frequency region 800–850 $cm^{-1}$ (marked accordingly in Figure 2) provides a fingerprint of the dimerization of the molecular cages [33].

### 2.2. Structural Properties at Ambient Pressure

The synchrotron X-ray powder diffraction patterns of the $(Li@C_{60})_2$ sample recorded at 300 and 3.4 K at ambient pressure are shown in Figure 3. An immediate observation is that they appear strikingly similar to those reported for as-prepared solid $(C_{59}N)_2$, in terms of both the position and the relative intensity of the diffraction peaks present [18,34]. As-prepared $(C_{59}N)_2$ crystallizes in a hexagonal lattice (space group $P6_3/mmc$) with lattice constants at ambient temperature, $a$ = 9.97 Å and $c$ = 16.18 Å [18]. In a remarkable analogy, the patterns of $(Li@C_{60})_2$ prepared by electrochemical reduction are also consistent with the material adopting an isostructural hexagonal closed-packed (*hcp*) arrangement of the fullerene units with a somewhat smaller $(c/a)$ anisotropy (space group $P6_3/mmc$, at 300 K: $a$ = 10.04374(10) Å and $c$ = 16.13451(19) Å; at 3.4 K: $a$ = 9.9560(3) Å and $c$ = 16.0317(5) Å). Excellent refinements of the diffraction profiles are achieved by the LeBail pattern–decomposition technique using the GSAS suite (at 300 K: $R_{wp}$ = 1.88%, $R_{exp}$ = 1.48%; at 3.4 K: $R_{wp}$ = 2.23%, $R_{exp}$ = 1.43%) (Figure 3).

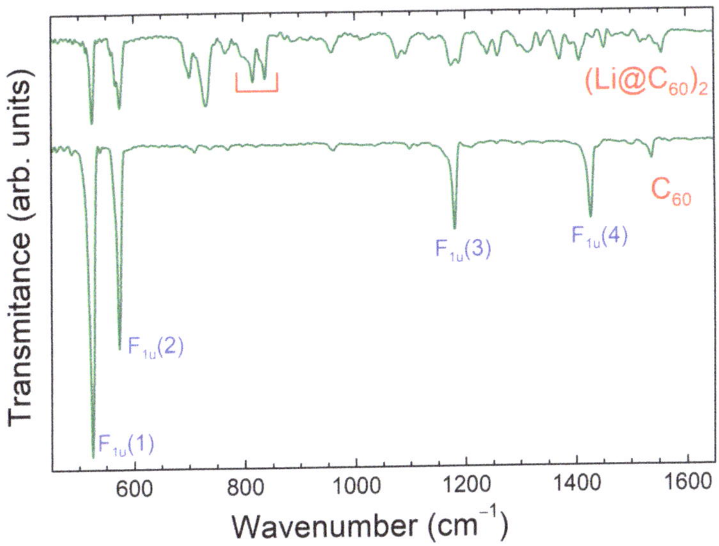


**Figure 2.** IR transmission spectrum of the $(Li@C_{60})_2$ sample prepared by electrochemical reduction and compared to that of solid $C_{60}$. Characteristic "dimer peaks" in the frequency region 800–850 $cm^{-1}$ are indicated. The mode assignment refers to the irreducible representations of the $C_{60}$ molecule in $I_h$ symmetry.

On the other hand, the measured profiles of the present $(Li@C_{60})_2$ sample contrast sharply with those of both $(Li@C_{60})_2$ prepared by chemical reduction route [17] and sublimed $(C_{59}N)_2$ [19]. The former displays a powder pattern, which can be indexed on a pseudo-cubic unit cell with a lattice size of ~14.15 Å (interfullerene separation ~10.01 Å) [17], whereas the latter provides clear evidence of a dimerized molecular structure and a lowering of crystal symmetry to monoclinic [19].

The apparent *hcp* crystal structure (Figure 3a inset) of $(Li@C_{60})_2$ prepared by electrochemical reduction is an archetypal packing arrangement of monomer units. Therefore, it is counterintuitive and contrasts sharply with the dimerization of the molecular units that has been unambiguously established by vibrational spectroscopy (*vide supra*). In the $P6_3/mmc$ space group, the $Li@C_{60}$ balls reside in the 2*d* positions, (2/3, 1/3, 1/4) and (1/3, 2/3, 3/4). As a result, their center-to-center distance at ambient temperature is 9.94 Å, implicit of the absence of interball bond formation. Due to the translational symmetry of the hexagonal unit cell, there are six nearest neighbors at this distance of 9.94 Å, together with six additional ones in the *ab* pane at 10.04 Å. These are close to the $C_{60}$—$C_{60}$ distances in monomeric $C_{60}$ (~10 Å) and are longer than those (~9.34 Å) in typical dimerized $C_{60}$ structures like that of $RbC_{60}$ [35–37].

The inconsistency with the vibrational results may be resolved if we introduce orientational disorder of the dimer units residing on the *hcp* sites. To this effect, we recall a much-celebrated analogous situation—namely, that of the Phase I of solid dihydrogen, $H_2$ [38]. At room temperature, this phase is stable between 5 and 190 GPa and comprises freely rotating dihydrogen molecules, forming an *hcp* crystal structure (space group $P6_3/mmc$). We therefore attempted Rietveld refinements of the diffraction data by modeling the pronounced orientational disorder of the fullerene moieties in terms of spherical shell scatterers placed at the 2*d* sites of $\overline{6}m2$ symmetry in the *hcp* lattice. In the course of the refinements, we found it necessary to also introduce scattering density at the 2*a* sites of $\overline{3}m$. symmetry, presumably reflecting the presence of residual $CH_2Cl_2$ solvent molecules intercalated during the electrochemical synthetic protocol. Rietveld refinements now proceeded smoothly, resulting in a stoichiometry of $(Li@C_{60})_2{\cdot}(CH_2Cl_2)_{0.81(2)}$ for the material. We note that an analogous situation was encountered in the Rietveld refinements of the diffraction data of as-prepared *hcp*-structured azafullerene solid, whose stoichiometry was established as $(C_{59}N)_2{\cdot}(CS_2)_{1.0}$ [34]. The refinement results are shown in Figure 4 and Table S2.

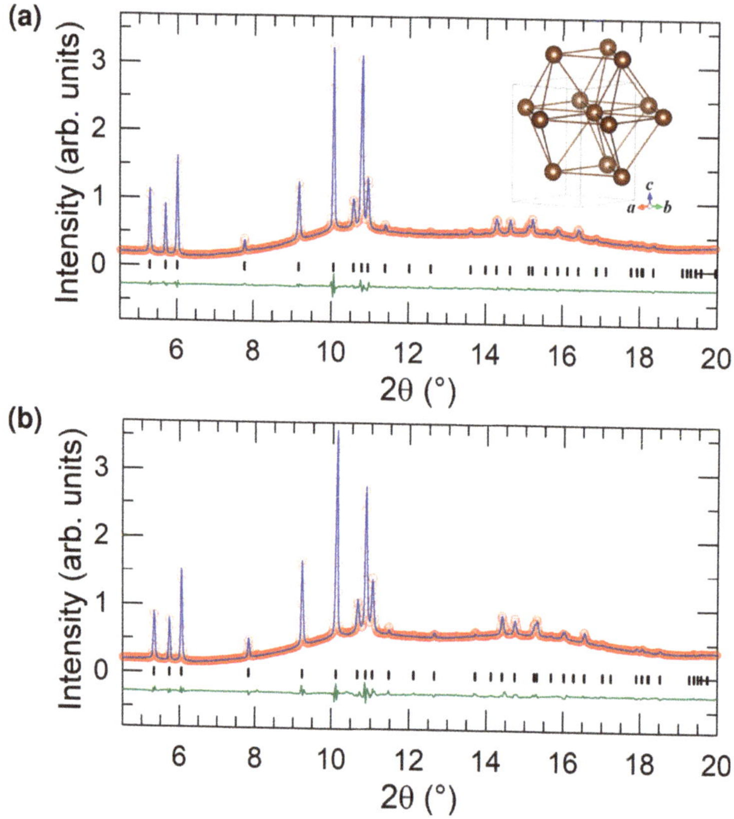


**Figure 3.** LeBail refinements (using the GSAS suite of programs) of the synchrotron X-ray (λ = 0.80031 Å) powder diffraction profiles of $(Li@C_{60})_2$ at (**a**) 300 K and (**b**) 3.4 K. The red open circles represent the measured intensity, and the blue solid line shows the calculated intensity. The lower green solid line shows the difference profile, and the tick marks show the reflection positions in space group $P6_3/mmc$. The inset in (**a**) shows a schematic diagram of the apparent hexagonal close-packing motif of solid spheres, consistent with the space group symmetry.

Figure 5a,b shows the temperature dependence of the unit–cell volume and the lattice parameters $a$ and $c$, as extracted from the LeBail analyses of the diffraction data. The smooth evolution in lattice size confirms the absence of any structural phase transition down to 3.4 K. The only apparent effect is a decrease in the lattice anisotropy, as measured by the $(c/a)$ ratio, which monotonically increases with decreasing temperature, approaching a value just larger than 1.61 at 3.4 K (Figure 5c). This is still smaller than the ideal value of 1.633 in perfect hexagonal close-packed architectures, but the decreased anisotropy on cooling may reflect a partial freezing out of the orientational disorder of the fullerene units.

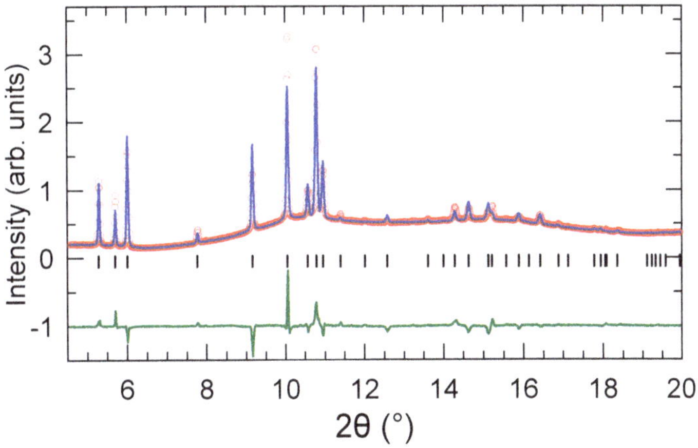


**Figure 4.** Rietveld analysis result (space group $P6_3/mmc$) using the Fullprof suite of programs—final observed (red circles) and calculated (blue solid line) synchrotron X-ray (λ = 0.80031 Å) powder diffraction profiles for the $(Li@C_{60})_2$ sample at 300 K. The lower solid line shows the difference profile, and the tick marks show the reflection positions. Refined parameters and agreement indices are provided in Supplementary Information, Table S2.

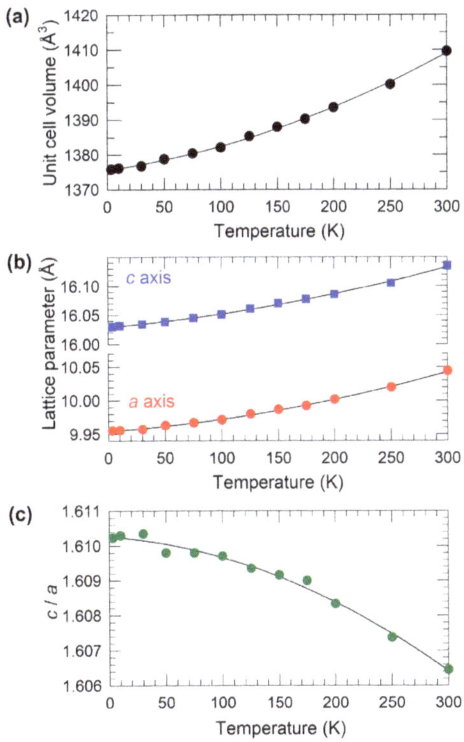


**Figure 5.** Temperature dependence of (**a**) the unit cell volume, (**b**) the hexagonal lattice constants, *a* and *c*, and (**c**) the (*c*/*a*) ratio in $(Li@C_{60})_2$ at ambient pressure. The lines through the points are guides-to-the-eye.

### 2.3. Structural Properties at High Pressure

Synchrotron X-ray powder diffraction data of $(Li@C_{60})_2$ were collected at ambient temperature and at selected pressures between 0.5 and 13.1 GPa. As the pressure increases, the diffraction profiles show gradual Bragg peak shifting towards higher 2θ angles consistent with unit cell contraction; this is accompanied by some peak broadening due to increasing inhomogeneities upon pressurization (Figure 6). No evidence of the occurrence of any structural phase transition is found in the pressure range of the present experiments. All diffraction patterns index with the same hexagonal unit cell (space group $P6_3/mmc$) found at all temperatures at ambient pressure. Representative LeBail refinements at base and highest pressures are included in Figure 6 (at 0.5 GPa: $a$ = 9.89824(18) Å and $c$ = 15.9333(3) Å; $R_{wp}$ = 0.35%, $R_{exp}$ = 4.60%; at 13.1 GPa: $a$ = 9.2495(5) Å and $c$ = 15.0437(11) Å; $R_{wp}$ = 0.29%, $R_{exp}$ = 4.94%). The evolution of the $(Li@C_{60})_2$ unit cell dimensions upon pressurization was investigated by LeBail refinements of the diffraction profiles (Figure 7a,b). Some spurious weak peaks, which were present in the diffraction datasets when the sample was introduced into the diamond anvil cell (DAC)—but were absent for the same sample batch outside the pressure cell—were excluded from the refinements. Figure 7c displays the change in the lattice anisotropy of the *hcp* structure, as monitored by the evolution of the $(c/a)$ ratio. Upon pressurization, $c/a$ increases monotonically at the beginning but eventually saturates above ~10 GPa towards a value of ~1.626, close to the ideal value of 1.633 expected for perfect *hcp* stacking of solid spheres. Comparable behavior was also observed before for the isostructural dimeric $(C_{59}N)_2$ solid, in which the $(c/a)$ ratio approached ~1.633 at ~6.5 GPa [18]. Such behavior implies that, at these elevated pressures, the intradimer and interdimer C-C distances in the compressed material have become comparable. Notably, this occurs in the same pressure range that the interdimer center-to-center separations in $(Li@C_{60})_2$ have decreased to values on the order of ~9.2–9.3 Å. Such close interball contacts have provided the signature of single C-C bond formation in fullerene-derived polymers [35–37] before.

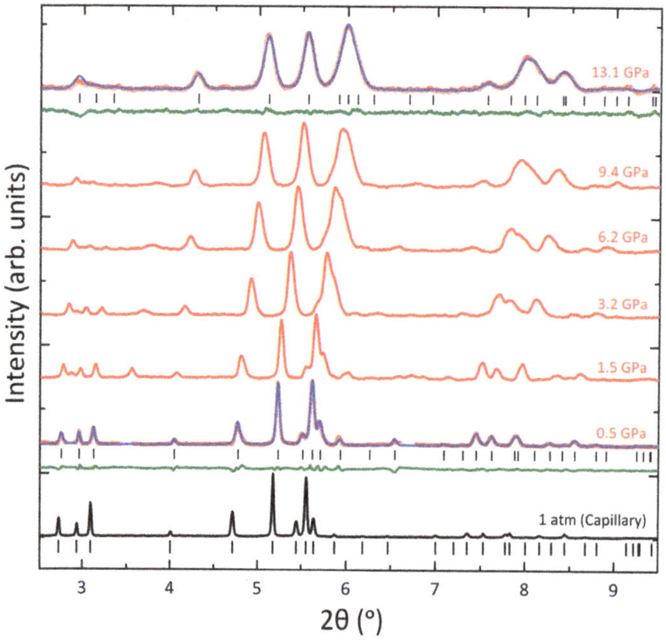


**Figure 6.** Representative high-pressure synchrotron X-ray powder diffraction profiles after background subtraction (λ = 0.4128 Å) of $(Li@C_{60})_2$ (red solid lines and open circles) at elevated pressures.

The corresponding diffraction profile at 1 atm (capillary) from Figure 5 is also displayed for comparison after conversion to the same wavelength (black solid line). The LeBail refinement results for base (0.5 GPa) and highest (13.1 GPa) pressure are also included (blue solid lines: calculated, red circles: observed). Tick marks label the reflections of the hexagonal lattice (space group $P6_3/mmc$), and the green solid lines are the difference profiles between observed and calculated profiles. Some spurious weak peaks present in the high-pressure data with the sample inside the diamond anvil cell (DAC) are excluded from the refinements.

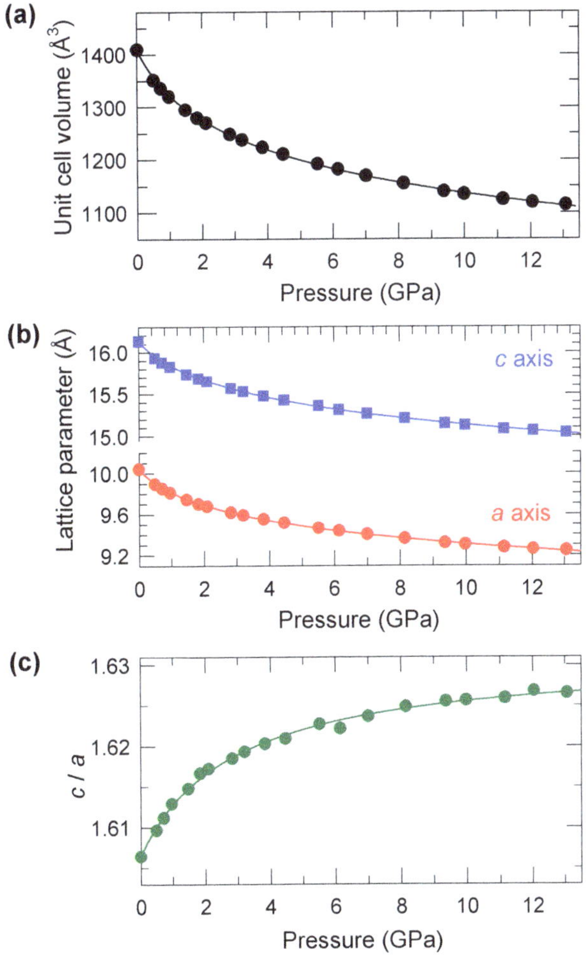


**Figure 7.** Pressure dependence of (**a**) the unit cell volume, (**b**) the hexagonal lattice constants, *a* and *c*, and (**c**) the (*c*/*a*) ratio in $(Li@C_{60})_2$ at ambient temperature. The solid lines in (**a**,**b**) are the results of least-squares fits to the semi-empirical 2nd-order Murnaghan equation-of-states (EoS) and its axial compressibility variants, respectively. The line through the points in (**c**) is a guide-to-the-eye.

The pressure dependence of the unit cell volume of $(Li@C_{60})_2$ was next investigated using the semi-empirical second-order Murnaghan equation-of-state (EoS):

$$P = \left(\frac{K_0}{K_0'}\right)\left[\left(\frac{V_0}{V}\right)^{K_0'} - 1\right] \quad (1)$$

where $K_0$ is the atmospheric pressure isothermal bulk modulus, $K_0'$ is its pressure derivative ($\equiv dK_0/dP)_{P=0}$, and $V_0$ is the unit–cell volume at zero pressure (approximated by the ambient pressure datum at 1 atm; here, $V_0$ = 1409.58 Å$^3$). The EoS fit to the $V(P)$ data (see Equation (S1) in the Supplementary Materials) of $(Li@C_{60})_2$ up to a pressure of 13.1 GPa results in values of $K_0$ = 10.8(3) GPa (leading to an initial volume compressibility, $\kappa = -d\ln V/dP$ = 0.093(3) GPa$^{-1}$) and of $K_0'$ = 11.4(2) (Figure 7a). The measured volume compressibility of $(Li@C_{60})_2$ is higher than those of $(C_{59}N)_2$ ($\kappa$ = 0.046(2) GPa$^{-1}$) and of the monomeric $C_{60}$ ($\kappa$ = 0.055 GPa$^{-1}$) and $(Li^+@C_{60})(PF_6)$ ($\kappa$ = 0.0571(4) GPa$^{-1}$) [15,18,39] solids, implying more flexible nearest-neighbor bonds and somewhat less tight crystal packing. Modified EoS equations, in which $K_0$ and its pressure derivative, $K_0'$, were replaced by $K_x$ and $K_x'$ ($x$ = $a$ or $c$) parameters, were used to fit the corresponding lattice axes data to extract initial axial compressibilities, $\beta_x$ ( = $1/(3K_x)$ GPa$^{-1}$, $x$ = $a$ or $c$) (Figure 7b). As anticipated by the earlier discussion on the pressure dependence of the $(c/a)$ ratio, we find that the compressibility along the $c$-axis ($\beta_c$ = 0.0086(2) GPa$^{-1}$) is smaller than that along the $a$-axis ($\beta_a$ = 0.0112(3) GPa$^{-1}$). Such a pressure response was also observed for $(C_{59}N)_2$ and was then interpreted as a consequence of the intradimer bonds being oriented along the (1/3, 2/3, 1/2) direction and therefore inclined at a larger angle to the basal plane than to the $c$-axis [18].

## 3. Materials and Methods

The $(Li^+@C_{60})(PF_6^-)$ sample used as a precursor in the present work was purchased from Idea International Corporation. Its electrochemical reduction to afford $(Li@C_{60})_2$ was performed according to the experimental protocol described in the literature before [16].

Raman spectra of $(Li@C_{60})_2$ were recorded in backscattering geometry using a LabRam HR (HORIBA, Kyoto, Japan) micro-Raman spectrometer, equipped with a Peltier cooled CCD detector. Laser lines at 633 and 785 nm were alternatively used for excitation, focused on the samples by means of a 100× objective at a power lower than 0.3 mW to avoid any laser-heating-induced effects.

Infrared (IR) transmission spectra of the samples under argon atmosphere were collected by means of a JASCO (Tokyo, Japan) FT-IR 6200 spectrometer.

Synchrotron X-ray powder diffraction measurements of $(Li@C_{60})_2$ at various temperatures between 3.4 and 300 K at ambient pressure were performed with the high-resolution powder diffractometer on beamline BL44B2 at SPring-8, Hyogo, Japan using a helium-flow cryostat [40]. The powder sample was sealed in a 0.5 mm diameter thin-wall glass capillary under He gas. The wavelength was 0.800312(1) Å. LeBail and Rietveld analyses were performed with the GSAS and Fullprof suites of refinement programs [41–43].

Synchrotron X-ray powder diffraction measurements of $(Li@C_{60})_2$ up to 13.1 GPa at ambient temperature were conducted at the BL10XU beamline, SPring-8, Hyogo, Japan. The powder X-ray diffraction data (monochromatized beam with $\lambda$ = 0.41277(5) Å, beam size = 40 µm) were collected on an image plate detector (Rigaku (Tokyo, Japan) R-AXIS IV$^{++}$, 300 × 300 mm$^2$, and 0.10 mm pixel size) while oscillating the sample [44]. Masking of the Bragg reflections of the DAC and integration of the 2D diffraction images to afford one-dimensional diffraction profiles were processed with the IPanalyzer software [45]. LeBail analysis of the diffraction profiles was performed with the GSAS suite of refinement programs [41,42]. A DAC [46,47], equipped with single-crystal diamonds of 500 µm culet diameter and with an 80 µm-thick stainless-steel gasket, was used for pressure generation. Powder samples were introduced into 250 µm diameter holes made at the center of the gasket, together with a ruby sphere and helium gas (solidification pressure ~12.1 GPa at room temperature, pressure uncertainty less than 0.5% at the highest pressure of the present experiment [48]) as pressure marker and pressure medium, respectively. The applied pressure was measured by the ruby–fluorescence method [49]. All measured pressure values are quoted to one decimal point.

## 4. Conclusions

In conclusion, we have presented a detailed structural and vibrational characterization of the dimerized endohedral metallofullerene, $(Li@C_{60})_2$, in the solid state. The material used in the present work was obtained by electrochemical reduction of the trifluoromethanesulfonyl imide precursor salt, $(Li^+@C_{60})(TFSI^-)$ [16]. Its highly crystalline nature permits facile identification of spectroscopic features related to interfullerene C-C bonding. Remarkably, the crystal structure of the material remains hexagonal close-packed (space group $P6_3/mmc$) both at low temperatures and ambient pressure and at high pressures and ambient temperature, concealing the effects of dimerization and the anticipated lowering in symmetry. High orientational disorder (together with the presence of some residual dichloromethane solvent) can account for the apparent high symmetry. A comparable orientational disorder scenario has been used to rationalize the *hcp* Phase I of solid dihydrogen ($H_2$) at high pressures.

**Supplementary Materials:** The following supporting information can be downloaded at: https://www.mdpi.com/article/10.3390/inorganics12040099/s1, Figure S1: Comparison of Raman spectra of $(Li@C_{60})_2$ obtained by chemical and electrochemical reduction; Table S1: Assignment of observed Raman peaks and their frequencies for $C_{60}$, $(Li@C_{60})(PF_6)$, $(C_{59}N)_2$, and $(Li@C_{60})_2$; Table S2: Refined parameters of the Rietveld analysis of the synchrotron X-ray powder diffraction data of $(Li@C_{60})_2$; Equation (S1): $V(P)$ form of the Murnaghan equation-of-state.

**Author Contributions:** Synthesis, M.V., T.N. and K.K. (Ken Kokubo); synchrotron X-ray diffraction—data collection and analysis, M.V., T.N., M.M., Y.T., K.K. (Kenichi Kato) and K.P.; high-pressure synchrotron X-ray diffraction—data collection and analysis, N.Y., K.M., S.K.-I., H.K. and K.P.; vibrational spectroscopy, J.A.; conceptualization and overall conclusions, J.A., Y.K. and K.P.; project supervision, K.P.; writing—review and editing, K.P. All authors have read and agreed to the published version of the manuscript.

**Funding:** This research was financially supported by Grants-in-Aid for Scientific Research (JSPS KAKENHI grant numbers JP18K05254, JP21H01907, JP22K18693, and JP23KJ1843) and by JST SPRING, grant number JPMJSP2139.

**Data Availability Statement:** All data supporting the reported results in this paper can be obtained from the authors.

**Acknowledgments:** We thank SPring-8 for access to synchrotron XRPD measurements (proposal no. 2017A4131, 2017A1363, 2017B1276, 2017B1358, 2018A1398, and 2022A1699).

**Conflicts of Interest:** The authors declare no conflicts of interest.

## References

1. Hirsch, A.; Brettreich, M. *Fullerenes: Chemistry and Reactions*; Wiley-VCH, 5 Wiley-VCH Verlag GmbH & Co. KgaA: Weinheim, Germany, 2005.
2. Takabayashi, Y.; Prassides, K. Unconventional high-$T_c$ superconductivity in fullerides. *Philos. Trans. R. Soc. A Math. Phys. Eng. Sci.* **2016**, *374*, 20150320. [CrossRef] [PubMed]
3. Zadik, R.H.; Takabayashi, Y.; Klupp, G.; Colman, R.H.; Ganin, A.Y.; Potocnik, A.; Jeglic, P.; Arcon, D.; Matus, P.; Kamaras, K.; et al. Optimized unconventional superconductivity in a molecular Jahn-Teller metal. *Sci. Adv.* **2015**, *1*, e1500059. [CrossRef]
4. Nomura, Y.; Sakai, S.; Capone, M.; Arita, R. Unified understanding of superconductivity and Mott transition in alkali-doped fullerides from first principles. *Sci. Adv.* **2015**, *1*, e1500568. [CrossRef] [PubMed]
5. Kasahara, Y.; Takeuchi, Y.; Zadik, R.H.; Takabayashi, Y.; Colman, R.H.; McDonald, R.D.; Rosseinsky, M.J.; Prassides, K.; Iwasa, Y. Upper critical field reaches 90 tesla near the Mott transition in fulleride superconductors. *Nat. Commun.* **2017**, *8*, 14467. [CrossRef] [PubMed]
6. Hummelen, J.C.; Knight, B.; Pavlovich, J.; Gonzalez, R.; Wudl, F. Isolation of the heterofullerene $C_{59}N$ as its dimer $(C_{59}N)_2$. *Science* **1995**, *269*, 1554–1556. [CrossRef] [PubMed]
7. Prassides, K.; Keshavarz-K, M.; Hummelen, J.C.; Andreoni, W.; Giannozzi, P.; Beer, E.; Bellavia, C.; Cristofolini, L.; Gonzalez, R.; Lappas, A.; et al. Isolation, structure, and electronic calculations of the heterofullerene salt $K_6C_{59}N$. *Science* **1996**, *271*, 1833–1835. [CrossRef]
8. Shinohara, H.; Tagmatarchis, N. *Endohedral Metallofullerenes: Fullereness with Metal Inside*; Wiley: Chichester, UK, 2015.
9. Akasaka, T.; Nagase, S. *Endofullerenes: A New Family of Carbon Clusters*; Springer: Dordrecht, The Netherlands, 2002.

10. Komatsu, K.; Murata, M.; Murata, Y. Encapsulation of molecular hydrogen in fullerene $C_{60}$ by organic synthesis. *Science* **2005**, *307*, 238–240. [CrossRef] [PubMed]
11. Bloodworth, S.; Whitby, R.J. Synthesis of endohedral fullerenes by molecular surgery. *Commun. Chem.* **2022**, *5*, 121. [CrossRef] [PubMed]
12. Krachmalnicoff, A.; Bounds, R.; Mamone, S.; Alom, S.; Concistrè, M.; Meier, B.; Kouřil, K.; Light, M.E.; Johnson, M.R.; Rols, S.; et al. The dipolar endofullerene HF@$C_{60}$. *Nat. Chem.* **2016**, *8*, 953–957. [CrossRef]
13. Aoyagi, S.; Nishibori, E.; Sawa, H.; Sugimoto, K.; Takata, M.; Miyata, Y.; Kitaura, R.; Shinohara, H.; Okada, H.; Sakai, T.; et al. A layered ionic crystal of polar $Li^+$@$C_{60}$ superatoms. *Nat. Chem.* **2010**, *2*, 678–683. [CrossRef]
14. Aoyagi, S.; Sado, Y.; Nishibori, E.; Sawa, H.; Okada, H.; Tobita, H.; Kasama, Y.; Kitaura, R.; Shinohara, H. Rock-salt-type crystal of thermally contracted $C_{60}$ with encapsulated lithium cation. *Angew. Chem.* **2012**, *124*, 3433–3437. [CrossRef]
15. Colman, R.H.; Okur, H.E.; Garbarino, G.; Ohishi, Y.; Aoyagi, S.; Shinohara, H.; Prassides, K. Pressure effects on the crystal structure of the cubic metallofullerene salt [Li@$C_{60}$][$PF_6$] to 12 GPa. *Mater. Today Commun.* **2022**, *31*, 103275. [CrossRef]
16. Ueno, H.; Aoyagi, S.; Yamazaki, Y.; Ohkubo, K.; Ikuma, N.; Okada, H.; Kato, T.; Matsuo, Y.; Fukuzumi, S.; Kokubo, K. Electrochemical Reduction of Cationic $Li^+$@$C_{60}$ to Neutral $Li^+$@$C_{60}{}^{\cdot -}$: Isolation and Characterisation of Endohedral [60] Fulleride. *Chem. Sci.* **2016**, *7*, 5770–5774. [CrossRef] [PubMed]
17. Okada, H.; Ueno, H.; Takabayashi, Y.; Nakagawa, T.; Vrankic, M.; Arvanitidis, J.; Kusamoto, T.; Prassides, K.; Matsuo, Y. Chemical reduction of $Li^+$@$C_{60}$ by decamethylferrocene to produce neutral $Li^+$@$C_{60}{}^{\bullet -}$. *Carbon* **2019**, *153*, 467–471. [CrossRef]
18. Brown, C.M.; Beer, E.; Bellavia, C.; Cristofolini, L.; González, R.; Hanfland, M.; Häusermann, D.; Keshavarz-K, M.; Kordatos, K.; Prassides, K.; et al. Effects of Pressure on the Azafullerene $(C_{59}N)_2$ Molecular Solid to 22 GPa. *J. Am. Chem. Soc.* **1996**, *118*, 8715–8716. [CrossRef]
19. Brown, C.M.; Cristofolini, L.; Kordatos, K.; Prassides, K.; Bellavia, C.; González, R.; Keshavarz-K, M.; Wudl, F.; Cheetham, A.K.; Zhang, J.P.; et al. On the Crystal Structure of Azafullerene $(C_{59}N)_2$. *Chem. Mater.* **1996**, *8*, 2548–2550. [CrossRef]
20. Momma, K.; Izumi, F. VESTA 3 for three-dimensional visualization of crystal, volumetric and morphology data. *J. Appl. Crystallogr.* **2011**, *44*, 1272–1276. [CrossRef]
21. Rao, A.M.; Zhou, P.; Wang, K.-A.; Hager, G.T.; Holden, J.M.; Wang, Y.; Lee, W.-T.; Bi, X.-X.; Ecklund, P.C.; Cornett, D.S.; et al. Photoinduced polymerization of solid $C_{60}$ films. *Science* **1993**, *259*, 955–957. [CrossRef]
22. Bethune, D.S.; Meijer, G.; Tang, W.C.; Rosen, H.J.; Golden, W.G.; Seki, H.; Brown, C.A.; de Vries, M.S. Vibrational Raman and infrared spectra of chromatographically separated $C_{60}$ and $C_{70}$ fullerene clusters. *Chem. Phys. Lett.* **1991**, *179*, 181–186. [CrossRef]
23. Davydov, V.A.; Kashevarova, L.S.; Rakhmanina, A.V.; Senyavin, V.M.; Ceolin, R.; Szwarc, H.; Allouchi, H.; Agafonov, V. Spectroscopic study of pressure-polymerized phases of $C_{60}$. *Phys. Rev. B* **2000**, *61*, 11936–11945. [CrossRef]
24. Kuzmany, H.; Plank, W.; Winter, J.; Dubay, O.; Tagmatarchis, N.; Prassides, K. Raman spectrum and stability of $(C_{59}N)_2$. *Phys. Rev. B* **1999**, *60*, 1005–1012. [CrossRef]
25. Plank, W.; Pichler, T.; Kuzmany, H.; Dubay, O.; Tagmatarchis, N.; Prassides, K. Resonance Raman excitation and electronic structure of the single bonded dimers $(C_{60}{}^-)_2$ and $(C_{59}N)_2$. *Eur. Phys. J. B* **2000**, *17*, 33–42. [CrossRef]
26. Winter, J.; Kuzmany, H. Potassium doped fullerene $K_xC_{60}$ with $x$ = 0, 1, 2, 3, 4 and 6. *Solid State Commun.* **1992**, *84*, 935–938. [CrossRef]
27. Kuzmany, H.; Matus, M.; Burger, B.; Winter, J. Raman scattering in $C_{60}$ fullerenes and fullerides. *Adv. Mater.* **1994**, *6*, 731–745. [CrossRef]
28. Kosaka, M.; Tanigaki, K.; Prassides, K.; Margadonna, S.; Lappas, A.; Brown, C.M.; Fitch, A.N. Superconductivity in $Li_xCsC_{60}$ fullerides. *Phys. Rev. B* **1999**, *59*, R6628–R6630. [CrossRef]
29. Jishi, R.A.; Dresselhaus, M.S. Mode softening and mode stiffening in $C_{60}$ doped with alkali metals. *Phys. Rev. B* **1992**, *45*, 6914–6918. [CrossRef] [PubMed]
30. Zhou, P.; Wang, K.A.; Wang, Y.; Eklund, P.C.; Dresselhaus, M.S.; Dresselhaus, G.; Jishi, R.A. Raman scattering in $C_{60}$ and alkali-metal-saturated $C_{60}$. *Phys. Rev. B* **1992**, *46*, 2595–2605. [CrossRef] [PubMed]
31. Meletov, K.P.; Velkos, G.; Arvanitidis, J.; Christofilos, D.; Kourouklis, G.A. Raman study of the photopolymer formation in the {Pt(dbdtc)$_2$}·$C_{60}$ fullerene complex and the decomposition kinetics of the photo-oligomers. *Chem. Phys. Lett.* **2017**, *68*, 124–129. [CrossRef]
32. Krause, M.; Dunsch, L.; Seifert, G.; Fowler, P.W.; Gromov, A.; Kratschmer, W.; Gutierrez, R.; Porezag, D.; Frauenheim, T.J. Vibrational signatures of fullerene oxides. *J. Chem. Soc. Faraday Trans.* **1998**, *94*, 2287–2294. [CrossRef]
33. Krause, M.; Baes-Fischlmair, S.; Pfeiffer, R.; Plank, W.; Pichler, T.; Kuzmany, H.; Tagmatarchis, N.; Prassides, K. Thermal Stability and high temperature graphitization of bisazafullerene $(C_{59}N)_2$ as studied by IR and Raman spectroscopy. *J. Phys. Chem. B* **2001**, *105*, 11964–11969. [CrossRef]
34. Prassides, K.; Wudl, F.; Andreoni, W. Solid Azafullerenes and Azafullerides. *Fullerene Sci. Technol.* **1997**, *5*, 801–812. [CrossRef]
35. Oszlányi, G.; Bortel, G.; Faigel, G.; Gránásy, L.; Bendele, G.M.; Stephens, P.W.; Forró, L. Single C-C Bond in $(C_{60})_2{}^{2-}$. *Phys. Rev. B* **1996**, *54*, 11849–11852. [CrossRef]
36. Prassides, K.; Vavekis, K.; Kordatos, K.; Tanigaki, K.; Bendele, G.M.; Stephens, P.W. Loss of cubic symmetry in low-temperature $Na_2RbC_{60}$. *J. Am. Chem. Soc.* **1997**, *119*, 834–835. [CrossRef]
37. Bendele, G.M.; Stephens, P.W.; Prassides, K.; Vavekis, K.; Kordatos, K.; Tanigaki, K. Effect of charge state on polymeric bonding geometry: The ground state of $Na_2RbC_{60}$. *Phys. Rev. Lett.* **1998**, *80*, 736–739. [CrossRef]

38. Ji, C.; Li, B.; Liu, W.; Smith, J.S.; Majumdar, A.; Luo, W.; Ahuja, R.; Shu, J.; Wang, J.; Sinogeikin, S.; et al. Ultrahigh-pressure isostructural electronic transitions in hydrogen. *Nature* **2019**, *573*, 558–562. [CrossRef]
39. Duclos, S.J.; Brister, K.; Haddon, R.C.; Kortan, A.R.; Thiel, F.A. Effects of Pressure and Stress on $C_{60}$ Fullerite to 20 GPa. *Nature* **1991**, *351*, 380–382. [CrossRef]
40. Kato, K.; Tanaka, H. Visualizing charge densities and electrostatic potentials in materials by synchrotron X-ray powder diffraction. *Adv. Phys. X* **2016**, *1*, 55–80. [CrossRef]
41. Larson, A.C.; von Dreele, R.B. General Structure Analysis System (GSAS). In *Los Alamos National Laboratory Report LAUR*; The Regents of the University of California: Oakland, CA, USA, 2000; pp. 86–748.
42. Toby, B.H. *EXPGUI*, a Graphical User Interface for *GSAS*. *J. Appl. Crystallogr.* **2001**, *34*, 210–213. [CrossRef]
43. Rodríguez-Carvajal, J. Recent Advances in Magnetic Structure Determination by Neutron Powder Diffraction. *Physica B Condens. Matter* **1993**, *192*, 55–69. [CrossRef]
44. Hirao, N.; Kawaguchi, S.I.; Hirose, K.; Shimizu, K.; Ohtani, E.; Ohishi, Y. New developments in high-pressure X-ray diffraction beamline for diamond anvil cell at SPring-8. *Matter Radiat. Extrem.* **2020**, *5*, 018403. [CrossRef]
45. Seto, Y.; Nishio-Hamane, D.; Nagai, T.; Sata, N. Development of a software suite on X-ray diffraction experiments. *Rev. High Press. Sci. Technol.* **2010**, *20*, 269–276. [CrossRef]
46. Sakai, T.; Yagi, T.; Ohfuji, H.; Irifune, T.; Ohishi, Y.; Hirao, N.; Suzuki, Y.; Kuroda, Y.; Asakawa, T.; Kanemura, T. High-pressure generation using double stage micro-paired diamond anvils shaped by focused ion beam. *Rev. Sci. Instrum.* **2015**, *86*, 033905. [CrossRef] [PubMed]
47. Sakai, T.; Yagi, T.; Irifune, T.; Kadobayashi, H.; Hirao, N.; Kunimoto, T.; Ohfuji, H.; Kawaguchi-Imada, S.; Ohishi, Y.; Tateno, S.; et al. High pressure generation using double-stage diamond anvil technique: Problems and equations of state of rhenium. *High Press. Res.* **2018**, *38*, 107–119. [CrossRef]
48. Zha, C.-S.; Mao, H.-K.; Hemley, R.J. Elasticity of MgO and a Primary Pressure Scale to 55 GPa. *Proc. Natl. Acad. Sci. USA* **2000**, *97*, 13494. [CrossRef]
49. Klotz, S.; Chervin, J.C.; Munsch, P.; Le Marchand, G. Hydrostatic limits of 11 pressure transmitting media. *J. Phys. D Appl. Phys.* **2009**, *42*, 075413. [CrossRef]

*inorganics*

*Article*

# Stability and Electronic Properties of 1D and 2D Ca@$C_{60}$ Oligomers and Polymers

**Yabei Wu [1], Zhonghao Zhou [2,*] and Zhiyong Wang [1,*]**

1 Key Laboratory of Advanced Light Conversion Materials and Biophotonics, Department of Chemistry, Renmin University of China, Beijing 100872, China
2 School of Materials Science and Engineering, Dalian Jiaotong University, Dalian 116028, China
* Correspondence: zhonghao_zhou@ruc.edu.cn (Z.Z.); zhiyongwang@ruc.edu.cn (Z.W.)

**Abstract:** The polymerization of fullerenes is a significant method for obtaining fullerene-based materials that possess intriguing properties. Metallofullerenes, as a notable type of fullerene derivatives, are also capable of undergoing polymerization, potentially resulting in the creation of metallofullerene polymers. However, there is currently limited knowledge regarding the polymerization process of metallofullerenes. In this study, we have selected Ca@$C_{60}$ as a representative compound to investigate the polymerization process of metallofullerenes. The objective of this research is to determine whether the polymerization process is energetically favorable and to examine how the electronic properties of the metallofullerene are altered throughout the polymerization process. Ca@$C_{60}$ is a unique metallofullerene molecule that exhibits insolubility in common fullerene solvents like toluene and carbon disulfide but is soluble in aniline. This behavior suggests a potential tendency for Ca@$C_{60}$ to form oligomers and polymers that resist dissolution. However, the structures and properties of polymerized Ca@$C_{60}$ remain unknown. We employed density functional theory calculations to investigate the stability and electronic properties of one-dimensional and two-dimensional Ca@$C_{60}$ oligomers and polymers. Our findings indicate that the coalescence of Ca@$C_{60}$ monomers is energetically favorable, with a significant contribution from van der Waals interactions between the fullerene cages. The polymerization process of Ca@$C_{60}$ also involves the formation of covalent linkages, including four-atom rings and C-C single bonds. The increase in the number of the Ca@$C_{60}$ units to three and four in the oligomer leads to a significant decrease in the HOMO-LUMO gap. In the two-dimensional polymerized Ca@$C_{60}$, the organization of the monomers closely resembles the spatial configuration of carbon atoms in graphene. With a direct bandgap of 0.22 eV, the polymerized Ca@$C_{60}$ holds potential for utilization in optoelectronic devices.

**Keywords:** fullerenes; metallofullerenes; density functional theory calculations

## 1. Introduction

Fullerenes, which typically consist of several tens of carbon atoms, are carbon allotropes with a cage-like structure. These structures possess conjugated electronic configurations and display intriguing optical, electrical and electrochemical properties [1–6]. $C_{60}$, the most prominent member of the fullerene family, has been extensively researched for several decades. Its derivatives have found wide-ranging applications in photovoltaic devices and have shown potential for disease treatment [7–10]. Alongside the $C_{60}$ monomer, there has been significant interest in studying the polymerized form of $C_{60}$. Reports suggest that when exposed to photo irradiation or high pressure treatment, $C_{60}$ molecules can polymerize into oligomers and polymers [11,12]. The properties of the $C_{60}$ polymers have been explored through theoretical and experimental studies [3,13–16]. During the polymerization process, the reaction between two neighboring $C_{60}$ molecules typically occurs via a [2 + 2] cycloaddition mechanism, leading to the formation of a cyclobutane ring that connects two fullerene cages. Subsequent to polymerization, the resulting products usually exhibit high insolubility in solvents that can dissolve $C_{60}$ [16]. The structural composition of the products derived from $C_{60}$ sample is determined by the reaction conditions. The formation of polymers and metastable materials occurs when the pressure exceeds several GPa and the temperature surpasses several hundred degrees Celsius. At higher pressures and temperatures, the $C_{60}$ cage can rupture, leading to the production of diamond-like materials [16,17]. Sabirov utilized density-functional theory to examine the polarizability of fullerene oligomers $(C_{60})_n$ (n = 1–5) [18]. The study findings demonstrated that the average polarizability of $(C_{60})_n$ is not linearly correlated with the number of fullerene cores in the molecules, suggesting a violation of additivity. This phenomenon, referred to as the exaltation of polarizability, arises from the interaction between the $\pi$-electronic systems of the fullerene cores in $(C_{60})_n$ and becomes increasingly significant as the distance between them increases. Quantum chemical calculations of fullerene oligomers are expensive because of the large size of the oligomers. Sabirov investigated energy, topology, and information theoretic descriptors, and obtained a ternary energy-topology-symmetry law. The findings indicate that topological indices are effective in identifying the most and least reactive atoms within the fullerene dimer structure [19].

Recently, the synthesis of monolayer and few-layer two-dimensional fullerene networks has been reported [20,21]. In the polymeric $C_{60}$ monolayer, the $C_{60}$ molecules are covalently bonded, showcasing excellent crystallinity and thermodynamic stability. Measurements of the electronic band structure of the polymeric $C_{60}$ monolayer indicate a bandgap of approximately 1.6 eV. The two-dimensional crystalline polymer of $C_{60}$ has been coined "graphullerene" and serves as a bridge between molecular and extended carbon materials. The synthesis of graphullerene involves the growth of single crystals of layered polymeric $Mg_4C_{60}$, followed by the removal of magnesium atoms [20,21]. Theoretical studies reveal that graphullerene is a meta-stable semiconductor with an indirect band gap [22].

Metallofullerenes, as a notable type of fullerene derivatives, may also have the ability to polymerize and form metallofullerene polymers. However, there is currently limited knowledge regarding the polymerization process of metallofullerenes. The aim of this study is to investigate the stability and electronic properties of polymerized metallofullerenes, a previously unexplored area of research. A variety of metal elements can be accommodated within the interior cavity of fullerene cages. Metallofullerene molecules have been extensively studied in terms of their synthesis, structures, and properties [23–33]. The arc-discharge technique is commonly employed for the synthesis of various metallofullerenes. While some metallofullerenes are soluble in organic solvents like toluene and carbon disulfide, others are not. One notable metallofullerene, $Ca@C_{60}$, is soluble in aniline but insoluble in toluene and carbon disulfide [34]. This can be attributed to its reactivity towards polymerization and its lack of solubility in non-polar organic solvents. The solubility in aniline may be due to the formation of a charge-transfer complex. Currently, the understanding of the structure and characteristics of polymerized $Ca@C_{60}$ is very limited. To address this gap, our study focuses on examining the stability and electronic properties of $Ca@C_{60}$ oligomers and polymers using density functional theory (DFT) calculations. We also analyze the interactions between individual $Ca@C_{60}$ monomers.

## 2. Results and Discussion

In a previous study by Kubozono and co-workers, endohedral metallofullerene $Ca@C_{60}$ was prepared and extracted [34]. The preparation involved arc-heating graphite rods containing CaO, followed by extraction with aniline under an air atmosphere. The laser desorption time-of-flight mass spectra of the extracted solutions showed prominent peaks attributed to $Ca@C_{60}$, demonstrating the stability of $Ca@C_{60}$. We conducted DFT calculations on $Ca@I_h{-}C_{60}$. Figure 1a illustrates the DFT-optimized structure of $Ca@C_{60}$. Upon encapsulation of a Ca atom, the molecule's symmetry is reduced from $I_h$ to $C_{2v}$. The Ca atom is positioned adjacent to a carbon-carbon bond, with a distance of 2.45 Å between the Ca atom and the carbon atom in the bond. In some metallofullerenes, the metal atom resides in close proximity to the center of a six-member carbon atom ring, while the coordination occurring at a carbon-carbon bond results in an increase in energy. Regarding Ca@$C_{60}$, it is observed that the Ca atom is situated near a carbon-carbon bond and remains in an off-center position. If the Ca atom were to shift towards the center of the molecule, there would be a noticeable increase in the overall energy.

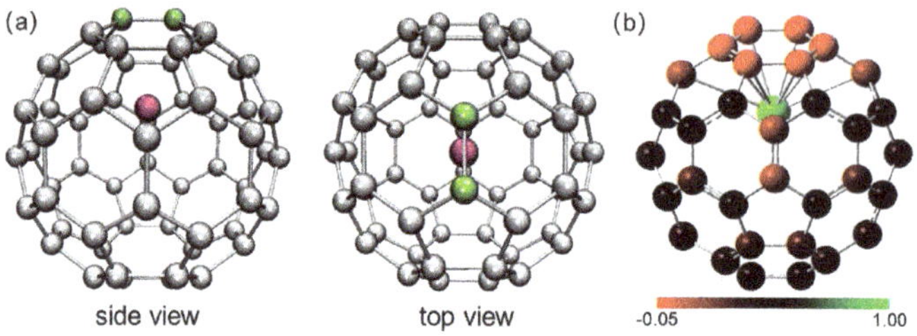


**Figure 1.** (**a**) Two views of the optimized structure of Ca@$C_{60}$. The carbon atoms adjacent to the Ca atom are indicated with a green color. (**b**) The charge distribution on the atoms of Ca@$C_{60}$. The green and red color indicates positive and negative charges, respectively. The black color indicates neutral atoms.

The HOMO-LUMO gap of Ca@$C_{60}$ is calculated to be 1.14 eV. To assess the charge distribution, we conducted a natural population analysis (NPA), revealing a NPA charge of 1.8 for the Ca atom. Considering that the Ca atom possesses two valence electrons, the calculated NPA charge signifies a transfer of two electrons to the fullerene cage. Therefore, the formal charge of Ca atom can be regarded as +2. The charge distribution of Ca@$C_{60}$ is illustrated in Figure 1b. It is evident that the negative charges on the fullerene cage are distributed unevenly. In particular, the carbon atoms in close proximity to the Ca atom exhibit a greater concentration of negative charges.

The $C_{60}$ cage contains two types of C-C bonds. One type corresponds to the hexagon-hexagon ring junction, known as the (6,6) bond, while the other corresponds to the pentagon-hexagon ring junction, known as the (5,6) bond. In our previous research, we investigated the reactivity of the (5,6) and (6,6) bonds in Ca@$C_{60}$ towards Diels-Alder cycloaddition reactions using density functional methods [35]. For the pristine $C_{60}$, the reactivity of the (6,6) bond is much higher than that of the (5,6) bond. We found that for the Diels-Alder mono-addition reaction, the unreactive (5,6) bond in $C_{60}$ can be effectively activated by encapsulating a Ca atom. The enhancement of reactivity is attributed to both electron transfer from the metal atom to the fullerene cage and the interaction between the metal cation and the cage. The functionalization of Ca@$C_{60}$ leads to significant increases in the HOMO-LUMO gap, which can improve the chemical stability of Ca@$C_{60}$. For the formation of oligomers of Ca@$C_{60}$, both the (5,6) and (6,6) bonds could serve as reaction sites. Consequently, various structures of Ca@$C_{60}$ oligomers were considered in this study.

It has been reported that the metallofullerene Li@$C_{60}$ can form dimers in solid states [36]. In the dimer of Li@$C_{60}$, the fullerene cages are connected by a single C-C bond. In our study on the dimer [Ca@$C_{60}$]$_2$, four possible isomers were considered. Isomer 1, referred to as [Ca@$C_{60}$]$_2$-1, consists of two fullerene cages connected by a single C-C bond (Figure 2a). Isomers 2–4, on the other hand, involve the linkage of the fullerene cages through a four-atom ring (Figure 2b–d). Isomers 2–4 can be seen as the result of [2 + 2] cycloaddition reactions between two Ca@$C_{60}$ molecules. Three products were considered from the [2 + 2] cycloaddition reactions between two Ca@$C_{60}$ molecules. The product formed by connecting two (6,6) bonds is labeled as [Ca@$C_{60}$]$_2$-2 (Figure 2b). The product formed by connecting two (5,6) bonds is labeled as [Ca@$C_{60}$]$_2$-3 (Figure 2c). Finally, the product formed by connecting a (6,6) bond and a (5,6) bond is labeled as [Ca@$C_{60}$]$_2$-4 (Figure 2d). Our DFT calculations revealed that [Ca@$C_{60}$]$_2$-3 has the lowest energy, indicating that the dimerization of Ca@$C_{60}$ prefers to occur at the (5,6) bond of the fullerene cage. This situation is different from the $C_{60}$ dimer. In the case of $C_{60}$ dimer, the fullerene cages are connected by cycloaddition reaction between two (6,6) bonds. Once metal atoms are encapsulated, the cage's reactivity is altered due to electron transfer from the metal atom to the fullerene cage. As previously discussed, the unreactive (5,6) bond in $C_{60}$ can be effectively activated by encapsulating a Ca atom in the context of the Diels-Alder mono-addition reaction. It is a common trend that the reactivity of the (5,6) bond towards Diels-Alder reactions is heightened in the case of metallofullerenes compared to empty fullerenes.

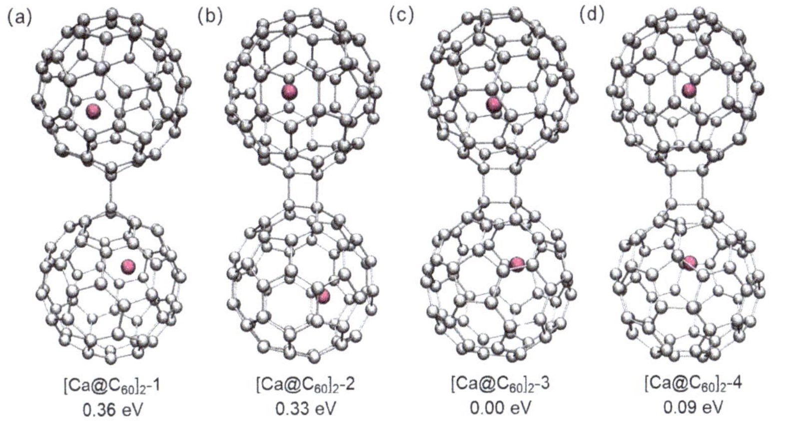


**Figure 2.** The optimized structures of the dimers $[Ca@C_{60}]_2$. The calculated relative energies are shown.

We examined five isomers of the trimers $[Ca@C_{60}]_3$. The first isomer, $[Ca@C_{60}]_3$-1, is formed by linking three $Ca@C_{60}$ molecules with C-C single bonds (Figure 3a). Isomers 2–4 can be regarded as the product of [2 + 2] cycloaddition reactions between three $Ca@C_{60}$ molecules. In $[Ca@C_{60}]_3$-2, only (6,6) bonds are involved in the cycloaddition reactions (Figure 3b), while in $[Ca@C_{60}]_3$-3, only (5,6) bonds are involved (Figure 3c). In $[Ca@C_{60}]_3$-4, the four-atom ring connecting the fullerene cages is formed by a (6,6) bond and a (5,6) bond (Figure 3d). In $[Ca@C_{60}]_3$-5, the three fullerene cages are arranged in a triangular shape and connected by a four-atom ring and two C-C single bonds (Figure 3e). The four-atom ring is formed by cycloaddition reaction between two (5,6) bonds. DFT calculations indicate that $[Ca@C_{60}]_3$-5 has the lowest energy, making the triangular trimer more thermodynamically stable than the linear isomers.

For the tetramers $[Ca@C_{60}]_4$, we considered six isomers. Isomers 1–4, which are linear molecules (Figure 4a–d), have a bonding pattern similar to isomers 1–4 of the trimers. The fifth isomer, $[Ca@C_{60}]_4$-5, has a rectangular shape, with four fullerene cages connected by two four-atom rings and two single bonds (Figure 4e). In the sixth isomer, $[Ca@C_{60}]_4$-6, the fullerene cages form a rhombus shape (Figure 4f). DFT calculations show that $[Ca@C_{60}]_4$-5 and $[Ca@C_{60}]_4$-6 have significantly lower energy compared to the other isomers. The arrangement of the fullerene cages in $[Ca@C_{60}]_4$-6 is similar to the reported two-dimensional monolayer $C_{60}$ polymer "graphullerene" [20–22]. The computed charge distribution of the atoms in $[Ca@C_{60}]_4$-5 and $[Ca@C_{60}]_4$-6 is depicted in Figure 4e,f. It is evident that the negative charges on the fullerene cage are concentrated in close proximity to the Ca atom. Consequently, there exists a robust electrostatic attraction between the negatively charged fullerene cage and the Ca cation.

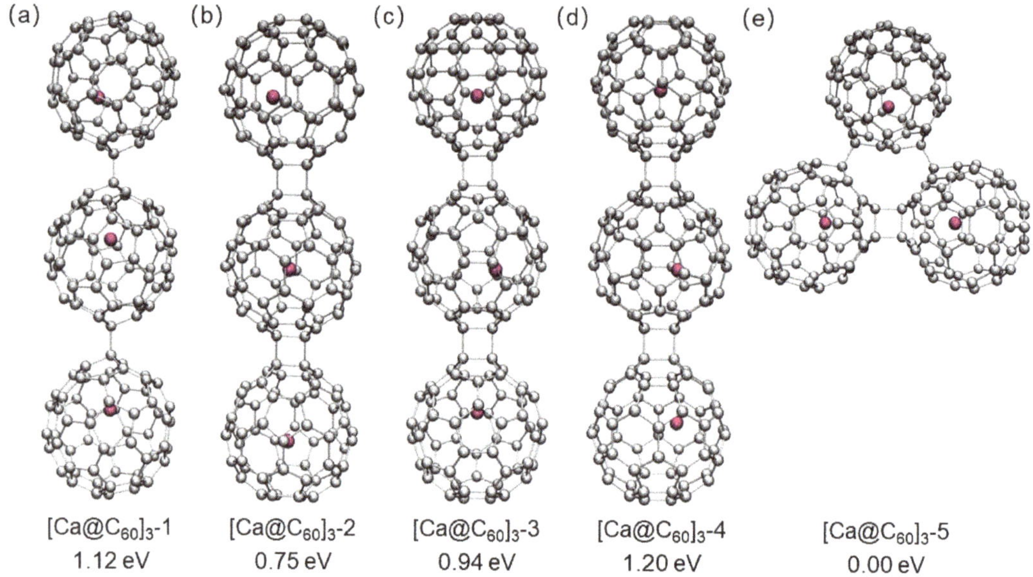


**Figure 3.** The optimized structures of the trimers $[Ca@C_{60}]_3$. The calculated relative energies are shown.

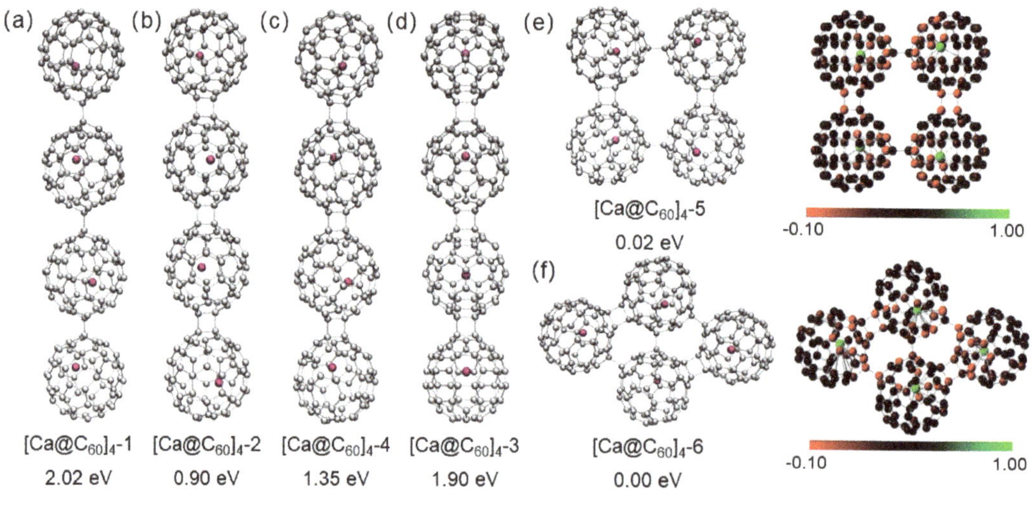


**Figure 4.** The optimized structures of the tetramers $[Ca@C_{60}]_4$. The calculated relative energies are shown.

Figure 5 depicts the molecular orbital diagrams for $Ca@C_{60}$, $[Ca@C_{60}]_2$-3, $[Ca@C_{60}]_3$-5 and $[Ca@C_{60}]_4$-6. DFT calculations reveal a significant alteration in the HOMO-LUMO gap of the oligomers $[Ca@C_{60}]_n$ compared to the monomer $Ca@C_{60}$. The HOMO-LUMO gaps for $Ca@C_{60}$, $[Ca@C_{60}]_2$-3, $[Ca@C_{60}]_3$-5 and $[Ca@C_{60}]_4$-6 are 1.14, 1.36, 0.60 and 0.27 eV, respectively. The dimer $[Ca@C_{60}]_2$-3 shows an enlarged HOMO-LUMO gap in comparison to $Ca@C_{60}$. For the trimer and tetramers, the HOMO-LUMO gap substantially decreases as the number of $Ca@C_{60}$ units increases. The HOMO-LUMO gaps for other isomers of the oligomers are summarized in Table 1. In the case of empty $C_{60}$, the HOMOs and LUMOs are degenerate, with the HOMOs being five-fold degenerate and the LUMOs being triply degenerate. However, when a Ca atom is encapsulated, two electrons occupy the LUMO of $C_{60}$, resulting in the HOMO of Ca@$C_{60}$ having a similar shape to the LUMO of $C_{60}$. Additionally, the LUMO and LUMO+1 of Ca@$C_{60}$, which are derived from the LUMOs of $C_{60}$, are almost degenerate. Similarly, the HOMO-6 to HOMO-1 of Ca@$C_{60}$, which are derived from the five-fold HOMOs of $C_{60}$, have very close energies.

In the case of the dimer [Ca@$C_{60}$]$_2$-3, both the HOMO and LUMO are observed to be spread across two fullerene cages. Additionally, the HOMO is also observed to be present at the C-C bridge that connects the fullerene cages. Moving on to the trimer [Ca@$C_{60}$]$_3$-5, it can be observed that the HOMO and LUMO are primarily distributed at the two fullerene cages that are connected by a four-atom ring. In the case of the tetramers [Ca@$C_{60}$]$_4$-6, the HOMO and LUMO are distributed across the two fullerene cages, which are separated by a significant distance.

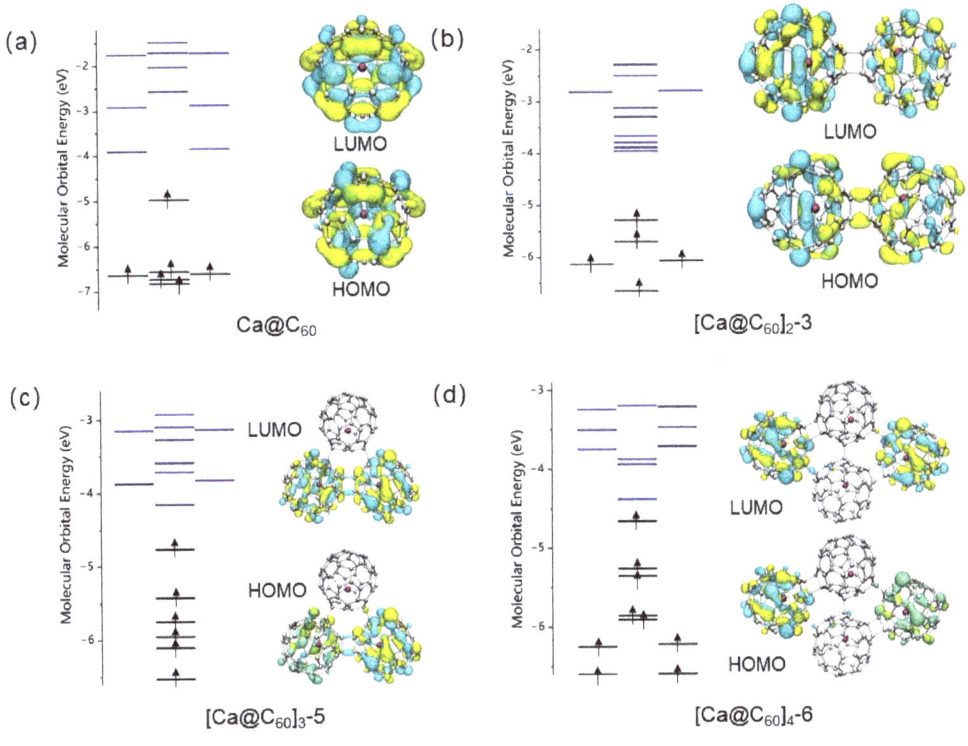


**Figure 5.** The calculated molecular orbital diagrams for (**a**) Ca@$C_{60}$, (**b**) [Ca@$C_{60}$]$_2$-3, (**c**) [Ca@$C_{60}$]$_3$-5 and (**d**) [Ca@$C_{60}$]$_4$-6. The occupied and unoccupied molecular orbitals are denoted in black and blue, respectively.

**Table 1.** The DFT-calculated HOMO-LUMO gaps (in eV) of Ca@$C_{60}$ and its oligomers.

| | **HOMO-LUMO Gap** |
|---|---|
| Ca@$C_{60}$ | 1.14 |
| 2-1 | 0.50 |
| 2-2 | 1.07 |
| 2-3 | 1.36 |
| 2-4 | 0.94 |

**Table 1.** *Cont.*

| | HOMO-LUMO Gap |
|---|---|
| 3-1 | 0.22 |
| 3-2 | 1.02 |
| 3-3 | 1.21 |
| 3-4 | 0.79 |
| 3-5 | 0.60 |
| 4-1 | 0.14 |
| 4-2 | 1.05 |
| 4-3 | 1.14 |
| 4-4 | 0.80 |
| 4-5 | 0.38 |
| 4-6 | 0.27 |

The UV-vis-NIR absorption spectra for Ca@$C_{60}$, [Ca@$C_{60}$]$_2$-3, and [Ca@$C_{60}$]$_3$-5 are presented in Figure 6. It is observed that both Ca@$C_{60}$ and [Ca@$C_{60}$]$_2$-3 exhibit an absorption onset at approximately 1000 nm, with a prominent absorption peak at around 800 nm. Notably, [Ca@$C_{60}$]$_3$-5 displays a broad absorption peak spanning the range of 1400–3000 nm, which can be attributed to its small HOMO-LUMO gap. This characteristic suggests potential applications in the fields of biomedicine and optoelectronic devices. On the other hand, the TDDFT calculations for [Ca@$C_{60}$]$_4$-6 were unsuccessful due to the high computational cost.

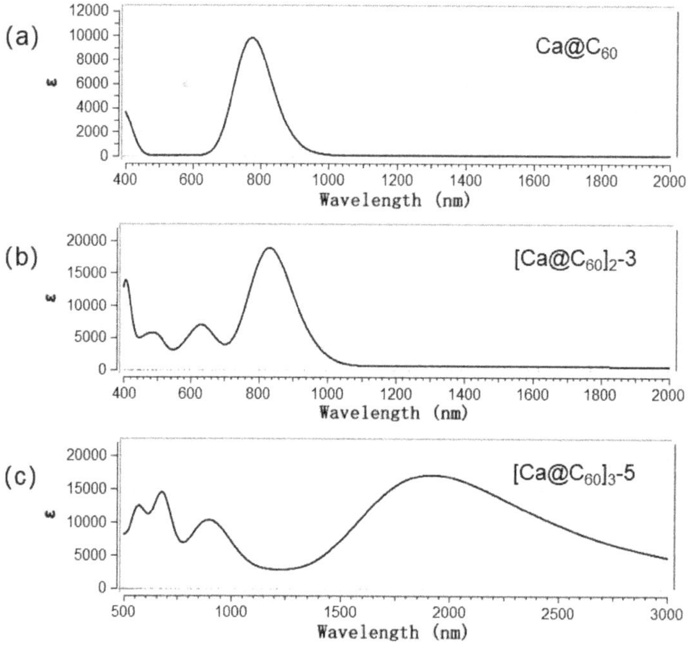


**Figure 6.** The calculated UV-vis-NIR absorption spectra for (**a**) Ca@$C_{60}$, (**b**) [Ca@$C_{60}$]$_2$-3 and (**c**) [Ca@$C_{60}$]$_3$-5 by using the TDDFT method.

On the basis of DFT calculations, we compared the energy between the oligomers [Ca@$C_{60}$]$_n$ and the monomer Ca@$C_{60}$. The results indicate that the formation of oligomers is an exothermic process. We calculated the relative energies for [Ca@$C_{60}$]$_n$ per Ca@$C_{60}$ unit using the monomer Ca@$C_{60}$ as a reference. The calculated values for [Ca@$C_{60}$]$_2$-3,

$[Ca@C_{60}]_3$-5,$[Ca@C_{60}]_4$-5 and $[Ca@C_{60}]_4$-6 per $Ca@C_{60}$ unit are −0.54, −0.67, −0.69 and −0.70 eV, respectively, revealing significantly lower energies compared to the monomer $Ca@C_{60}$. Remarkably, the relative energy decreases with an increase in the number of $Ca@C_{60}$ units. Consequently, the process of polymerizing $Ca@C_{60}$ exhibits a favorable energy profile.

The stability of the oligomers $[Ca@C_{60}]_n$ can be attributed to two factors. Firstly, the covalent bonding between the fullerene cages plays a crucial role. Secondly, the van der Waals interaction between the fullerene cages also contributes to the stability. To illustrate this, we examined the dimer $[Ca@C_{60}]_2$-3 as a representative compound and calculated the dispersion energy between the two fullerene units. This was achieved by performing energy decomposition using dispersion corrected density functional theory (DFT-D3). The calculated dispersion energy was found to be 0.49 eV, indicating that the dispersion energy significantly contributes to the dimer formation. To visualize the weak interactions between the fullerene cages, we conducted an analysis using the Independent Gradient Model based on Hirshfeld partition (IGMH) for $[Ca@C_{60}]_2$-3, $[Ca@C_{60}]_3$-5, and $[Ca@C_{60}]_4$-6. The results are presented in Figure 7. The isosurface of $\delta g$ indicates the region where dispersion interaction is present. It is evident that dispersion interaction exists between the fullerene cages.

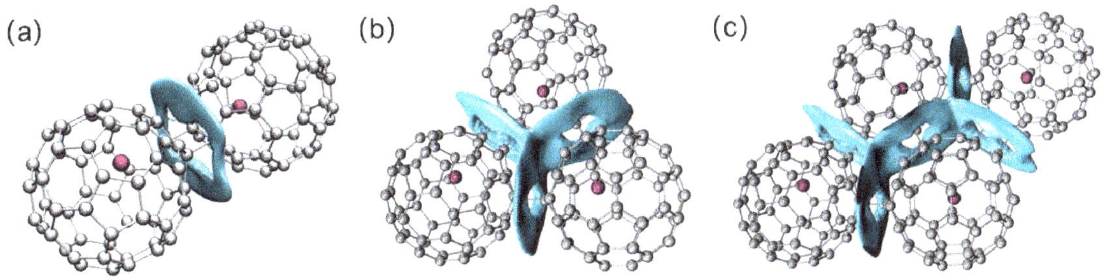


**Figure 7.** IGMH-$\delta g$ isosurface (isovalue = 0.001 a.u.) showing the dispersion interaction between the $Ca@C_{60}$ fragments for (**a**) $[Ca@C_{60}]_2$-3, (**b**) $[Ca@C_{60}]_3$-5 and (**c**) $[Ca@C_{60}]_4$-6.

For the two-dimensional polymers $[Ca@C_{60}]_\infty$, we considered two possible structures. The first one, $[Ca@C_{60}]_\infty$-1, is constructed by extension of the $Ca@C_{60}$ units in $[Ca@C_{60}]_4$-6. The second one, $[Ca@C_{60}]_\infty$-2, is constructed by extension of the $Ca@C_{60}$ units in $[Ca@C_{60}]_4$-5. Figure 8 displays the optimized structures of these polymers. In $[Ca@C_{60}]_\infty$-1, each fullerene cage is connected to six adjacent cages, resembling a "graphullerene" structure [20–22]. In $[Ca@C_{60}]_\infty$-2, each fullerene cage connects with four adjacent cages. DFT calculations indicate that $[Ca@C_{60}]_\infty$-1 is thermodynamically more stable than $[Ca@C_{60}]_\infty$-2. The energy difference for the unit cell containing two $Ca@C_{60}$ units is 1.08 eV. Figure 9 shows the band structure for $[Ca@C_{60}]_\infty$-1, with a direct bandgap of 0.22 eV. Notably, this bandgap is significantly smaller in comparison to that of "graphullerene" [20,22]. This reduction in bandgap can be attributed to the transfer of electrons from Ca atoms to the fullerene cage. As a result, a portion of the conduction band of "graphullerene" is occupied by these transferred electrons. Consequently, the polymer $[Ca@C_{60}]_\infty$-1 exhibits distinct electronic characteristics in comparison to "graphullerene". The band structure of $[Ca@C_{60}]_\infty$-1 reveals that the wave vectors corresponding to the Conduction band minimum (CBM) and Valance band maximum (VBM) are both situated at the high symmetry point G in the Brillouin zone. The electron transition mode displays vertical transitions, while the band structure itself demonstrates direct bandgap characteristics. Notably, since the CBM and VBM coincide at the same momentum point, the transition from the valence band to the conduction band in $[Ca@C_{60}]_\infty$-1, upon absorption of photons by electrons, only requires adherence to the energy conservation constraint, without necessitating additional phonon interaction. In comparison with indirect bandgap materials, $[Ca@C_{60}]_\infty$-1 exhibits superior quantum efficiency, rendering it a promising contender for optoelectronic devices.

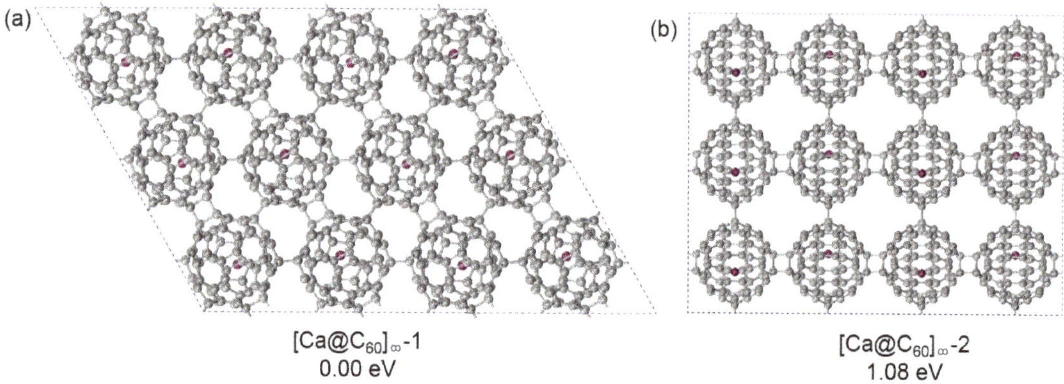


**Figure 8.** The optimized structures of $[Ca@C_{60}]_\infty$-1 and $[Ca@C_{60}]_\infty$-2. The calculated relative energies for a unit cell are shown.

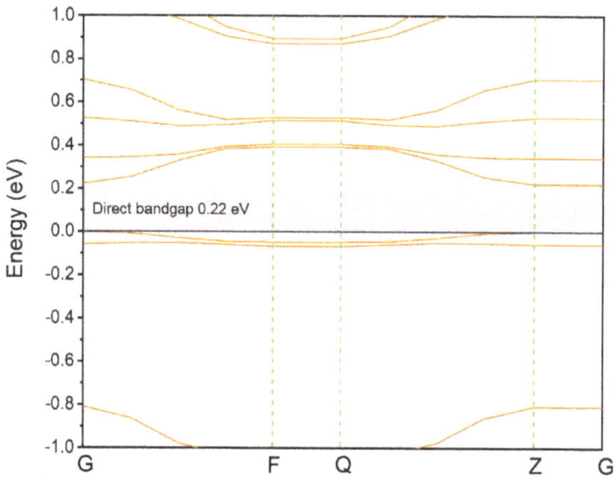


**Figure 9.** The calculated band structure of $[Ca@C_{60}]_\infty$-1.

## 3. Computational Details

DFT calculations were conducted on $Ca@C_{60}$ and its oligomers and polymers. The oligomers included the dimer$[Ca@C_{60}]_2$, the trimer $[Ca@C_{60}]_3$ and the tetramer $[Ca@C_{60}]_4$. The polymers were referred to as $[Ca@C_{60}]_\infty$. Gaussian16 version A03 [37] was used for the calculations on the oligomers, with the PBE0 functional employed [38]. To account for the significant van der Waals interaction between the fullerene cages, the dispersion correction D3(BJ) was applied [39]. The geometric optimization utilized the def2SVP basis set for the carbon and calcium atoms [40], while the def2-TZVP basis set was used for single-point energy calculations [41]. The counterpoise method of Boys and Bernardi was employed to correct the basis set superposition error (BSSE) [42]. The interaction between $Ca@C_{60}$ monomers was investigated using the MULTIWFN program [43], specifically the analysis of independent gradient model based on Hirshfeld partition (IGMH) [44,45]. The visualization of the calculation results was achieved using the Visual Molecular Dynamics (VMD, version 1.9.3) software [46].

We utilized the Dmol3 package [47,48] to perform calculations on the polymers $[Ca@C_{60}]_\infty$ within periodic models. The generalized gradient approximation (GGA) with the Perdew-Burke-Ernzerhof (PBE) functional [49] was applied to simulate the electron exchange correlation interaction. Electron-ion interactions were described using the DFT Semi-core Pseudopotentials (DSPP) [50]. The wave function is expanded by Double Numerical plus polarization (DNP) basis set, and the Brillouin zone is used by $2 \times 4 \times 1$ K-point grid for $[Ca@C_{60}]_\infty-1$, $4 \times 2 \times 1$ for $[Ca@C_{60}]_\infty-2$. The van der Waals interaction was described using the Tkatchenko-Scheffler method [51]. In order to prevent interlayer interaction between the same research objects, a 15 Å vacuum layer was applied.

## 4. Conclusions

We conducted a comprehensive investigation into the stability and electronic properties of oligomers and polymers of $Ca@C_{60}$ through DFT calculations. In the most stable $Ca@C_{60}$ dimer, the fullerene cages are connected through a four-atom ring, with each fullerene cage utilizing a (5,6) bond to form the four-atom ring. The trimer of $Ca@C_{60}$ exhibits a most stable isomer with a triangular structure. For $Ca@C_{60}$ tetramers, two thermodynamically stable isomers exist—one with a rectangular shape and the other with a rhombus shape. Our findings indicate a decreasing HOMO-LUMO gap as the number of $Ca@C_{60}$ units increases to three and four in the oligomers. Furthermore, our study reveals that the oligomer formation is an exothermic process, with van der Waals interactions between fullerene cages playing a crucial role. In the case of the two-dimensional polymer of $Ca@C_{60}$, we observe structural similarities with graphullerene. This polymer functions as a semiconductor, featuring a direct bandgap of 0.22 eV. Given the solubility of the $Ca@C_{60}$ monomer in aniline, future experimental studies on the polymerization of $Ca@C_{60}$ are deemed feasible.

**Author Contributions:** Conceptualization, Z.W.; methodology, Y.W. and Z.Z.; software, Y.W. and Z.Z.; validation, Y.W. and Z.Z.; formal analysis, Y.W. and Z.Z.; investigation, Y.W. and Z.Z.; resources, Z.W.; data curation, Z.Z. and Z.W.; writing—original draft preparation, Z.W.; writing—review and editing, Z.W.; visualization, Y.W.; supervision, Z.W.; project administration, Z.W.; funding acquisition, Z.W. All authors have read and agreed to the published version of the manuscript.

**Funding:** This work was supported by the Beijing Municipal Natural Science Foundation (grant no. 2212030) and the National Natural Science Foundation of China (grant no. 22175199).

**Institutional Review Board Statement:** Not applicable.

**Informed Consent Statement:** Not applicable.

**Data Availability Statement:** The data presented in this study are available in this article.

**Acknowledgments:** We are grateful for the support of the Public Computing Cloud at Renmin University of China.

**Conflicts of Interest:** The authors declare no conflict of interest.

## References

1. Lindle, J.; Pong, R.; Bartoli, F.; Kafafi, Z. Nonlinear optical properties of the fullerenes $C_{60}$ and $C_{70}$ at 1.064 µm. *Phys. Rev. B* **1993**, *48*, 9447. [CrossRef] [PubMed]
2. Sun, Y.P.; Lawson, G.E.; Riggs, J.E.; Ma, B.; Wang, N.; Moton, D.K. Photophysical and nonlinear optical properties of [60] fullerene derivatives. *J. Phys. Chem. A* **1998**, *102*, 5520–5528. [CrossRef]
3. Makarova, T. Electrical and optical properties of pristine and polymerized fullerenes. *Semiconductors* **2001**, *35*, 243–278. [CrossRef]
4. Sachdeva, S.; Singh, D.; Tripathi, S. Optical and electrical properties of fullerene $C_{70}$ for solar cell applications. *Opt. Mater.* **2020**, *101*, 109717. [CrossRef]
5. Guo, K.; Li, N.; Bao, L.; Lu, X. Fullerenes and derivatives as electrocatalysts: Promises and challenges. *Green Energy Environ.* **2022**, *9*, 7–27. [CrossRef]
6. Huang, C.; Yang, Y.; Li, M.; Qi, X.; Pan, C.; Guo, K.; Bao, L.; Lu, X. Ultrahigh Capacity from Complexation-Enabled Aluminum-Ion Batteries with $C_{70}$ as the Cathode. *Adv. Mater.* **2023**, 2306244. . [CrossRef]
7. Li, C.Z.; Yip, H.L.; Jen, A.K.Y. Functional fullerenes for organic photovoltaics. *J. Mater. Chem.* **2012**, *22*, 4161–4177. [CrossRef]

8. Collavini, S.; Delgado, J.L. Fullerenes: The stars of photovoltaics. *Sustain. Energy Fuels* **2018**, *2*, 2480–2493. [CrossRef]
9. Castro, E.; Garcia, A.H.; Zavala, G.; Echegoyen, L. Fullerenes in biology and medicine. *J. Mater. Chem. B* **2017**, *5*, 6523–6535. [CrossRef]
10. Su, S.; Zhen, M.; Zhou, C.; Cao, X.; Sun, Z.; Xu, Y.; Li, L.; Jia, W.; Wu, Z.; Wang, C. Efficiently Inhibiting Systemic Inflammatory Cascades by Fullerenes for Retarding HFD-Fueled Atherosclerosis. *Adv. Healthc. Mater.* **2023**, *12*, 2202161. [CrossRef]
11. Burger, B.; Winter, J.; Kuzmany, H. Dimer and cluster formation in $C_{60}$ photoreaction. *Z. Phys. B Condens. Matter* **1996**, *101*, 227–233. [CrossRef]
12. Blank, V.D.; Buga, S.G.; Dubitsky, G.A.; Serebryanaya, N.R.; Popov, M.Y.; Sundqvist, B. High-pressure polymerized phases of $C_{60}$. *Carbon* **1998**, *36*, 319–343. [CrossRef]
13. Okada, S.; Saito, S. Electronic structure and energetics of pressure-induced two-dimensional $C_{60}$ polymers. *Phys. Rev. B* **1999**, *59*, 1930. [CrossRef]
14. Xu, C.H.; Scuseria, G.E. Theoretical predictions for a two-dimensional rhombohedral phase of solid $C_{60}$. *Phys. Rev. Lett.* **1995**, *74*, 274. [CrossRef] [PubMed]
15. Rao, A.; Eklund, P.; Venkateswaran, U.; Tucker, J.; Duncan, M.; Bendele, G.; Stephens, P.; Hodeau, J.L.; Marques, L.; Nunez-Regueiro, M.; et al. Properties of $C_{60}$ polymerized under high pressure and temperature. *Appl. Phys. A* **1997**, *64*, 231–2239. [CrossRef]
16. Giacalone, F.; Martin, N. Fullerene polymers: Synthesis and properties. *Chem. Rev.* **2006**, *106*, 5136–5190. [CrossRef]
17. Blank, V.; Buga, S.; Serebryanaya, N.; Denisov, V.; Dubitsky, G.; Ivlev, A.; Mavrin, B.; Popov, M.Y. Ultrahard and superhard carbon phases produced from $C_{60}$ by heating at high pressure: Structural and Raman studies. *Phys. Lett. A* **1995**, *205*, 208–216. [CrossRef]
18. Sabirov, D.S. Polarizability of $C_{60}$ fullerene dimer and oligomers: The unexpected enhancement and its use for rational design of fullerene-based nanostructures with adjustable properties. *RSC Adv.* **2013**, *3*, 19430–19439. [CrossRef]
19. Sabirov, D.S.; Ori, O.; Tukhbatullina, A.A.; Shepelevich, I.S. Covalently bonded fullerene nano-aggregates $(C_{60})n$: Digitalizing their energy–topology–symmetry. *Symmetry* **2021**, *13*, 1899. [CrossRef]
20. Hou, L.; Cui, X.; Guan, B.; Wang, S.; Li, R.; Liu, Y.; Zhu, D.; Zheng, J. Synthesis of a monolayer fullerene network. *Nature* **2022**, *606*, 507–510. [CrossRef]
21. Meirzadeh, E.; Evans, A.M.; Rezaee, M.; Milich, M.; Dionne, C.J.; Darlington, T.P.; Bao, S.T.; Bartholomew, A.K.; Handa, T.; Rizzo, D.J.; et al. A few-layer covalent network of fullerenes. *Nature* **2023**, *613*, 71–76. [CrossRef]
22. Argaman, U.; Makov, G. Structure and properties of graphullerene: A semiconducting two-dimensional $C_{60}$ crystal. *Npj Comput. Mater.* **2023**, *9*, 211. [CrossRef]
23. Shinohara, H. Endohedral metallofullerenes. *Rep. Prog. Phys.* **2000**, *63*, 843. [CrossRef]
24. Chaur, M.N.; Melin, F.; Ortiz, A.L.; Echegoyen, L. Chemical, electrochemical, and structural properties of endohedral metallofullerenes. *Angew. Chem. Int. Ed.* **2009**, *48*, 7514–7538. [CrossRef] [PubMed]
25. Cong, H.; Yu, B.; Akasaka, T.; Lu, X. Endohedral metallofullerenes: An unconventional core–shell coordination union. *Coord. Chem. Rev.* **2013**, *257*, 2880–2898. [CrossRef]
26. Popov, A.A.; Yang, S.; Dunsch, L. Endohedral fullerenes. *Chem. Rev.* **2013**, *113*, 5989–6113. [CrossRef] [PubMed]
27. Wang, T.; Wang, C. Endohedral Metallofullerenes Based on Spherical $I_h$-$C_{80}$ Cage: Molecular Structures and Paramagnetic Properties. *Acc. Chem. Res.* **2014**, *47*, 450–458. [CrossRef] [PubMed]
28. Yang, S.; Wei, T.; Jin, F. When metal clusters meet carbon cages: Endohedral clusterfullerenes. *Chem. Soc. Rev.* **2017**, *46*, 5005–5058. [CrossRef] [PubMed]
29. Jin, P.; Li, Y.; Magagula, S.; Chen, Z. Exohedral functionalization of endohedral metallofullerenes: Interplay between inside and outside. *Coord. Chem. Rev.* **2019**, *388*, 406–439. [CrossRef]
30. Shen, W.; Hu, S.; Lu, X. Endohedral metallofullerenes: New structures and unseen phenomena. *Chem.- Eur. J.* **2020**, *26*, 5748–5757. [CrossRef]
31. Li, M.; Zhao, R.; Dang, J.; Zhao, X. Theoretical study on the stabilities, electronic structures, and reaction and formation mechanisms of fullerenes and endohedral metallofullerenes. *Coord. Chem. Rev.* **2022**, *471*, 214762. [CrossRef]
32. Shen, W.; Bao, L.; Lu, X. Endohedral Metallofullerenes: An Ideal Platform of Sub-Nano Chemistry. *Chin. J. Chem.* **2022**, *40*, 275–284. [CrossRef]
33. Hu, Z.; Wang, Y.; Ullah, A.; Gutiérrez-Finol, G.M.; Bedoya-Pinto, A.; Gargiani, P.; Shi, D.; Yang, S.; Shi, Z.; Gaita-Ariño, A.; et al. High-temperature magnetic blocking in a monometallic dysprosium azafullerene single-molecule magnet. *Chem* **2023**, *9*, 3613–3622. [CrossRef]
34. Kubozono, Y.; Noto, T.; Ohta, T.; Maeda, H.; Kashino, S.; Emura, S.; Ukita, S.; Sogabe, T. Extractions of Ca@$C_{60}$ and Sr@$C_{60}$ with aniline. *Chem. Lett.* **1996**, *25*, 453–454. [CrossRef]
35. Wu, Y.; Jiang, Y.; Deng, J.; Wang, Z. Capturing unconventional metallofullerene M@C60 through activation of the unreactive [5,6] bond toward Diels–Alder reaction. *Phys. Chem. Chem. Phys.* **2020**, *22*, 24249–24256. [CrossRef] [PubMed]
36. Ueno, H.; Aoyagi, S.; Yamazaki, Y.; Ohkubo, K.; Ikuma, N.; Okada, H.; Kato, T.; Matsuo, Y.; Fukuzumi S.; Kokubo K. Electrochemical reduction of cationic $Li^+$@$C_{60}$ to neutral Li+@$C_{60}^{\cdot -}$: isolation and characterisation of endohedral [60]fulleride. *Chem. Sci.* **2016**, *7*, 5770–5774. [CrossRef] [PubMed]
37. Frisch, M.J.; Trucks, G.W.; Schlegel, H.B.; Scuseria, G.E.; Robb, M.A.; Cheeseman, J.R.; Scalmani, G.; Barone, V.; Petersson, G.A.; Nakatsuji, H.; et al. *Gaussian 16*; Revision A.03; Gaussian Inc.: Wallingford, CT, USA, 2016.

38. Adamo, C.; Barone, V. Toward reliable density functional methods without adjustable parameters: The PBE0 model. *J. Chem. Phys.* **1999**, *110*, 6158–6170. [CrossRef]
39. Grimme, S.; Ehrlich, S.; Goerigk, L. Effect of the damping function in dispersion corrected density functional theory. *J. Comput. Chem.* **2011**, *32*, 1456–1465. [CrossRef]
40. Weigend, F.; Ahlrichs, R. Balanced basis sets of split valence, triple zeta valence and quadruple zeta valence quality for H to Rn: Design and assessment of accuracy. *Phys. Chem. Chem. Phys.* **2005**, *7*, 3297–3305. [CrossRef]
41. Schäfer, A.; Huber, C.; Ahlrichs, R. Fully Optimized Contracted Gaussian Basis Sets of Triple Zeta Valence Quality for Atoms Li to Kr. *J. Chem. Phys.* **1994**, *100*, 5829–5835. [CrossRef]
42. Boys, S.F.; Bernardi, F. The calculation of small molecular interactions by the differences of separate total energies. Some procedures with reduced errors. *Mol. Phys.* **1970**, *19*, 553–566. [CrossRef]
43. Lu, T.; Chen, F. Multiwfn: A multifunctional wavefunction analyzer. *J. Comput. Chem.* **2012**, *33*, 580–592. [CrossRef]
44. Lefebvre, C.; Rubez, G.; Khartabil, H.; Boisson, J.C.; Contreras-García, J.; Hénon, E. Accurately extracting the signature of intermolecular interactions present in the NCI plot of the reduced density gradient versus electron density. *Phys. Chem. Chem. Phys.* **2017**, *19*, 17928–17936. [CrossRef] [PubMed]
45. Lu, T.; Chen, Q. Independent gradient model based on Hirshfeld partition: A new method for visual study of interactions in chemical systems. *J. Comput. Chem.* **2022**, *43*, 539–555. [CrossRef] [PubMed]
46. Humphrey, W.; Dalke, A.; Schulten, K. VMD: Visual Molecular Dynamics. *J. Mol. Graph.* **1996**, *14*, 33–38. [CrossRef]
47. Delley, B. An all-electron numerical method for solving the local density functional for polyatomic molecules. *J. Chem. Phys.* **1990**, *92*, 508–517. [CrossRef]
48. Delley, B. From molecules to solids with the DMol 3 approach. *J. Chem. Phys.* **2000**, *113*, 7756–7764. [CrossRef]
49. Perdew, J.P.; Burke, K.; Ernzerhof, M. Generalized gradient approximation made simple. *Phys. Rev. Lett.* **1996**, *77*, 3865. [CrossRef]
50. Hamann, D.; Schlüter, M.; Chiang, C. Norm-conserving pseudopotentials. *Phys. Rev. Lett.* **1979**, *43*, 1494. [CrossRef]
51. Tkatchenko, A.; Scheffler, M. Accurate Molecular Van Der Waals Interactions from Ground-State Electron Density and Free-Atom Reference Data. *Phys. Rev. Lett.* **2009**, *102*, 073005. [CrossRef]

inorganics

MDPI

Article

# Geometries, Electronic Structures, Bonding Properties, and Stability Strategy of Endohedral Metallofullerenes $TM@C_{28}$ (TM = $Sc^-$, $Y^-$, $La^-$, Ti, Zr, Hf, $V^+$, $Nb^+$, $Ta^+$)

**Dong Liu, Yuan Shui and Tao Yang ***

MOE Key Laboratory for Non-Equilibrium Synthesis and Modulation of Condensed Matter, School of Physics, Xi'an Jiaotong University, Xi'an 710049, China; dliu_sd@stu.xjtu.edu.cn (D.L.); f24270216f@stu.xjtu.edu.cn (Y.S.)
* Correspondence: taoyang1@xjtu.edu.cn

**Abstract:** We performed quantum chemical calculations on the geometries, electronic structures, bonding properties, and stability strategy of endohedral metallofullerenes $TM@C_{28}$ (TM = $Sc^-$, $Y^-$, $La^-$, Ti, Zr, Hf, $V^+$, $Nb^+$, $Ta^+$). Our calculations revealed that there are three different lowest-energy structures with $C_{2v}$, $C_{3v}$, and $T_d$ symmetries for $TM@C_{28}$. The HOMO–LUMO gap of all these structures ranges from 1.35 eV to 2.31 eV, in which $[V@C_{28}]^+$ has the lowest HOMO–LUMO gap of 1.35 eV. The molecular orbitals are mainly composed of fullerene cage orbitals and slightly encapsulated metal orbitals. The bonding analysis on the metal–cage interactions reveals they are dominated by the Coulomb term $\Delta E_{elstat}$ and the orbital interaction term $\Delta E_{orb}$, in which the orbital interaction term $\Delta E_{orb}$ contributes more than the Coulomb term $\Delta E_{elstat}$. The addition of one or two $CF_3$ groups to $[V@C_{28}]^+$ could increase the HOMO–LUMO gap and further increase the stability of $[V@C_{28}]^+$.

**Keywords:** fullerenes; quantum-chemical calculations; chemical bonding; stability

## 1. Introduction

Fullerenes, structurally unique closed hollow carbon cages, are composed of exactly 12 pentagons and a certain number of hexagons. Among them, buckminsterfullerene $C_{60}$ stands out as the most symmetrical and chemically inert member [1–3]. The smallest fullerene, $C_{20}$, presents a singular dodecahedral structure but is theoretically predicted to be highly reactive [4]. The reason why these small fullerenes are highly reactive is their high curvature. Thus, there are two main strategies to stabilize these fullerenes. On the one hand, there is the encapsulation of metal inside fullerenes, which gives rise to endohedral metallofullerenes (EMFs), which could lead to a charge transfer from metal to fullerenes and stabilize the high curvature motifs [5]. On the other hand, exohedral derivatization on high curvature motifs could release the curvature directly and make them transform from $sp^2$ hybridization to $sp^3$ hybridization [6,7].

EMFs, which encapsulate metal atoms or clusters within fullerene cages [5,8,9], represent an innovative class of metal–carbon clusters. These EMFs combine the structural merits of fullerene frameworks with the photoelectromagnetic advantages of metals. Particularly, $C_{28}$ has been observed to exhibit one remarkable peak in the domain of supersonic carbon cluster beam experiments [4,10]. However, the isolation of $C_{28}$ has been challenging due to its open-shell electronic nature. EMFs not only improve the stability of reactive fullerenes such as $C_{28}$ [11,12] but also introduce novel functionalities. For instance, the encapsulation of tetravalent metals like titanium (Ti), zirconium (Zr), and uranium (U) culminates in the formation of stable endohedral metallofullerenes, namely $Ti@C_{28}$, $Zr@C_{28}$, and $U@C_{28}$ [4,13–17]. $U^{2+}@C_{28}$ has been theoretically predicted to adhere to the 32-electron principle [18], of which central f-block metals like uranium have their s-, p-, d-, and f-type valence shells completely filled and exhibit remarkable stability. $U@C_{28}$ also follows the 32-electron rule and accommodates two valence electrons within the peripheral cage structure [19].

The synthesis of EMFs involves various experimental methods [20], including electric arc discharge [21], which generates EMFs by vaporizing graphite rods and metal oxides in an inert atmosphere, and laser ablation [22], where a high-power laser vaporizes a carbon-rich target and metal oxides to facilitate EMF formation. Chemical vapor deposition (CVD) methods [23,24], involving the thermal decomposition of hydrocarbon gases on metal catalysts, contribute to fullerene production. Additionally, the surgery approach [25] is often employed to open fullerene cages, put transition metals inside, and close fullerene cages.

The presence of a fullerene can also modify the physical and chemical properties of encapsulated atoms. Recently, one theoretical study showed that in $^{7}Be@C_{28}$, the electron capture decay rate of $^{7}Be$ could become slower because of the metal–cage interactions [26]. Wang and Gao revealed that a hyperhydrogenated water species, $H_4O$, can be easily formed using $H_2$ and $H_2O$ if they are put inside small-sized fullerenes like $C_{28}$ as nanoreactors [27]. Spano and coworkers theoretically studied the $Cr@C_{28}$ and found there are two stable sites of Cr metal in $C_{28}$, leading to two different measurable currents in single-molecule devices [28]. The bonding, electron affinity, optical, and magnetic properties of $M@C_{28}$ (M = Ti, Zr, Hf, U) have been investigated theoretically [29]. However, the element effect on the geometries and electronic structures is rarely discussed. Moreover, the chemical bonding between the encapsulated metal and fullerene cage $C_{28}$ has been explored.

In the present study, we studied the geometries, electronic structures, bonding properties, and stability strategy of endohedral metallofullerenes $TM@C_{28}$ (TM = $Sc^-$, $Y^-$, $La^-$, Ti, Zr, Hf, $V^+$, $Nb^+$, $Ta^+$) by using quantum chemical calculations. The chemical bonding between metal and $C_{28}$ was studied using the energy decomposition analysis. The stability strategy of the highly reactive EMFs is discussed using trifluoromethyl groups.

## 2. Computational Methods

The structural search was started from two fullerene cages, including $C_{28}$-$D_2$(1) and $C_{28}$-$T_d$(2). The positions of transition metal atoms were generated randomly within a geometric sphere centered on the fullerene cages and with a radius smaller than that of the fullerene cage. In total, over 1000 initial structures were randomly generated. Clustering of these structures was conducted using the ISOSTAT module within the Molclus program [30] with two criteria: (a) an energy difference of less than 0.25 kcal/mol and (b) a geometric deviation of less than 0.1 Å. The geometric deviation was defined by calculating a distance matrix based on the atomic coordinates of the structures and converting its nondiagonal elements into a one-dimensional array. The calculations start from the semiempirical GFN-xTB method [31,32] and follow density functional theory (DFT) calculations at the PBE-D3(BJ)/def2-SVP level [33]. The final optimizations were performed at the PBE-D3(BJ)/def2-TZVPPD level [34–41]. The wave function stability and frequency analysis were subsequently verified, confirming these structures as local minima on the potential energy surface. The atomic charge distribution was computed using natural bond orbital (NBO7.0) [42] at the PBE-D3(BJ)/def2-TZVPPD level to gain electron distribution and bonding interactions. NBO analysis offers a comprehensive view of molecular bonding by translating electron densities into more chemically intuitive patterns like bonds and lone pairs, thus aiding in the understanding of intrinsic molecular interactions. All these DFT calculations were performed using Gaussian 16 software [43].

The energy decomposition analysis (EDA) developed by Ziegler and Rauk [44] is one of the most powerful methods to study the nature of chemical bonds. Here, the bonding nature between the metal dimer and fullerene cage was studied via EDA. Scalar relativistic effects and zeroth-order regular approximation (ZORA) [45] were considered. In EDA calculations, the molecule AB divides into two fragments, A and B, and the kernel is put on the intrinsic interaction between fragments that hold a specified electronic reference state in the frozen geometry [46,47]. The EDA approach decomposes the instantaneous interaction energy $\Delta E_{int}$ of the A–B bond into four energy terms, of which the first three are the main components:

$$\Delta E_{int} = \Delta E_{elstat} + \Delta E_{Pauli} + \Delta E_{orb} + \Delta E_{disp} \quad (1)$$

The term $\Delta E_{elstat}$ is the *quasi*-classical electrostatic interaction between the unperturbed charge distributions of the prepared atoms and is usually an attractive term. The Pauli repulsion $\Delta E_{Pauli}$ is the energy change in the transformation from the superposition of the unperturbed electron densities $\rho_A + \rho_B$ of the isolated fragments to the wave function $\Psi^0 = N\hat{A}[\Psi_A \Psi_B]$, which properly obeys the Pauli principle through explicit antisymmetrization ($\hat{A}$ operator) and renormalization (N = constant) of the product wavefunction. $\Delta E_{Pauli}$ comprises the destabilizing interactions between electrons of the same spin on either fragment. The orbital interaction $\Delta E_{orb}$ derives from charge transfer, electron-pair bonding, and polarization effects. The EDA calculations were performed with the ADF2019 [48,49] package using the BP86 [50,51] functional with Grimme's dispersion correction D3(BJ) [33] and TZ2P+ basis set [52]. The zeroth-order regular approximation (ZORA) Hamiltonian was adopted to include the relativistic effect [45]. All these EDA calculations were performed based on the PBE-D3(BJ)/def2-TZVPPD-optimized structures.

## 3. Results and Discussion

There are two isomers of $C_{28}$, including $D_2(1)$ and $T_d(2)$. Our calculations show that all these metals prefer to be encapsulated inside the $T_d(2)$, giving rise to TM@$T_d(2)$-$C_{28}$. Figure 1 presents the calculated structural models of optimized TM@$T_d(2)$-$C_{28}$ (TM = $Sc^-$, $Y^-$, $La^-$, Ti, Zr, Hf, $V^+$, $Nb^+$, $Ta^+$). All these structures have a singlet as the ground state. However, the optimized structures are quite different for the different metals and slightly depend on the charge state. In the case of $[Y@C_{28}]^-$, $[La@C_{28}]^-$, and $[Hf@C_{28}]$, the metal could stay in the center of the fullerene cage, leading to the $T_d$ symmetry of the whole endohedral metallofullerenes. The metal in TM@$C_{28}$ (TM = $Sc^-$, Ti, $V^+$, $Ta^+$) is positioned off-center and close to the carbon atom along the $C_3$ axis, giving rise to the $C_{3v}$ symmetry of this EMF. In contrast, $[Zr@C_{28}]$ and $[Nb@C_{28}]^+$ exhibit the lowest symmetry ($C_{2v}$) due to the off-center positioning of the metal. Despite the Zr and Ti metals having the same d-electron configuration, Zr@$C_{28}$ ($C_{2v}$) and Ti@$C_{28}$ ($C_{3v}$) exhibit different symmetries and structures.

The Root Mean Square Deviation (RMSD) serves as an essential quantitative index for analyzing alterations in the symmetry and structural integrity of molecular frameworks. In Figure 2, visualized with the aid of VMD software [53], a structure from each symmetry class is presented, with concomitant structures co-aligned through strategic translation and rotation. Subsequent calculations of RMSD values between disparate structures (expressed in angstrom units) were conducted. RMSD values in Figure 2a,b primarily stem from the locomotion of the encapsulated transition metal atoms, while in Figure 2c, the RMSD is attributable to the dimensional augmentation of the carbon cage. Consequently, in the lower portion of Figure 2c, a line representation is adopted, where the central atoms of varying systems are overlaid in a spatial context, with lines in diverse hues representing the different structural configurations. This representational approach succinctly delineates the subtle disparities in atomic positioning and cage structural modifications resulting from the incorporation of distinct transition metals.

The RMSD values for the $C_{2v}$ symmetry systems with Zr and $Nb^+$ encapsulation are notably low (0.000 and 0.076, respectively), suggesting minor structural deviations due to the movement of the transition metals along the $C_2$ axis. For the $C_{3v}$ symmetry systems encompassing $V^+$, $Sc^-$, $Ta^+$, and Ti, there is an observed increment in RMSD values (0.000, 0.144, 0.164, and 0.181, respectively), indicating structural adaptations associated with the distribution of transition metals along different $C_3$ axes. In the $T_d$ symmetry systems containing $Y^+$, Hf, and $La^-$, the RMSD changes (0.000, 3.950, and 3.981, respectively) denote significant structural changes, principally due to the global expansion or contraction of the carbon cage. This behavior could be a direct consequence of the larger atomic radii of these encapsulated metals, which may enforce a more dramatic spatial rearrangement of the carbon atoms, altering the fullerene cage dimensions. These observed RMSD variations between different metals underscore the influence of atomic size, electronic configuration, and oxidation state on the resulting symmetry and structural modifications within the fullerene cage. The complexity of these interactions suggests a nuanced interplay between metal properties and fullerene adaptability.

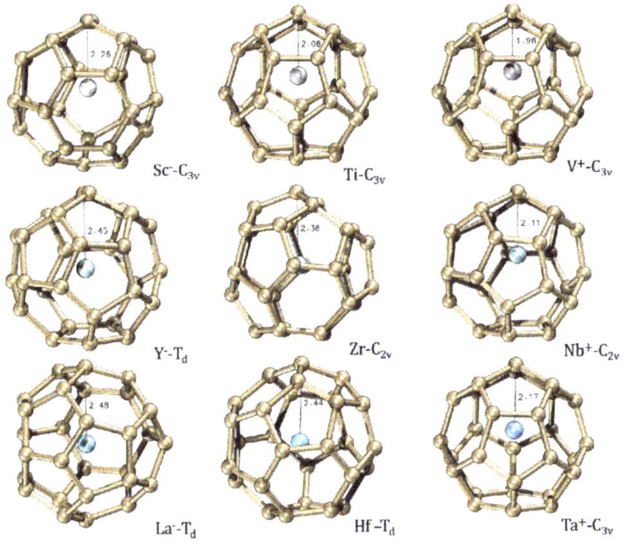


**Figure 1.** The optimized structures of TM@$C_{28}$ (TM= $Sc^{-}$, $Y^{-}$, $La^{-}$, Ti, Zr, Hf, $V^{+}$, $Nb^{+}$, $Ta^{+}$) with the shortest TM-C bond length.

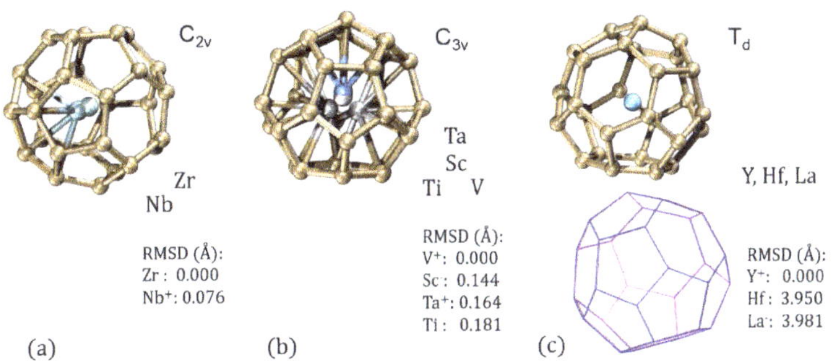


**Figure 2.** Visualization of transition metal-doped $C_{28}$ fullerenes exhibiting varied symmetry classes, analyzed using VMD software [53]. Panel (**a**) illustrates the $C_{2v}$ symmetric structure with Zr and $Nb^{+}$ encapsulation, panel (**b**) displays the $C_{3v}$ symmetric structure with $V^{+}$, $Sc^{-}$, $Ta^{+}$, and Ti encapsulation, and panel (**c**) represents the $T_d$ symmetric structure with $Y^{+}$, Hf, and $La^{-}$ encapsulation. The calculated RMSD values (in angstroms) highlight the structural deviations due to the specific movements of the embedded atoms and the expansion or contraction of the carbon cage. Panel (**c**) also features a line representation for clarity in structural differentiation, with superimposed central atoms from different systems and color-coded lines to distinguish between the structures.

The RMSD values derived from structural analysis provide a quantitative reflection of the divergence in geometric structures among different endohedral metallofullerenes. This divergence can be partially rationalized by applying the Hard and Soft Acids and Bases (HSAB) principle [54], which suggests a preferential interaction between "hard" acids and "hard" bases, as well as "soft" acids with "soft" bases. The HSAB principle is predicated on the notion that the chemical behavior of atoms and molecules during bond formation is influenced by their electronic properties, specifically the polarizability and the relative size of their orbitals.

Within the context of the periodic table, these properties are not constant; they demonstrate notable variation. For instance, as transition metal changes from titanium (Ti) to hafnium (Hf), there is an observed increase in the size of the metal ions. According to HSAB theory, this increment in size correlates with an increase in softness, which implies a greater capacity for deformation or polarization upon interaction with a base, which is the fullerene cage in this case. The softness of the metal ion can enhance the degree of distortion or expansion of the cage structure, as softer ions can form more malleable and adaptable bonds, accommodating themselves within the confines of the fullerene lattice.

Furthermore, this interaction is not merely a function of size or electronic softness; other chemical and physical parameters also play important roles. Steric effects, for example, pertain to the spatial arrangement of atoms and can drastically affect how a metal ion is incorporated into the fullerene cage. The specific nature of the metal–fullerene interaction, dictated by the unique chemistry of each metal, further contributes to the resultant geometric configuration. Additionally, the overall stability of the resulting complex is an aggregate effect of these individual interactions and properties.

It is also worth noting that while elements such as Ti, Zr, and Hf may share similar d-electron configurations, which suggests comparable chemical reactivity, the subtle differences in their ionic radii, softness, and other chemical properties can lead to distinct geometric structures within the fullerene framework. The HSAB concept provides a framework to understand these variations. A holistic understanding of the geometric forms adopted by these complexes necessitates an integrated approach, considering not only HSAB theory but also delving into quantum mechanical explanations, electron density distribution, and the dynamics of bond formation and breaking. Only by encompassing these multifaceted chemical and physical factors can one begin to fully elucidate the reasons behind the observed structural diversity in these endohedral metallofullerenes.

Table 1 presents the optimized shortest TM−C bond lengths of all nine EMFs. The TM−C distances range from 1.96 Å to 2.48 Å. It is not surprising that the shortest TM−C distance for the same group metal increases monotonically. For instance, the shortest TM−C distance for TM@$T_d$(2)-$C_{28}$ (TM = $Sc^-$, $Y^-$, $La^-$) increases from 2.26 to 2.48 Å. On the other hand, the TM−C distance decreases for the same period element. Particularly, the shortest V−C distance is only 1.96 Å. The calculations on the singlet–triplet splitting energy ($\Delta E_{S\text{-}T}$) revealed that all those values are larger than 1.04 eV, suggesting that the ground state for all these EMFs is a singlet.

**Table 1.** Symmetry (Sym.), the singlet–triplet splitting energy ($\Delta E_{S\text{-}T}$), the shortest TM−C bond length ($d_{TM-C}$) of TM@$C_{28}$ (TM = $Sc^-$, $Y^-$, $La^-$, Ti, Zr, Hf, $V^+$, $Nb^+$, $Ta^+$).

| EMFs | Sym. | $\Delta E_{S\text{-}T}$ (eV) | $d_{TM-C}$ (Å) |
|---|---|---|---|
| $[Sc@C_{28}]^-$ | ca. $C_{3v}$ | 1.87 | 2.26 |
| $[Y@C_{28}]^-$ | ca. $T_d$ | 1.98 | 2.45 |
| $[La@C_{28}]^-$ | ca. $T_d$ | 1.93 | 2.48 |
| $[Ti@C_{28}]$ | ca. $C_{3v}$ | 1.56 | 2.06 |
| $[Zr@C_{28}]$ | ca. $C_{2v}$ | 2.01 | 2.38 |
| $[Hf@C_{28}]$ | ca. $T_d$ | 2.03 | 2.44 |
| $[V@C_{28}]^+$ | ca. $C_{3v}$ | 1.04 | 1.96 |
| $[Nb@C_{28}]^+$ | ca. $C_{2v}$ | 1.86 | 2.11 |
| $[Ta@C_{28}]^+$ | ca. $C_{3v}$ | 1.64 | 2.17 |

To assess the kinetic stabilities of the EMFs, Figure 3 presents the gap between the highest occupied molecular orbitals (HOMO) and the lowest unoccupied molecular orbitals (LUMO). Except Ti@$C_{28}$, $[V@C_{28}]^+$, and $[Ta@C_{28}]^+$, all the other EMFs hold a large HOMO–LUMO gap above 2.0 eV, indicating that these EMFs could be isolated experimentally.

Particularly, both Zr@$C_{28}$ and Hf@$C_{28}$ have the highest HOMO–LUMO gap of 2.31 eV. On the other hand, the HOMO–LUMO gap of $[V@C_{28}]^+$ is as small as 1.35 eV, suggesting that this EMF is reactive and might not be isolated easily. Particularly, the HOMO–LUMO gap results are in agreement with the above the singlet–triplet splitting energy $\Delta E_{S\text{-}T}$ results; the larger the HOMO–LUMO gap, the larger $\Delta E_{S\text{-}T}$. Figure 4 presents the frontier molecular orbitals (FMOs) of three EMFs with different symmetries. Cage orbitals which are from the 2p atomic orbitals obviously contribute to all these FMOs. On the other hand, the metal orbitals show slight contributions. Particularly, the metal *d* orbitals show some contributions to the HOMO-1 and HOMO-2 of $[Ta@C_{28}]^+$ and HOMO of $[La@C_{28}]^-$.

To further analyze the electronic structures of TM@$C_{28}$, an NBO analysis was performed, and the results of the metals are shown in Table 2. The charges of metals range from −0.18 to +1.81. Although TM@$C_{28}$ (TM = $Sc^-$, $Y^-$, $La^-$) is negatively charged, the charge is mainly localized on the cage rather than on the metal, which is in line with the above molecular orbitals results. The highly positive charge (+1.34) of Y metal in $[Y@C_{28}]^-$ may come from the fact that the NBO analysis does not treat (n)p functions as valence orbitals of transition metals [55]. For the other six EMFs, the metal is positively charged, with the positive charge ranging from 0.53 to 1.81.

The formation energy of EMFs, which is defined by the energy difference from total energy minus the sum of neutral fullerene and neutral/charged transition metals, is important to assess the stability of EMFs. The formation energy of TM@$C_{28}$ was calculated and is shown in Table 2. All the formation energy falls in the range of −5.91 to −10.63 eV, indicating that these EMFs could form in the experiments. Particularly, Ti@$C_{28}$, which was observed experimentally, has a formation energy of −8.18 eV.

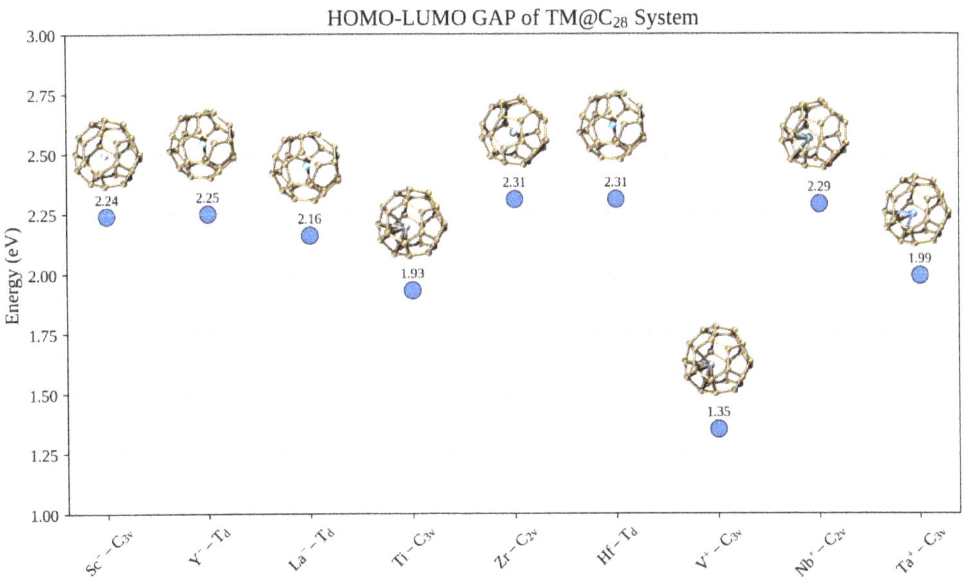


**Figure 3.** The HOMO–LUMO gap of TM@$C_{28}$ (TM = $Sc^-$, $Y^-$, $La^-$, Ti, Zr, Hf, $V^+$, $Nb^+$, $Ta^+$).

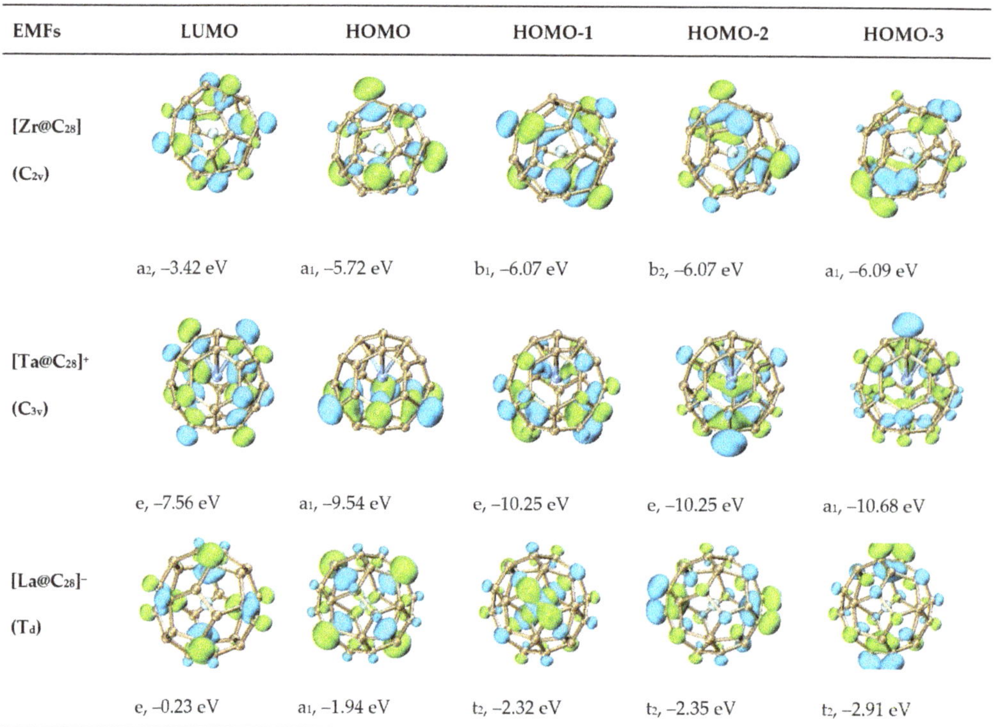


**Figure 4.** Selected molecular orbitals (isovalue = 0.05) of [TM@$C_{28}$] at the PBE-D3(BJ)/def2-TZVPPD level.

**Table 2.** NBO partial charges, the formation energy from the carbon cage $C_{28}$ and transition metal atoms, and the first excitation energies and corresponding oscillator strengths of TM@$C_{28}$ (TM = $Sc^-$, $Y^-$, $La^-$, Ti, Zr, Hf, $V^+$, $Nb^+$, $Ta^+$). The T and Q represent the triplet and quintet states, respectively.

| EMFs (Singlet) | Fragment | Natural Charge (TM) | Formation Energy (eV) | Excitation Energies (eV) | Oscillator Strengths (nm) |
|---|---|---|---|---|---|
| $[Sc@C_{28}]^-$ | $Sc^-$(T) + $C_{28}$(Q) | −0.18 | −10.63 | 2.31 | 537.7 |
| $[Y@C_{28}]^-$ | $Y^-$(T) + $C_{28}$(Q) | 1.34 | −10.13 | 2.30 | 539.5 |
| $[La@C_{28}]^-$ | $La^-$(T) + $C_{28}$(Q) | 0.19 | −5.91 | 2.22 | 559.0 |
| $[Ti@C_{28}]$ | Ti(Q) + $C_{28}$(Q) | 0.53 | −8.18 | 2.05 | 605.8 |
| $[Zr@C_{28}]$ | Zr(Q) + $C_{28}$(Q) | 1.68 | −9.02 | 2.37 | 523.7 |
| $[Hf@C_{28}]$ | Hf(T) + $C_{28}$(Q) | 1.64 | −8.66 | 2.37 | 523.9 |
| $[V@C_{28}]^+$ | $V^+$(Q) + $C_{28}$(Q) | 1.38 | −6.66 | 1.48 | 837.6 |
| $[Nb@C_{28}]^+$ | $Nb^+$(Q) + $C_{28}$(Q) | 1.33 | −8.00 | 2.38 | 521.5 |
| $[Ta@C_{28}]^+$ | $Ta^+$(Q) + $C_{28}$ (Q) | 1.81 | −8.47 | 2.13 | 582.9 |

The excitation energies and corresponding oscillator strengths, which are related to the electron excitation from the HOMO to LUMO, were calculated using time-dependent density functional theory (TDDFT), as shown in Table 2. The excitation energies range from 1.48 to 2.38 eV, with oscillator strengths ranging from 837.56 to 521.49 nm. The smaller the excitation energy, the larger the oscillator strength. The excitation energy also agrees well with the above HOMO–LUMO gap. The $[V@C_{28}]^+$ needs the smallest excitation energy.

Although the electronic structures of TM@$C_{28}$ like Ti@$C_{28}$ have been discussed in previous papers [13–16], the bonding nature between the metals and $C_{28}$ has not been analyzed.

Recently, we found that the EDA analysis could describe the bonding nature of EMFs [56] and Zintl clusters [57,58] well. Thus, to study the bonding nature between the encapsulated metal and fullerenes $C_{28}$, we performed the EDA analysis on TM@$C_{28}$. Because the EDA results strongly depend on how to divide a whole EMF into two interacting fragments, we first carried out the EDA analysis on the TM@$C_{28}$ by using different models. According to the criterion that the bonding model that has the smallest absolute orbital interactions $|\Delta E_{orb}|$ is usually determined as the best bonding model, the EDA results with the best bonding models are shown in Table 3, with the remaining models in the Supporting Information. As shown in Table 3, the bonding energy ranges from −331.9 kcal/mol to −851.5 kcal/mol, suggesting that all these EMFs are stable. The attractive interaction terms, including the Coulomb term $\Delta E_{elstat}$ and the orbital interaction term $\Delta E_{orb}$, compensate for the repulsive terms $\Delta E_{Pauli}$. In general, the Coulomb term $\Delta E_{elstat}$ contributes less than the orbital interaction term $\Delta E_{orb}$. Particularly, for [Nb@$C_{28}$]$^+$, the contribution from the orbital interaction term $\Delta E_{orb}$ to the total attractive energy is as high as 76.9%. The dispersion interactions contribute slightly to the total bonding strength with a range of −2.8 to −6.9 kcal/mol.

**Table 3.** EDA results (in kcal/mol) of the [TM@$C_{28}$] with $TM^{m+}$ and $C_{28}{}^{n-}$ as interacting fragments at the BP86-D3(BJ)/TZ2P+ level. The interaction energies ($\Delta E_{int}$), Pauli repulsion ($\Delta E_{Pauli}$), electrostatic interactions ($\Delta E_{elstat}$), orbital interactions ($\Delta E_{orb}$), and dispersion ($\Delta E_{disp}$) are presented with the percentages showing the contribution of orbital interactions or electrostatic attractions to the total attractive interactions.

| **EMFs** | **[Sc@$C_{28}$]$^-$** | **[Y@$C_{28}$]$^-$** | **[La@$C_{28}$]$^-$** |
|---|---|---|---|
| Fragments | $Sc^+$ ($4s^03d^2$) + [$C_{28}$]$^{2-}$ | $Y^+$ ($5s^04d^2$) + [$C_{28}$]$^{2-}$ | $La^+$ ($6s^05d^2$) + [$C_{28}$]$^{2-}$ |
| $\Delta E_{int}$ | −552.3 | −361.4 | −308.7 |
| $\Delta E_{Pauli}$ | +472.3 | +752.4 | +1160.2 |
| $\Delta E_{elstat}$ [a] | −363.7 (35.6%) | −522.8 (47.1%) | −703.7 (48.1%) |
| $\Delta E_{orb}$ [a] | −658.1 (64.4%) | −588.3 (52.9%) | −760.8 (51.9%) |
| $\Delta E_{disp}$ | −2.8 | −2.8 | −4.3 |
| **EMFs** | **[Ti@$C_{28}$]** | **[Zr@$C_{28}$]** | **[Hf@$C_{28}$]** |
| Fragments | $Ti^+$ ($4s^13d^2$) + [$C_{28}$]$^-$ | $Zr^+$ ($5s^14d^2$) + [$C_{28}$]$^-$ | $Hf^+$ ($6s^15d^2$) + [$C_{28}$]$^-$ |
| $\Delta E_{int}$ | −423.0 | −331.9 | −559.6 |
| $\Delta E_{Pauli}$ | +908.5 | +848.4 | +1053.6 |
| $\Delta E_{elstat}$ [a] | −464.5 (35.0%) | −485.7 (41.3%) | −585.9 (36.4%) |
| $\Delta E_{orb}$ [a] | −861.3 (65.0%) | −689.4 (58.7%) | −1024.4 (63.6%) |
| $\Delta E_{disp}$ | −5.7 | −5.2 | −2.8 |
| **EMFs** | **[V@$C_{28}$]$^+$** | **[Nb@$C_{28}$]$^+$** | **[Ta@$C_{28}$]$^+$** |
| Fragments | $V^{2+}$ ($4s^03d^3$) + [$C_{28}$]$^-$ | $Nb^{2+}$ ($5s^04d^3$) + [$C_{28}$]$^-$ | $Ta^{2+}$ ($6s^05d^3$) + [$C_{28}$]$^-$ |
| $\Delta E_{int}$ | −851.5 | −508.0 | −622.1 |
| $\Delta E_{Pauli}$ | +831.1 | +1003.9 | +690.4 |
| $\Delta E_{elstat}$ [a] | −463.1 (27.6%) | −348.2 (23.1%) | −388.8 (29.7%) |
| $\Delta E_{orb}$ [a] | −1214.1 (72.4%) | −1156.9 (76.9%) | −918.8 (70.3%) |
| $\Delta E_{disp}$ | −5.4 | −6.9 | −4.9 |

[a] The percentages corresponding to the total attractive interactions ($\Delta E_{elstat}$ + $\Delta E_{orb}$) are in parentheses.

The HOMO–LUMO gap of [V@$C_{28}$]$^+$ is as small as 1.35 eV, which is less than the average HOMO–LUMO gap (2.09 eV) in the nine EMFs, suggesting the relatively lower kinetic stability of [V@$C_{28}$]$^+$. Previous studies have shown that electronegative organic groups like trifluoromethyl can enhance the stability of EMFs by increasing the HOMO–LUMO gap [59,60]. In these previous studies, La@$C_{60}$, which has an open shell structure and is difficult to isolate experimentally, is theoretically predicted to be able to be stabilized by introducing $CF_3$ groups [56]. Then, this theoretical proposal was verified by the experimental single-crystal structure of the $CF_3$-adduct of La@$C_{60}$ [57]. In the present study, we studied the addition of the trifluoromethyl group to $[V@C_{28}]^+$ to enhance its stability and facilitate its future experimental isolation.

We focused on the single addition and double addition of the trifluoromethyl group to $[V@C_{28}]^+$. For the lowest-energy $[V@C_{28}]^+$ ($C_{3v}$), we tried all 28 configurations ($C_{28}^1$) of single $CF_3$ addition and 378 configurations ($C_{28}^2$) of two $CF_3$ additions at the PBE-D3(BJ)/def2-SVP level. After optimizations, only 7 structures and 51 structures exist for $[V@C_{28}]^+$-$CF_3$ and $[V@C_{28}]^+$-$(CF_3)_2$, respectively. Figure 5 shows the seven possible adducts of $[V@C_{28}]^+$-$CF_3$ with the relative reaction energies and HOMO–LUMO gaps, and the remaining results are shown in Supporting Information. The two lowest-energy structures of $[V@C_{28}]^+$-$CF_3$ have almost the same reaction energy of 68.3 kcal/mol and 69.2 kcal/mol, respectively. The HOMO–LUMO gaps of these two structures are 1.72 and 1.84 eV, obviously higher than that of $[V@C_{28}]^+$.

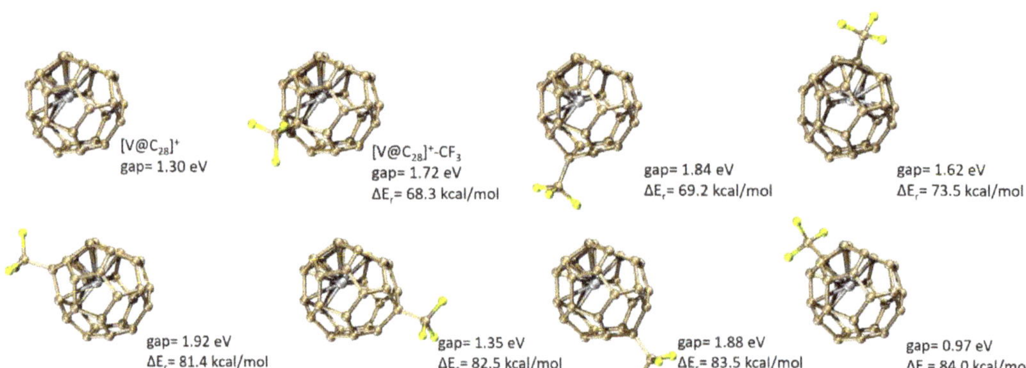


**Figure 5.** The HOMO–LUMO gaps and reaction energies of $[V@C_{28}]^+$ and $[V@C_{28}]^+$-$CF_3$ isomers at the PBE-D3(BJ)/def2-SVP level.

Further addition of one more $CF_3$ group to $[V@C_{28}]^+$-$CF_3$ gives rise to lots of possible $[V@C_{28}]^+$-$(CF_3)_2$ isomers. Figure 6 only presents four isomers with relative energy between $[V@C_{28}]^+$ and two $CF_3$ groups. The lowest-energy isomer has the two $CF_3$ group additions on the two carbon atoms connected by one C—C bond with a reaction energy of 152.5 kcal/mol. The HOMO–LUMO gap then increases to 2.18 eV. The reaction energy of the second isomer is only 1.20 kcal/mol higher than that of the first one, but its HOMO–LUMO gap is only 1.78 eV, indicating its lower kinetic stability. Therefore, it is clear from the above discussion that the addition of $CF_3$ groups to $[V@C_{28}]^+$ will increase its stability, which will assist future experimental isolation of $[V@C_{28}]^+$-$(CF_3)_2$.

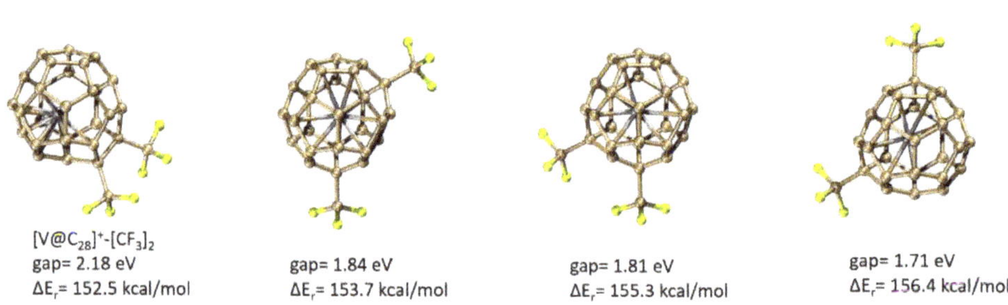


**Figure 6.** The HOMO–LUMO gaps and reaction energies of low-energy $[V@C_{28}]^+$-$(CF_3)_2$ isomers at the PBE-D3(BJ)/def2-SVP level.

## 4. Conclusions

By using quantum chemical calculations, we studied the geometries, electronic structures, bonding properties, and stability strategy of endohedral metallofullerenes $TM@C_{28}$ ($TM = Sc^-, Y^-, La^-$, Ti, Zr, Hf, $V^+, Nb^+, Ta^+$). $TM@C_{28}$ has three different structures with $C_{2v}$, $C_{3v}$, and $T_d$ symmetries. The HOMO–LUMO gap ranges from 1.35 eV to 2.31 eV, with $[V@C_{28}]^+$ having the lowest HOMO–LUMO gap of 1.35 eV. The trend of singlet–triplet splitting energy and first excitation energy is in line with the HOMO–LUMO gap. The frontier molecular orbitals analysis revealed that the molecular orbitals are mainly from fullerene cage orbitals, with a slight contribution from the encapsulated metals. The bonding interactions between encapsulated metal and fullerene cages are dominated by the Coulomb term $\Delta E_{elstat}$ and the orbital interaction term $\Delta E_{orb}$, in which the orbital interaction term $\Delta E_{orb}$ contributes more than the Coulomb term $\Delta E_{elstat}$. The addition of one or two $CF_3$ groups to $[V@C_{28}]^+$ could increase the HOMO–LUMO gap, giving rise to the possible isolation of $[V@C_{28}]^+$.

**Supplementary Materials:** The following supporting information can be downloaded at https://www.mdpi.com/article/10.3390/inorganics12020040/s1, Figure S1: The HOMO-LUMO GAP after attaching a CF3 group to [V@C28]+ and the energy differences between different isomers, at the PBE-D3(BJ)/def2-SVP level.; Table S1: The atomic (ionic) energies (in eV) calculated using PBE-D3(BJ)/def2-TZVPPD.; Figure S2: After the addition of two CF3 groups, the HOMO-LUMO GAP and the energy relationships among different isomers, at the PBE-D3(BJ)/def2-SVP level; Table S2: The EDA results (in kcal/mol) of the $[TM@C_{28}]$ with $TM^{m+}$ and $C_{28}^{n-}$ as interacting fragments at the BP86-D3(BJ)/TZ2P+ level.

**Author Contributions:** For the author contributions in our manuscript: D.L. was responsible for the majority of the computations, analysis, and original draft preparation. Y.S. contributed to parts of the data discussion and analysis. T.Y., as the corresponding author, provided project management, funding acquisition, supervision, and was involved in the review and editing of the manuscript. All authors have read and agreed to the published version of the manuscript.

**Funding:** This research was funded by the National Natural Science Foundation of China, grant number 12274337.

**Data Availability Statement:** We emphasize open data sharing and hereby provide the following data availability statement: All research data have been meticulously recorded and are stored in a trusted data repository. Some essential data are attached in the Supplementary Materials of this paper for reference. We encourage interested researchers to contact us for further details on data availability.

**Acknowledgments:** The authors gratefully acknowledge the financial support from the National Natural Science Foundation of China (Grant No. 12274337).

**Conflicts of Interest:** The authors declare no conflict of interest.

## References

1. Kroto, H.W.; Heath, J.R.; O'Brien, S.C.; Curl, R.F.; Smalley, R.E. C60: Buckminsterfullerene. *Nature* **1985**, *318*, 162–163. [CrossRef]
2. Fowler, P.W.; Manolopoulos, D.E. *An Atlas of Fullerenes*; Courier Corporation: North Chelmsford, MA, USA, 2007.
3. Lu, X.; Akasaka, T.; Slanina, Z. *Handbook of Fullerene Science and Technology*; Springer Nature: Berlin, Germany, 2022.
4. Guo, T.; Diener, M.; Chai, Y.; Alford, M.; Haufler, R.; McClure, S.; Ohno, T.; Weaver, J.; Scuseria, G.; Smalley, R. Uranium Stabilization of C28: A Tetravalent Fullerene. *Science* **1992**, *257*, 1661–1664. [CrossRef] [PubMed]
5. Popov, A.A.; Yang, S.; Dunsch, L. Endohedral Fullerenes. *Chem. Rev.* **2013**, *113*, 5989–6113. [CrossRef] [PubMed]
6. Tan, Y.-Z.; Xie, S.-Y.; Huang, R.-B.; Zheng, L.-S. The Stabilization of Fused-Pentagon Fullerene Molecules. *Nat. Chem.* **2009**, *1*, 450–460. [CrossRef] [PubMed]
7. Xie, S.-Y.; Gao, F.; Lu, X.; Huang, R.-B.; Wang, C.-R.; Zhang, X.; Liu, M.-L.; Deng, S.-L.; Zheng, L.-S. Capturing the Labile Fullerene[50] as C50Cl10. *Science* **2004**, *304*, 699. [CrossRef] [PubMed]
8. Akasaka, T.; Nagase, S. *Endofullerenes: A New Family of Carbon Clusters*; Springer Science & Business Media: Berlin, Germany, 2002; Volume 3.
9. Akasaka, T.; Wudl, F.; Nagase, S. *Chemistry of Nanocarbons*; John Wiley & Sons: Hoboken, NJ, USA, 2010.
10. Kroto, H. The Stability of the Fullerenes C n, with N = 24, 28, 32, 36, 50, 60 and 70. *Nature* **1987**, *329*, 529–531. [CrossRef]

11. Cai, W.; Chen, C.-H.; Chen, N.; Echegoyen, L. Fullerenes as Nanocontainers That Stabilize Unique Actinide Species inside: Structures, Formation, and Reactivity. *Acc. Chem. Res.* **2019**, *52*, 1824–1833. [CrossRef]
12. Shen, W.; Hu, S.; Lu, X. Endohedral Metallofullerenes: New Structures and Unseen Phenomena. *Chem. Eur. J.* **2020**, *26*, 5748–5757. [CrossRef]
13. Dunlap, B.I.; Haeberlen, O.D.; Roesch, N. Asymmetric Localization of Titanium in Carbon Molecule (C28). *J. Phys. Chem.* **1992**, *96*, 9095–9097. [CrossRef]
14. Muñoz-Castro, A.; King, R.B. Formation of Spherical Aromatic Endohedral Metallic Fullerenes. Evaluation of Magnetic Properties of M@ C28 (M = Ti, Zr, and Hf) from DFT Calculations. *Inorg. Chem.* **2017**, *56*, 15251–15258. [CrossRef]
15. Mulet-Gas, M.; Abella, L.; Dunk, P.W.; Rodríguez-Fortea, A.; Kroto, H.W.; Poblet, J.M. Small Endohedral Metallofullerenes: Exploration of the Structure and Growth Mechanism in the Ti@ C 2n (2 N = 26–50) Family. *Chem. Sci.* **2015**, *6*, 675–686. [CrossRef]
16. Dunk, P.W.; Kaiser, N.K.; Mulet-Gas, M.; Rodríguez-Fortea, A.; Poblet, J.M.; Shinohara, H.; Hendrickson, C.L.; Marshall, A.G.; Kroto, H.W. The Smallest Stable Fullerene, M@ C28 (M = Ti, Zr, U): Stabilization and Growth from Carbon Vapor. *J. Am. Chem. Soc.* **2012**, *134*, 9380–9389. [CrossRef]
17. Zhao, K.; Pitzer, R.M. Electronic Structure of C28, Pa@ C28, and U@ C28. *J. Phys. Chem.* **1996**, *100*, 4798–4802. [CrossRef]
18. Dognon, J.-P.; Clavaguéra, C.; Pyykkö, P. A Predicted Organometallic Series Following a 32-Electron Principle: An@ C28 (An = Th, Pa+, U2+, Pu4+). *J. Am. Chem. Soc.* **2009**, *131*, 238–243. [CrossRef]
19. Dai, X.; Gao, Y.; Jiang, W.; Lei, Y.; Wang, Z. U@ C 28: The Electronic Structure Induced by the 32-Electron Principle. *Phys. Chem. Chem. Phys.* **2015**, *17*, 23308–23311. [CrossRef]
20. Mojica, M.; Alonso, J.A.; Méndez, F. Synthesis of Fullerenes. *J. Phys. Org. Chem.* **2013**, *26*, 526–539. [CrossRef]
21. Alekseev, N.I.; Filippov, B.M.; Basargin, I.V.; Sedov, A.I. Investigation of Semicommercial Arc Plants for Fullerene-Production. *J. Eng. Phys. Thermophys.* **2011**, *84*, 1087–1098. [CrossRef]
22. Ishigaki, T.; Suzuki, S.; Kataura, H.; Krätschmer, W.; Achiba, Y. Characterization of Fullerenes and Carbon Nanoparticles Generated with a Laser-Furnace Technique. *Appl. Phys. A* **2000**, *70*, 121–124. [CrossRef]
23. Nitze, F.; Abou-Hamad, E.; Wågberg, T. Carbon Nanotubes and Helical Carbon Nanofibers Grown by Chemical Vapour Deposition on C60 Fullerene Supported Pd Nanoparticles. *Carbon* **2011**, *49*, 1101–1107. [CrossRef]
24. Manawi, Y.M.; Ihsanullah; Samara, A.; Al-Ansari, T.; Atieh, M.A. A Review of Carbon Nanomaterials' Synthesis via the Chemical Vapor Deposition (CVD) Method. *Materials* **2018**, *11*, 822. [CrossRef] [PubMed]
25. Murata, M.; Murata, Y.; Komatsu, K. Surgery of Fullerenes. *Chem. Commun.* **2008**, *46*, 6083–6094. [CrossRef]
26. Li, Z.; Yang, T. Modifying the Electron Capture Decay Rate of Be 7 by Using Small Fullerenes. *Phys. Rev. C* **2023**, *108*, 054604. [CrossRef]
27. Wang, E.; Gao, Y. H2+ H2O→ H4O: Synthesizing Hyperhydrogenated Water in Small-Sized Fullerenes? *J. Phys. Chem. A* **2023**, *127*, 1190–1195. [CrossRef]
28. Spano, C.E.; Mo, F.; Ardesi, Y.; Roch, M.R.; Piccinini, G.; Graziano, M. Electronic Transport Study of Bistable Cr@C28 Single Molecule Device for High-Density Data Storage Applications. In Proceedings of the 8th World Congress on New Technologies (NewTech'22), Prague, Czech Republic, 3–5 August 2022. [CrossRef]
29. Muñoz-Castro, A.; Bruce King, R. Evaluation of Bonding, Electron Affinity, and Optical Properties of M@ C28 (M = Zr, Hf, Th, and U): Role of d-and f-Orbitals in Endohedral Fullerenes from Relativistic DFT Calculations. *J. Comput. Chem.* **2017**, *38*, 44–50. [CrossRef]
30. Lu, T. Molclus Program 1.12. Available online: http://www.keinsci.com/research/molclus.html (accessed on 1 September 2023).
31. Grimme, S.; Bannwarth, C.; Shushkov, P. A Robust and Accurate Tight-Binding Quantum Chemical Method for Structures, Vibrational Frequencies, and Noncovalent Interactions of Large Molecular Systems Parametrized for All Spd-Block Elements (Z = 1–86). *J. Chem. Theory Comput.* **2017**, *13*, 1989–2009. [CrossRef] [PubMed]
32. Bannwarth, C.; Ehlert, S.; Grimme, S. GFN2-xTB—An Accurate and Broadly Parametrized Self-Consistent Tight-Binding Quantum Chemical Method with Multipole Electrostatics and Density-Dependent Dispersion Contributions. *J. Chem. Theory Comput.* **2019**, *15*, 1652–1671. [CrossRef] [PubMed]
33. Grimme, S.; Antony, J.; Ehrlich, S.; Krieg, H. A Consistent and Accurate Ab Initio Parametrization of Density Functional Dispersion Correction (DFT-D) for the 94 Elements H-Pu. *J. Chem. Phys.* **2010**, *132*, 154104. [CrossRef] [PubMed]
34. Pritchard, B.P.; Altarawy, D.; Didier, B.; Gibson, T.D.; Windus, T.L. New Basis Set Exchange: An Open, up-to-Date Resource for the Molecular Sciences Community. *J. Chem. Inf. Model.* **2019**, *59*, 4814–4820. [CrossRef] [PubMed]
35. Feller, D. The Role of Databases in Support of Computational Chemistry Calculations. *J. Comput. Chem.* **1996**, *17*, 1571–1586. [CrossRef]
36. Schuchardt, K.L.; Didier, B.T.; Elsethagen, T.; Sun, L.; Gurumoorthi, V.; Chase, J.; Li, J.; Windus, T.L. Basis Set Exchange: A Community Database for Computational Sciences. *J. Chem. Inf. Model.* **2007**, *47*, 1045–1052. [CrossRef] [PubMed]
37. Weigend, F.; Ahlrichs, R. Balanced Basis Sets of Split Valence, Triple Zeta Valence and Quadruple Zeta Valence Quality for H to Rn: Design and Assessment of Accuracy. *Phys. Chem. Chem. Phys.* **2005**, *7*, 3297–3305. [CrossRef] [PubMed]
38. Dolg, M.; Stoll, H.; Preuss, H. A Combination of Quasirelativistic Pseudopotential and Ligand Field Calculations for Lanthanoid Compounds. *Theor. Chim. Acta* **1993**, *85*, 441–450. [CrossRef]
39. Rappoport, D.; Furche, F. Property-Optimized Gaussian Basis Sets for Molecular Response Calculations. *J. Chem. Phys.* **2010**, *133*, 134105. [CrossRef] [PubMed]

40. Dolg, M.; Stoll, H.; Savin, A.; Preuss, H. Energy-Adjusted Pseudopotentials for the Rare Earth Elements. *Theor. Chim. Acta* **1989**, *75*, 173–194. [CrossRef]
41. Andrae, D.; Haeussermann, U.; Dolg, M.; Stoll, H.; Preuss, H. Energy-Adjusted Ab Initio Pseudopotentials for the Second and Third Row Transition Elements. *Theor. Chim. Acta* **1990**, *77*, 123–141. [CrossRef]
42. Glendening, E.; Badenhoop, J.; Reed, A.; Carpenter, J.; Bohmann, J.; Morales, C.; Karafiloglou, P.; Landis, C.; Weinhold, F. NBO7. *Theor. Chem. Inst. Univ. Wis. Madison WI USA* **2018**. [CrossRef]
43. Frisch, M.; Trucks, G.; Schlegel, H.; Scuseria, G.; Robb, M.; Cheeseman, J.; Scalmani, G.; Barone, V.; Petersson, G.; Nakatsuji, H.; et al. *Gaussian 16, Revision A. 03*; Gaussian Inc.: Wallingford, CT, USA, 2016.
44. Ziegler, T.; Rauk, A. On the Calculation of Bonding Energies by the Hartree Fock Slater Method: I. The Transition State Method. *Theor. Chim. Acta* **1977**, *46*, 1–10. [CrossRef]
45. van Lenthe, E.; Baerends, E.-J.; Snijders, J.G. Relativistic Total Energy Using Regular Approximations. *J. Chem. Phys.* **1994**, *101*, 9783–9792. [CrossRef]
46. Zhao, L.; Hermann, M.; Schwarz, W.E.; Frenking, G. The Lewis Electron-Pair Bonding Model: Modern Energy Decomposition Analysis. *Nat. Rev. Chem.* **2019**, *3*, 48–63. [CrossRef]
47. Zhao, L.; Schwarz, W.E.; Frenking, G. The Lewis Electron-Pair Bonding Model: The Physical Background, One Century Later. *Nat. Rev. Chem.* **2019**, *3*, 35–47. [CrossRef]
48. Te Velde, G.T.; Bickelhaupt, F.M.; Baerends, E.J.; Fonseca Guerra, C.; van Gisbergen, S.J.; Snijders, J.G.; Ziegler, T. Chemistry with ADF. *J. Comput. Chem.* **2001**, *22*, 931–967. [CrossRef]
49. Baerends, E.J.; Ziegler, T.; Autschbach, J.; Bashford, D.; Bérces, A.; Bickelhaupt, F.; Bo, C.; Boerrigter, P.; Cavallo, L.; Chong, D. *ADF 2019*; SCM, Theoretical Chemistry, Vrije Universiteit: Amsterdam, The Netherlands, 2019.
50. Perdew, J.P. Density-Functional Approximation for the Correlation Energy of the Inhomogeneous Electron Gas. *Phys. Rev. B* **1986**, *33*, 8822. [CrossRef]
51. Becke, A.D. Density-Functional Exchange-Energy Approximation with Correct Asymptotic Behavior. *Phys. Rev. A* **1988**, *38*, 3098. [CrossRef]
52. Van Lenthe, E.; Baerends, E.J. Optimized Slater-Type Basis Sets for the Elements 1–118. *J. Comput. Chem.* **2003**, *24*, 1142–1156. [CrossRef]
53. Humphrey, W.; Dalke, A.; Schulten, K. VMD: Visual Molecular Dynamics. *J. Mol. Graph.* **1996**, *14*, 33–38. [CrossRef] [PubMed]
54. Chattaraj, P.K.; Lee, H.; Parr, R.G. HSAB Principle. *J. Am. Chem. Soc.* **1991**, *113*, 1855–1856. [CrossRef]
55. Chi, C.; Pan, S.; Meng, L.; Luo, M.; Zhao, L.; Zhou, M.; Frenking, G. Alkali Metal Covalent Bonding in Nickel Carbonyl Complexes ENi $(CO)_3^-$. *Angew. Chem.* **2019**, *131*, 1746–1752. [CrossRef]
56. Shui, Y.; Pei, G.; Zhao, P.; Xiong, M.; Li, S.; Ehara, M.; Yang, T. Understanding Electronic Structures, Chemical Bonding, and Fluxional Behavior of Lu2@ C2n (2n = 76–88) by a Theoretical Study. *J. Chem. Phys.* **2022**, *157*, 184306. [CrossRef] [PubMed]
57. Liu, D.; Xu, S.; Pei, G.; Xu, J.; Zhao, X.; Kong, C.; Yang, Z.; Yang, T. Geometries, Electronic Structures, and Bonding Properties of Endohedral Group-14 Zintl Clusters TM@ E10 (TM = Fe, Co, Ni; E = Ge, Sn, Pb). *J. Comput. Chem.* **2022**, *43*, 828–838. [CrossRef] [PubMed]
58. Zhao, X.; Pei, G.; Xu, S.; Kong, C.; Yang, Z.; Yang, T. Endohedral Group-14-Element Clusters TM@ E 9 (TM = Co, Ni, Cu; E = Ge, Sn, Pb) and Their Low-Dimensional Nanostructures: A First-Principles Study. *Phys. Chem. Chem. Phys.* **2021**, *23*, 20654–20665. [CrossRef] [PubMed]
59. Guan, J.; Tománek, D. Can CF3-Functionalized La@ C60 Be Isolated Experimentally and Become Superconducting? *Nano Lett.* **2017**, *17*, 3402–3408. [CrossRef] [PubMed]
60. Nakagawa, A.; Nishino, M.; Niwa, H.; Ishino, K.; Wang, Z.; Omachi, H.; Furukawa, K.; Yamaguchi, T.; Kato, T.; Bandow, S.; et al. Crystalline Functionalized Endohedral C60 Metallofullerides. *Nat. Commun.* **2018**, *9*, 3073. [CrossRef] [PubMed]

*inorganics*

MDPI

*Article*

# Synthesis and Characterization of a Novel Non-Isolated-Pentagon-Rule Isomer of $Th@C_{76}$:$Th@C_1(17418)$-$C_{76}$

**Yunpeng Xia [1,†], Yi Shen [1,†], Yang-Rong Yao [2], Qingyu Meng [1] and Ning Chen [1,*]**

1 College of Chemistry, Chemical Engineering and Materials Science, and State Key Laboratory of Radiation Medicine and Protection, Soochow University, Suzhou 215123, China
2 CAS Key Laboratory of Materials for Energy Conversion, Anhui Laboratory of Advanced Photon Science and Technology, Department of Materials Science and Engineering, University of Science and Technology of China, Hefei 230026, China
* Correspondence: chenning@suda.edu.cn
† These authors contributed equally to this work.

**Abstract:** A novel Non-Isolated-Pentagon-Rule (non-IPR) isomer of thorium-based endohedral mono-metallofullerenes (mono-EMFs), $Th@C_1(17418)$-$C_{76}$, was successfully synthesized and characterized using MALDI-TOF mass spectroscopy, single-crystal X-ray diffraction, UV-vis-NIR spectroscopy, and Raman spectroscopy. The molecular structure of this non-IPR isomer was determined unambiguously as $Th@C_1(17418)$-$C_{76}$ using a single-crystal X-ray diffraction analysis. The crystallographic results further revealed that the optimal Th site resided at the intersection of two adjacent pentagons, similar to that of $U@C_1(17418)$-$C_{76}$. Additionally, the UV-vis-NIR spectra of $Th@C_1(17418)$-$C_{76}$ exhibited distinct differences compared to the previously reported $U@C_1(17418)$-$C_{76}$, highlighting the distinctive electronic structure of actinium-based endohedral metallofullerenes (EMFs). The Raman spectrum of $Th@C_1(17418)$-$C_{76}$ exhibited similarities to that previously reported for thorium-based EMFs, indicating the analogous strong metal–cage interactions of thorium-based EMFs.

**Keywords:** endohedral metallofullerenes; mono-metallofullerenes; thorium; X-ray crystallography

## 1. Introduction

Endohedral metallofullerenes (EMFs) are a specialized type of carbon nanomaterial featured by the encapsulation of metal atoms or metal clusters within the carbon cage of the fullerenes [1–3]. In 1991, Smalley et al. achieved the isolation of the first EMF, $La@C_{82}$, which initiated the exploration of the structure and properties of EMFs [4]. Subsequently, numerous types of mono-EMFs and endohedral clusterfullerenes have been synthesized and reported by researchers [5–8]. Depending on the differences between endohedral species, EMFs can be further classified as mono-metallofullerenes, di-metallofullerenes, tri-metallofullerenes, and clusterfullerenes [9–13]. The electron transfer and host–guest interactions between the endohedral metal units and fullerene carbon cages contribute to the intriguing electronic structures and physicochemical properties of these EMFs, thereby presenting immense potential in various cutting-edge fields, including biomedicine, organic photovoltaic devices, and single-molecule devices [14–17].

Mono-EMFs represent the most extensively studied type of EMF, with the majority of reported mono-EMFs being based on lanthanide metals. These metals are embedded within the carbon cage of the fullerene in the form of $M@C_{2n}$ (M = Ce [18], Pr [19], Nd [20], Sm [21], Eu [22], Gd [23], Tb [24], Dy [25], Ho [26], Er [27], Yb [28], and Lu [29]). In recent years, research on actinium-based mono-EMFs has also been reported, primarily focused on uranium (U) and thorium (Th). Among them, the characterization of $Th@C_{3v}(8)$-$C_{82}$ [30] marked the beginning of the investigation into actinium-based mono-EMFs. In monometallic thorium-based EMFs, thorium usually exhibits an oxidation state of a +4 oxidation state. For example, Th@$T_d$(19151)-$C_{76}$ [31], Th@$D_{5h}$(6)-$C_{80}$ [32], Th@$C_{2v}$(9)-$C_{82}$, Th@$C_2$(5)-$C_{82}$ [33], and Th@$C_1$(11)-$C_{86}$ [34] have been discovered in previous research. All of them demonstrate a four-electron charge transfer from the metal atom to the fullerene cage. Conversely, uranium's oxidation state displays a remarkable flexibility. In U@$C_{2v}$(9)-$C_{82}$, the endohedral uranium transfers three electrons to the $C_{82}$ carbon cage, while in U@$C_2$(5)-$C_{82}$, the endohedral uranium transfers four electrons to the $C_{82}$ fullerene cage [35]. Because of the contributions from f-orbital electrons, the unique characteristics and variability of oxidation states in actinide metals lead to more complex electron configurations in actinide-based EMFs compared to lanthanide-based EMFs. Actinide-based EMFs also exhibit stronger metal–cage interactions than those of lanthanide-based EMFs, and actinide metal ions have a more significant influence on the properties of actinide-based EMFs than previously observed for lanthanide-based EMFs. For example, previously reported mono-EMFs based on uranium, known as U@$C_1$(28324)-$C_{80}$, and mono-EMFs based on thorium, labeled as Th@$C_1$(28324) $C_{80}$, possess the same fullerene carbon cage, and both of them undergo a four-electron charge transfer process. They demonstrate distinct electronic properties, as evidenced by significant differences in their UV-vis-NIR spectra [36].

Based on extensive research on actinium-based mono-EMFs, it is clear that the remarkable charge transfer and strong engagement between actinide metal ions and fullerene carbon cages are essential for stabilizing previously unstable fullerene cage structures. Thus, in addition to the experimentally verified isomers, there are numerous isomers that have only been predicted and researched theoretically in actinium-based mono-EMFs. Consequently, further experimental investigations are of paramount importance to exploring actinium-based mono-EMFs, as they will reveal the distinct electronic properties and carbon cage structures that differentiate them from traditional lanthanide-based mono-EMFs.

Recently, two independent computational studies proposed two possible isomeric structures of Th@$C_{76}$, one is referred to as Th@$T_d$(19151)-$C_{76}$, while the other is referred to as Th@$C_1$(17418)-$C_{76}$. [37,38]. Nevertheless, only the existence of Th@$T_d$(19151)-$C_{76}$ [31] has been experimentally confirmed. Further experimental investigation of Th@$C_1$(17418)-$C_{76}$ would contribute to a deeper understanding of the unique carbon cage structures, electronic properties, and host–guest interactions exhibited by actinide-based EMFs.

In this study, we present the synthesis and characterization of a novel non-IPR isomer of thorium-based mono-EMFs, Th@$C_1$(17418)-$C_{76}$. Th@$C_1$(17418)-$C_{76}$ was successfully synthesized by using a modified Krätschmer–Huffman direct-current DC arc-discharge method. It was thoroughly characterized through various techniques, including MALDI-TOF mass spectroscopy, single-crystal X-ray diffraction, UV-vis-NIR spectroscopy, and Raman spectroscopy. The crystallographic results revealed that the optimal Th site resided at the intersection of two adjacent pentagons, and the presence of a single Th ion played a crucial role in stabilizing the non-IPR $C_{76}$ fullerene cage.

## 2. Results and Discussion

### 2.1. Synthesis and Isolation of Th@$C_{76}$

Thorium-based EMFs were produced using a modified Krätschmer–Huffman direct-current DC arc-discharge method. Hollow graphite rods, packed with $ThO_2$ and graphite powder (molar ratio of Th: C = 1:24), were vaporized in an arcing reactor chamber under a 200 Torr helium atmosphere. The resulting soot was then collected and extracted with $CS_2$ for 12 h. Multistage high-performance liquid chromatography (HPLC) separation processes were employed to isolate and purify the Th@$C_{76}$ (Figure S1). The purity of the Th@$C_{76}$ was confirmed by the observation of single peaks with HPLC and positive-ion-mode matrix-assisted laser desorption-ionization time-of-flight (MALDI-TOF) mass spectrometry, as exhibited in Figure 1. The mass spectra of the isomer of Th@$C_{76}$ exhibited a peak at 1143.975 $m/z$, and the experimental isotopic distribution of the sample agrees well with the empirical projection.

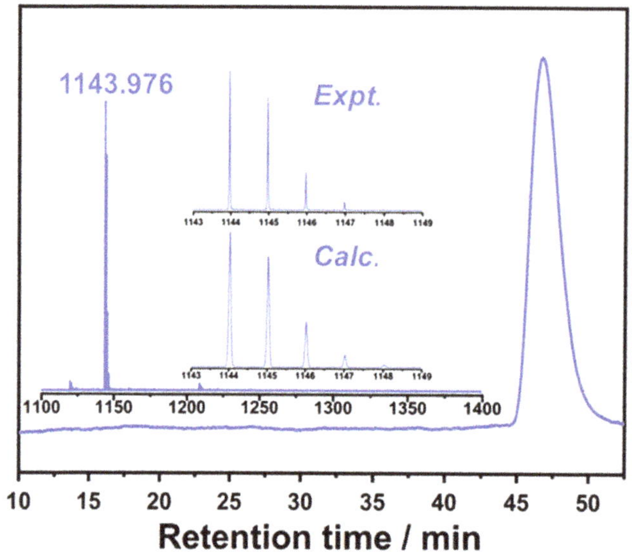


**Figure 1.** HPLC chromatogram of purified Th@$C_{76}$ on a Buckyprep column using toluene as the eluent, with a 4 mL/min flow rate. The mass spectra of Th@$C_{76}$ and associated empirical and hypothetical isotopic profiles of the sample are shown in the inset.

*2.2. Molecular Structure of Th@$C_1$(17418)-$C_{76}$*

The molecular configuration of Th@$C_{76}$ was unequivocally ascertained through single-crystal X-ray diffraction analyses. Th@$C_{76}$ and $Ni^{II}$-OEP (OEP = 2, 3, 7, 8, 12, 13, 17, 18-octaethylporphyrin dianion) were co-crystallized by slowly diffusing the benzene solution of $Ni^{II}$-OEP into the carbon disulfide solution of the purified compounds. The structure of Th@$C_{76}$ was resolved and refined in the $C2/c$ (No. 15) space group. The experimental results clearly showed that this new isomer of Th@$C_{76}$ showed a different structure from the reported Th@$T_d$(19151)-$C_{76}$ [31]. It shares the same carbon cage structure with U@$C_1$(17418)-$C_{76}$, which possesses a low-symmetry $C_1$(17418)-$C_{76}$ cage and one adjacent pentagon pair [36]. Interestingly, crystallographic investigations of the $C_1$(17418)-$C_{76}$ carbon cage have primarily been limited to actinide metallofullerenes, and no previous studies have reported the observation of or noticed the same cage in lanthanide-based EMFs. This indicates that a stronger interaction between U or Th and fullerene cages may facilitate the stabilization of these unconventional fullerene cage isomers. On the other hand, Th@$C_1$(17418)-$C_{76}$ was also predicted to be the most advantageous product in experiments at elevated temperatures due to its dominant molar fraction [38]. This result verifies the theoretical predications.

Figure 2 shows the molecular structure of Th@$C_1$(17418)-$C_{76}$ and its relationship to the co-crystallized $Ni^{II}$(OEP) molecule. The Th@$C_1$(17418)-$C_{76}$ moiety is close to the adjacent $Ni^{II}$(OEP), with a short Ni-to-cage carbon distance (Ni1-C12) of 2.875 Å, demonstrating a characteristic interplay between the carbon cage and $Ni^{II}$(OEP). There are five crystallographic sites for the Th atoms with occupancies of 0.50 for Th1, 0.30 for Th2, 0.14 for Th3, 0.03 for Th4, and 0.025 for Th5 (see Figure S2). Th1 (occupancy 0.50) is situated at the intersection of two adjacent pentagons with the shortest Th−C distance in the range of 2.359−2.449 Å (Table S2), as shown in Figure 3. The shortest Th−C distances of the Th@$C_1$(17418)-$C_{76}$ are similar to those in Th@$T_d$(19151)-$C_{76}$, demonstrating similar interactions between the Th atom and the carbon cage [31]. In comparison to two examples of lanthanide-based non-IPR monometallic $C_{76}$ (M@$C_{2v}$(19138)-$C_{76}$, M = Eu [39], Sm [40]), although the major metal sites are all located at the intersection of two adjacent pentagons, there are large differences in the shortest Th–cage distances. This discrepancy suggests that the interaction between the metal and cage in Th@$C_{76}$ could potentially possess a greater strength when compared to its lanthanide counterparts.

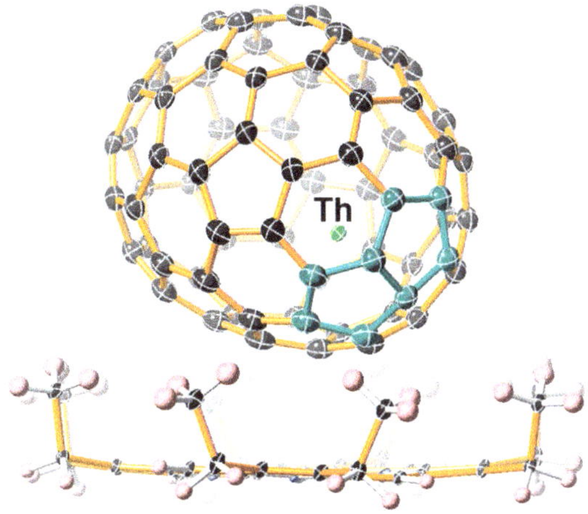


**Figure 2.** ORTEP drawings showing the relationship between the EMF and porphyrin moieties for Th@$C_1$(17418)-$C_{76}$·[$Ni^{II}$(OEP)] with 15% thermal ellipsoids. Only the major cage orientation and predominant thorium sites are shown.

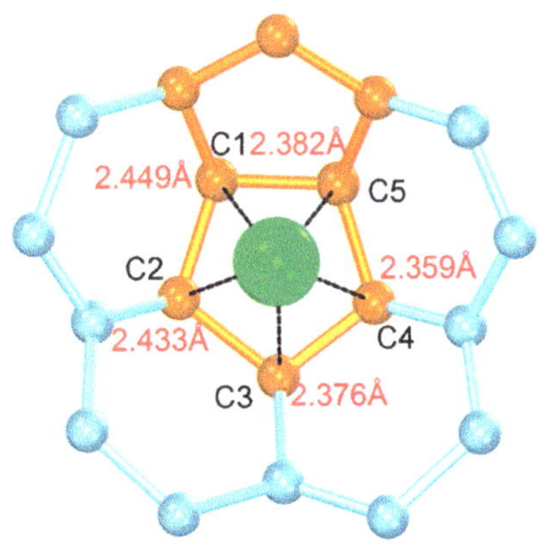


**Figure 3.** Perspective drawings showing the interaction of the major thorium closest cage portion in @$C_1$(17418)-$C_{76}$.

### 2.3. Spectroscopic Characterizations

UV-vis-NIR absorption is a useful technique that can help us to determine the electronic structures of EMFs. Normally, EMFs sharing the same cage and charge transfer exhibit almost identical absorption spectra [1]. The UV-vis-NIR absorption spectrum was utilized in this study to characterize the pure sample dissolved in a $CS_2$ solution, as depicted in Figure 4. This allowed for further analysis. Th@$C_1$(17418)-$C_{76}$ exhibits five characteristic peaks at 490, 602, 755, 930, and 1078 nm. The absorption onset occurs at approximately 1230 nm, indicating an optical band of 1.01 V. The absorption spectrum of U@$C_1$(17418)-$C_{76}$ [36], which shares the same cage isomer and charge transfer, though showing some similarity to that of the Th@$C_1$(17418)-$C_{76}$, only exhibits one absorption at 593 nm (Figure 4a). This result shows a notable influence of the actinide metal on the electronic structures of M@$C_1$(17418)-$C_{76}$(M = U, Th). A similar phenomenon can also be observed in M@$C_1$(28324)-$C_{80}$ [36] (M = U, Th), revealing that the endohedral metal itself also plays a significant role in the electronic structure of the EMFs. On the other hand, the spectrum of Th@$T_d$(19151)-$C_{76}$ [31] shows different characteristic peaks compared to those of the Th@$C_1$(17418)-$C_{76}$ (Figure 4b). This agrees with the crystallographic results that the Th@$C_{76}$ identified in this work is a novel isomer. To delve deeper into the electronic structure of Th@$C_1$(17418)-$C_{76}$, we conducted electrochemical tests on the compound. The cyclic voltammogram of Th@$C_1$(17418)-$C_{76}$ presents two oxidation peaks and three reduction peaks. However, the cyclic voltammogram of U@ $C_1$(17418)-$C_{76}$ only presents one oxidation peak and three reduction peaks [36]. In addition, compared to the first oxidation potential of U@$C_1$(17418)-$C_{76}$ (0.14 V), the first oxidation potential of Th@$C_1$(17418)-$C_{76}$ (0.37 V) shows a significant positive shift. This result shows the notable influence of the actinide metal on the electronic structures of M@$C_1$(17418)-$C_{76}$(M = U, Th).

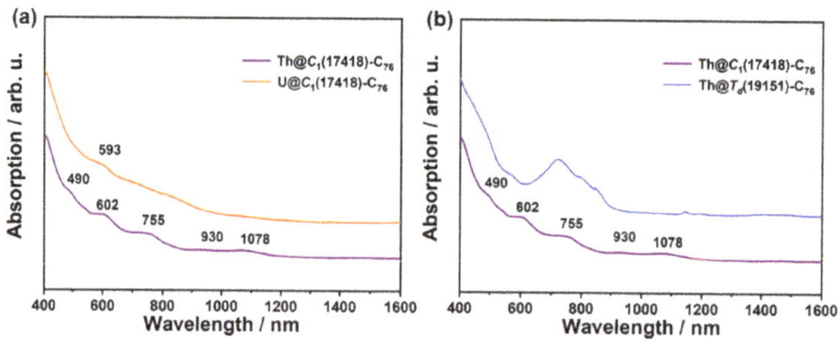


**Figure 4.** UV-vis-NIR absorption spectra of Th@$C_1$(17418)-$C_{76}$ and (**a**) U@$C_1$(17418)-$C_{76}$ and (**b**) Th@$T_d$(19151)-$C_{76}$ in $CS_2$.

Th@$C_1$(17418)-$C_{76}$ was also characterized using low-energy Raman spectroscopy, as shown in Figure 5. Sharp peaks at 149, 218, and 479 $cm^{-1}$ are observed for the Raman spectrum of Th@$C_1$(17418)-$C_{76}$. Similar frequencies also can be observed in Th@$D_{5h}$(6)-$C_{80}$ [32] (150 and 219 $cm^{-1}$), Th@$C_{2v}$(9)-$C_{82}$ (151 and 216 $cm^{-1}$), and Th@$C_2$(5)-$C_{82}$ [33] (152 and 215 $cm^{-1}$), indicating that thorium-based mono-EMFs have similar metal–cage interactions. Though the UV-vis-NIR absorption spectra of Th@$C_1$(17418)-$C_{76}$ differ from those of U@$C_1$(17418)-$C_{76}$, the vibrational frequencies on the Raman spectrum of U@$C_1$(17418)-$C_{76}$ are at 153, 219, and 359 $cm^{-1}$ [36], showing notable similarities to those of the Th@$C_1$(17418)-$C_{76}$. This phenomenon can also be observed in Th@$C_1$(28134)-$C_{80}$ and U@$C_1$(28134)-$C_{80}$ [36], which also share the same carbon cage. This may indicate that the cage isomers play a major role on the metal–cage interaction of the corresponding EMFs.

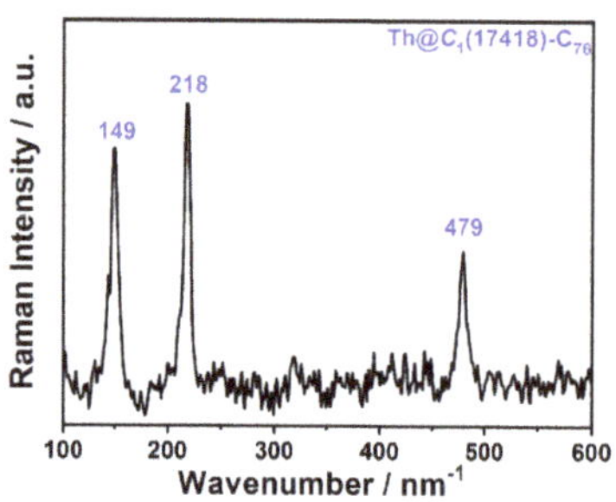


**Figure 5.** Low-energy Raman spectra of Th@$C_1$(17418)-$C_{76}$ at 633 nm excitation.

## 3. Materials and Methods

Materials. The chemicals employed in this research were $Th_2O_3$ (99.9%, School of Radiation Medicine and Protection of Soochow University., Suzhou, China), $CS_2$ (99%, Sinopharm Chemical Reagent Co., Ltd., Shanghai, China), toluene (99%, Sinopharm Chemical Reagent Co., Ltd.), hexane ethanol (99%, Sinopharm Chemical Reagent Co., Ltd.), hexane (99%, Sigma Aldrich Co., St. Louis, MO, USA), benzene (99%, Sigma Aldrich Co.), trichloromethane (99%, Sinopharm Chemical Reagent Co., Ltd.), ferrocene (99.8%, Beijing Bailing Wei Technology Co., Ltd., Beijing, China), Octaethylnickel porphyrin (99.8%, Beijing Bailing Wei Technology Co., Ltd.), tetrabutylammonium hexafluorophosphate (electrochemical-grade, Shanghai Aladdin Biochemical Technology Co., Ltd., Shanghai, China), 1,2-dichlorobenzene (99%, Shanghai Aladdin Biochemical Technology Co., Ltd.), and graphite rods and graphite powder (Shanghai Fengyi Carbon Co., Shanghai, China).

Synthesis of Th@$C_1$(17418)-$C_{76}$. Th@$C_1$(17418)-$C_{76}$ was synthesized using the direct-current arc discharge method. $ThO_2$ powders and graphite powder were mixed in an atomic molar ratio of Th:C = 1:24 (equivalent to a mass ratio of $ThO_2$:C = 1:1.09). The resulting mixture was then uniformly filled into a hollow graphite rod and compacted to achieve dense packing. The filled graphite rod was placed into a tube furnace and subjected to a 12 h reduction process at 1000 °C under a nitrogen atmosphere, aiming to convert the metal oxide into a metal carbide.

After the reduction process, the compaction graphite rod containing the reduced material was utilized as the anode, while a solid graphite rod served as the cathode within a vacuum arc furnace. The furnace was evacuated to a pressure of 4 Pa and then filled with 200 Torr of helium gas. The distance between the electrodes was maintained at approximately 1 cm, and the current was set to 90 A using an electric welding machine. Plasma evaporation took place during the arc discharge process. The anode was manually rotated with the left hand to achieve the optimal evaporation positioning, while the cathode was rotated with the right hand to push the solid graphite rod towards the consumed mixture graphite rod. When the mixture rod was nearly depleted, the cathode was retracted rapidly to prevent any damage to the metallic target on the anode. Subsequently, the power supply of the electric welding machine was switched off. Once the furnace temperature cooled down to room temperature, air was introduced into the furnace to reach atmospheric pressure. The furnace cover was opened, and the carbon ash accumulated on the furnace wall was collected using tools like a test tube brush. The collected carbon ash was soaked in a solution of carbon disulfide and left to stand for 12 h. Once the first immersion was completed, the obtained carbon ash was immersed twice with the same operation, so that as much of the obtained fullerene sample as possible was extracted. The extraction of fullerene samples needs to be carried out in a glove box with specific filtration devices.

The carbon disulfide solution was separated from the solid carbon residues using a vacuum filtration device equipped with a 0.45 μm organic membrane, resulting in a carbon disulfide filtrate. The carbon disulfide solvent was then evaporated using rotary evaporation until no liquid remained, and an appropriate quantity of toluene solution was added to dissolve the residue. The resulting reddish brown solution underwent another round of filtration to eliminate the solid particles of fullerenes, resulting in a saturated crude extract solution of fullerenes, which could be further utilized for High-Performance Liquid Chromatography (HPLC) separation in subsequent steps.

Isolation of Th@$C_1$(17418)-$C_{76}$. The separation and purification of Th@$C_1$(17418)-$C_{76}$ was achieved using multi-stage HPLC procedures. The HPLC was performed with toluene as the mobile phase and a UV detector with a detection wavelength of 310 nm. A 10 mL chromatographic injector was used to inject samples of fullerenes dissolved in toluene into a preparative HPLC system for separation. Prior to injection, a disposable syringe filter was employed to filter the fullerene sample, mitigating the interference caused by solid particles during the chromatographic separation. Distinct chromatographic columns were employed for the separation at various stages of the process.

Multiple HPLC columns, including a Buckyprep-M column (25 × 250 mm, Cosmosil, Nacalai Tesque Inc., Kyoto, Japan), a Buckprep column (10 × 250 mm, Cosmosil, Nacalai Tesque, Kyoto, Japan), and a 5PBB column (10 × 250 mm, Cosmosil, Nacalai Tesque, Kyoto, Japan), were utilized in the procedures. The purity of the target fractions was monitored by using MALDI-TOF during the separation process. In our mass spectrometry analysis, numerous signals corresponding to monometallic fullerenes based on thorium (Th), including Th@$C_{76}$, Th@$C_{80}$, Th@$C_{82}$, Th@$C_{84}$, Th@$C_{86}$, and Th@$C_{88}$, were observed. Among them, in particular, the isomer of Th@$C_{76}$ reported in this study has not been experimentally documented before. Therefore, in subsequent separation steps, our focus was solely on the purification and characterization of Th@$C_{76}$.

Th@$C_1$(17418)-$C_{76}$ was subjected to purification utilizing a four-stage high-performance liquid chromatography (HPLC) process, as shown in Figure S1. Initially, the compound was passed through a Buckyprep-Mcolumn (25 mm × 250 mm, Cosmosil, Nacalai Tesque) at a flow rate of 10 mL/min during the first stage of separation. Subsequently, the fraction eluted between 26 and 27.5 min (indicated by the color blue) was collected and reinjected into the Buckyprep column (10 mm × 250 mm, Cosmosil, Nacalai Tesque) for the second stage of separation, utilizing toluene as the eluent with a flow rate of 4 mL/min. The fraction containing Th@$C_1$(17418)-$C_{76}$ was obtained between 46 and 49.5 min (highlighted in red) and isolated accordingly. Moving on to the third stage of the separation, the 5PBB column (10 mm × 250 mm, Cosmosil, Nacalai Tesque) was employed, delivering a flow rate of 4 mL/min. The desired fraction, marked in orange, appeared within the 65 to 73 min range and specifically contained Th@$C_1$(17418)-$C_{76}$. Finally, the last stage involved passing the sample through the Buckyprep column (10 mm × 250 mm, Cosmosil, Nacalai Tesque) at a flow rate of 4 mL/min, bringing forth the achievement of pure Th@$C_1$(17418)-$C_{76}$. Notably, the confirmation of the extracted Th@$C_1$(17418)-$C_{76}$'s purity was impeccably established through the utilization of MALDI-TOF in positively charged mode, as visually exemplified in Figure 1.

Spectroscopic Studies of Th@$C_1$(17418)-$C_{76}$. The positive-ion mode matrix-assisted laser desorption ionization time-of-flight (MALDI-TOF) (Bruker, Karlsruhe, Germany) was employed for the mass characterization. A rotary evaporator was used to spin dry the fullerene sample under investigation. The substance was dissolved in carbon disulfide and the sample was transferred onto a mass spectrometry target using a pipette. A mass spectrometry analysis proceeded after the solvent evaporated. The purified Th@$C_1$(17418)-$C_{76}$ fullerene sample was dissolved in $CS_2$ and the carbon disulfide solution of the fullerene was transferred into a 2 mL quartz cuvette. A control experiment was prepared by adding blank carbon disulfide to another quartz cuvette. The Cary 5000 UV-vis-NIR spectrophotometer (Agilent, Santa Clara, CA, USA) was used to measure the ultraviolet-visible-near-infrared absorption wavelengths of the sample in the detection range of 400–1600 nm. After completing the measurements, the carbon disulfide solution was collected from the sample in a glass vial and the quartz cuvette was rinsed with clean carbon disulfide solution. The Raman spectra were obtained using a Horiba Lab RAM HR Evolution Raman spectrometer using a laser at 633 nm.

Electrochemical Studies of Th@$C_1$(17418)-$C_{76}$. The cyclic voltammetry was performed using a conventional three-electrode system consisting of a glassy carbon electrode as the working electrode, a silver wire as the reference electrode, and a platinum wire as the counter electrode. A 0.05 M $(n\text{-}Bu)_4NPF_6$ solution was prepared by adding ortho-dichlorobenzene to the electrolyte, which contained tetrabutylammonium hexafluorophosphate. Subsequently, the purified sample of monometallic fullerene was dissolved in the solution for a cyclic voltammetry analysis using a CHI-660E electrochemical workstation. The reference material used for calibration was ferrocene, and the scan rate was set at 100 mV/s.

X-ray Crystallographic Study. Single crystals of Th@$C_{76}$ were obtained, employing a solvent diffusion method characterized in a gradual process. A supersaturated solution of benzene (0.2 mL), encompassing $Ni^{II}$(OEP) (0.18 mg), was meticulously diffused into an impeccably concocted carbon disulfide solution (0.1 mL) of Th@$C_1$(17418)-$C_{76}$ (0.1 mg). Prior to the crystal growth procedure, the fabricated fullerene sample underwent a comprehensive purification and cleaning step utilizing n-hexane. Filtering and cleaning devices were obtained by using a rubber-tipped dropper. The filter paper was folded to obtain a filter core approximately 1 cm in size, precisely filling the neck of the pipette. The fullerene sample requiring filtration and cleaning was dissolved in carbon disulfide, followed by drying with nitrogen gas. A small amount of hexane was added, and ultrasonication was applied to suspend the fullerene sample in the hexane solution. Using a pipette, the mixture of hexane and fullerene samples was transferred to a prepared filtration pipette, with the rate of filtration controlled at one drop per second. As fullerene samples are insoluble in hexane, they remained on the filter core. The filter core was then rinsed with carbon disulfide, and the cleaning solution was collected, resulting in a fully cleaned fullerene sample.

In explicit detail, the carbon disulfide solution of Th@$C_1$(17418)-$C_{76}$ and the benzene solution of $Ni^{II}$(OEP) were introduced into a precisely maintained 5 mm inner diameter nuclear magnetic resonance (NMR) tube, adhering to an exact volumetric ratio of 1:2. Prior to crystal growth, the NMR tube necessitated a series of sequential cleansing steps. The NMR tube was thoroughly filled with solutions of soap, ethanol, n-hexane, and carbon disulfide in a successive manner. The subsequent cleansing process involved subjecting the NMR tube to ultrasonic irradiation for a duration exceeding 30 min, with three cycles of ultrasonic treatment for each solution. To ensure a meticulously controlled and hermetically sealed environment, the aperture of the NMR tube was securely enclosed using a specialized sealing film. This meticulously prepared assembly was subsequently transferred to a refrigerated environment, maintaining a constant temperature of 4 °C, and left undisturbed, allowing for a designated duration of one month.

After the designated duration, we used an optical microscope to observe the NMR tubes, black prism-shaped crystals, exhibiting a remarkable specular gleam, precipitated within the NMR tube. The meticulous selection of appropriately sized and shaped black prism-shaped crystals was conducted to facilitate a subsequent single-crystal X-ray diffraction analysis. These selected crystals were subjected to thorough testing and characterization employing a cutting-edge single-crystal diffractometer; the model was the Bruker D8 Venture with metaljet. The single-crystal X-ray data of Th@$C_1$(17418)-$C_{76}$ were collected at 120 K, respectively, on a diffractometer (APEX II; Bruker Analytik GmbH, Karlsruhe, Germany) equipped with a CCD collector. The multiscan method was used for absorption correction. The structures were solved using direct methods and refined on $F^2$ using full-matrix least-squares using the SHELXL2014 crystallographic software packages. Hydrogen atoms were inserted at calculated positions and constrained with isotropic thermal parameters.

Crystal data for Th@$C_1$(17418)-$C_{76}$·[$Ni^{II}$(OEP)]·2$C_6H_6$: $Mr = 1892.47$, 0.1 mm × 0.08 mm × 0.06 mm, monoclinic, $C2/c$ (No. 15), $a = 44.778(2)$ Å, $b = 14.9844(11)$ Å, $c = 25.431(22)$ Å, $\alpha = 90°$, $\beta = 121.405(3)°$, $\gamma = 90°$, $V = 14564(22)$ Å$^3$, $Z = 8$, $\rho_{calcd} = 1.756$ g cm$^{-3}$, $\mu$(Ga K$\alpha$) = 6.129 mm$^{-1}$, $\theta = 2.756$–58.099, $T = 120.0$ K, $R_1 = 0.1339$, $wR_2 = 0.2547$ for all data; $R_1 = 0.0901$, $wR_2 = 00.2256$ for 10155 reflections ($I > 2.0\sigma(I)$) with 1220 parameters. Goodness-of-fit indicator1.074. The Maximum residual electron density was 1.429 e Å$^{-3}$.

## 4. Conclusions

In conclusion, a novel non-IPR isomer of thorium-based mono-EMFs, the synthesis and characterization of Th@$C_1$(17418)-$C_{76}$ were accomplished through mass spectroscopy, single-crystal X-ray diffraction, Raman spectroscopy, and UV-vis-NIR spectroscopy. The crystallographic characterization unambiguously demonstrated that this isomer possesses a pair of adjacent pentagons, sharing the same carbon cage as the one previously reported for U@$C_1$(17418)-$C_{76}$. The major metal sites in U@$C_1$(17418)-$C_{76}$ and Th@$C_1$(17418)-$C_{76}$ are all situated at the intersection of two adjacent pentagons. Currently, the stabilization of the $C_1$(17418)-$C_{76}$ cage is only achieved by the encapsulation of U and Th, suggesting that the unique 5f electrons present in actinide metals enable them to stabilize uncommon carbon cage structures. In addition, despite similarity of their Raman vibrational spectra, the UV-vis-NIR absorption spectra of U@$C_1$(17418)-$C_{76}$ and Th@$C_1$(17418)-$C_{76}$ exhibit notable differences. This work introduces a novel addition to the group of thorium-based mono-EMFs, enhancing the comprehension of the distinctive carbon cage structure, electronic properties, and interactions exhibited by actinium-based EMFs.

**Supplementary Materials:** The following supporting information can be downloaded at: https://www.mdpi.com/2304-6740/11/11/422/s1, Figure S1: HPLC isolation procedures of Th@$C_1$(17418)-$C_{76}$ and the pure sample's MALDI-TOF mass spectrum.; Figure S2: Ball and stick representation of disordered Th sites inside Th@$C_1$(17418)-$C_{76}$, five positions (Th1, Th2, Th3, Th4, and Th5) are observed.; Figure S3: Cyclic voltammograms of Th@$C_1$(17418)-$C_{76}$ in *o*-dichlorobenzene.; Table S1: Occupancies of disordered thorium sites in Th@$C_1$(17418)-$C_{76}$.; Table S2: Closest Th-Cage distances (Å) in Th@$C_1$(17418)-$C_{76}$.; Table S3: Crystallographic information of Th@$C_1$(17418)-$C_{76}$.

**Author Contributions:** Conceptualization, Y.X.; funding acquisition, N.C.; investigation, Y.X., Y.S. and Y.-R.Y.; project administration, N.C. and Q.M.; supervision, N.C.; writing—original draft, Y.X.; writing—review and editing, N.C. and Y.X. All authors have read and agreed to the published version of the manuscript.

**Funding:** This research was funded by the National Science Foundation China (NSFC 52172051 and 22301288) and the Anhui Provincial Natural Science Foundation (2308085MB31).

**Data Availability Statement:** The crystallographic data can be obtained free of charge from the Cambridge Crystallographic Data Centre with the CCDC numbers 2295545 via www.ccdc.cam.ac.uk/data_request/cif (accessed on 18 September 2023). The other data are available upon reasonable request from the corresponding authors.

**Conflicts of Interest:** The authors declare no conflict of interest.

## References

1. Popov, A.A.; Yang, S.; Dunsch, L. Endohedral fullerenes. *Chem. Rev.* **2013**, *113*, 5989–6113. [CrossRef]
2. Yang, S.; Wei, T.; Jin, F. When metal clusters meet carbon cages: Endohedral clusterfullerenes. *Chem. Soc. Rev.* **2017**, *46*, 5005–5058. [CrossRef]
3. Bao, L.; Peng, P.; Lu, X. Bonding inside and outside Fullerene Cages. *Acc. Chem. Res.* **2018**, *51*, 810–815. [CrossRef] [PubMed]
4. Chai, Y.; Guo, T.; Jin, C.M.; Haufler, R.E.; Chibante, L.P.F.; Fure, J.; Wang, L.H.; Alford, J.M.; Smalley, R.E. Fullerenes with Metals Inside. *J. Phys. Chem.* **1991**, *95*, 7564–7568. [CrossRef]
5. Slanina, Z.; Nagase, S. $Sc_3N@C_{80}$: Computations on the Two-Isomer Equilibrium at High Temperatures. *Chem. Phys. Chem.* **2005**, *6*, 2060–2063. [CrossRef] [PubMed]
6. Wang, T.-S.; Feng, L.; Wu, J.-Y.; Xu, W.; Xiang, J.-F.; Tan, K.; Ma, Y.-H.; Zheng, J.-P.; Jiang, L.; Lu, X.; et al. Planar Quinary Cluster inside a Fullerene Cage: Synthesis and Structural Characterizations of $Sc_3NC@C_{80}$-$I_h$. *J. Am. Chem. Soc.* **2010**, *132*, 16362–16364. [CrossRef] [PubMed]
7. Kodama, T.; Fujii, R.; Miyake, Y.; Sakaguchi, K.; Nishikawa, H.; Ikemoto, I.; Kikuchi, K.; Achiba, Y. Structural study of four $Ca@C_{82}$ isomers by 13C NMR spectroscopy. *Chem. Phys. Lett.* **2003**, *377*, 197–200. [CrossRef]
8. Xu, Z.; Nakane, T.; Shinohara, H. Production and Isolation of $Ca@C_{82}$ (I–IV) and $Ca@C_{84}$ (I,II) Metallofullerenes. *J. Am. Chem. Soc.* **1996**, *118*, 11309–11310. [CrossRef]
9. Xu, W.; Feng, L.; Calvaresi, M.; Liu, J.; Liu, Y.; Niu, B.; Shi, Z.; Lian, Y.; Zerbetto, F. An experimentally observed trimetallofullerene $Sm_3@I_h$-$C_{80}$: Encapsulation of three metal atoms in a cage without a nonmetallic mediator. *J. Am. Chem. Soc.* **2013**, *135*, 4187–4190. [CrossRef]
10. Kurihara, H.; Lu, X.; Iiduka, Y.; Mizorogi, N.; Slanina, Z.; Tsuchiya, T.; Akasaka, T.; Nagase, S. $Sc_2C_2@C_{80}$ Rather than $Sc_2@C_{82}$: Templated Formation of Unexpected $C_{2v}$(5)-$C_{80}$ and Temperature-Dependent Dynamic Motion of Internal $Sc_2C_2$ Cluster. *J. Am. Chem. Soc.* **2011**, *133*, 2382–2385. [CrossRef]
11. Hu, S.; Liu, T.; Shen, W.; Slanina, Z.; Akasaka, T.; Xie, Y.; Uhlik, F.; Huang, W.; Lu, X. Isolation and Structural Characterization of Er@$C_{2v}$(9)-$C_{82}$ and Er@$C_s$(6)-$C_{82}$: Regioselective Dimerization of a Pristine Endohedral Metallofullerene Induced by Cage Symmetry. *Inorg. Chem.* **2019**, *58*, 2177–2182. [CrossRef] [PubMed]
12. Shen, W.; Bao, L.; Hu, S.; Yang, L.; Jin, P.; Xie, Y.; Akasaka, T.; Lu, X. Crystallographic characterization of $Lu_2C_{2n}$ (2n = 76–90): Cluster selection by cage size. *Chem. Sci.* **2019**, *10*, 829–836. [CrossRef] [PubMed]
13. Shen, W.; Bao, L.; Wu, Y.; Pan, C.; Zhao, S.; Fang, H.; Xie, Y.; Jin, P.; Peng, P.; Li, F.-F.; et al. $Lu_2@C_{2n}$ (2n = 82, 84, 86): Crystallographic Evidence of Direct Lu–Lu Bonding between Two Divalent Lutetium Ions Inside Fullerene Cages. *J. Am. Chem. Soc.* **2017**, *139*, 9979–9984. [CrossRef] [PubMed]

14. Ross, R.B.; Cardona, C.M.; Guldi, D.M.; Sankaranarayanan, S.G.; Reese, M.O.; Kopidakis, N.; Peet, J.; Walker, B.; Bazan, G.C.; Van Keuren, E.; et al. Endohedral fullerenes for organic photovoltaic devices. *Nat. Mater.* **2009**, *8*, 208–212. [CrossRef]
15. Liu, Y.; Chen, C.; Qian, P.; Lu, X.; Sun, B.; Zhang, X.; Wang, L.; Gao, X.; Li, H.; Chen, Z.; et al. Gd-metallofullerenol nanomaterial as non-toxic breast cancer stem cell-specific inhibitor. *Nat. Commun.* **2015**, *6*, 5988. [CrossRef]
16. Zhang, K.; Wang, C.; Zhang, M.; Bai, Z.; Xie, F.F.; Tan, Y.Z.; Guo, Y.; Hu, K.J.; Cao, L.; Zhang, S.; et al. A Gd@$C_{82}$ single-molecule electret. *Nat. Nanotech.* **2020**, *15*, 1019–1024. [CrossRef] [PubMed]
17. Okamura, N.; Yoshida, K.; Sakata, S.; Hirakawa, K. Electron transport in endohedral metallofullerene Ce@$C_{82}$ single-molecule transistors. *Appl. Phys. Lett.* **2015**, *106*, 043108. [CrossRef]
18. Wakahara, T.; Kobayashi, J.; Yamada, M.; Maeda, Y.; Tsuchiya, T.; Okamura, M.; Akasaka, T.; Waelchli, M.; Kobayashi, K.; Nagase, S.; et al. Characterization of Ce@$C_{82}$ and its anion. *J. Am. Chem. Soc.* **2004**, *126*, 4883–4887. [CrossRef]
19. Akasaka, T.; Okubo, S.; Kondo, M.; Maeda, Y.; Wakahara, T.; Kato, T.; Suzuki, T.; Yamamoto, K.; Kobayashi, K.; Nagase, S. Isolation and characterization of two Pr@$C_{82}$ isomers. *Chem. Phys. Lett.* **2000**, *319*, 153–156. [CrossRef]
20. Ding, J.; Lin, N.; Weng, L.-T.; Cue, N.; Yang, S. Isolation and characterization of a new metallofullerene Nd@$C_{82}$. *Chem. Phys. Lett.* **1996**, *261*, 92–97. [CrossRef]
21. Hu, Z.; Hao, Y.; Slanina, Z.; Gu, Z.; Shi, Z.; Uhlík, F.; Zhao, Y.; Feng, L. Popular $C_{82}$ Fullerene Cage Encapsulating a Divalent Metal Ion $Sm^{2+}$: Structure and Electrochemistry. *Inorg. Chem.* **2015**, *54*, 2103–2108. [CrossRef] [PubMed]
22. Sun, B.Y.; Sugai, T.; Nishibori, E.; Iwata, K.; Sakata, M.; Takata, M.; Shinohara, H. An Anomalous Endohedral Structure of Eu@$C_{82}$ Metallofullerenes. *Angew. Chem. Int. Ed.* **2005**, *44*, 4568–4571. [CrossRef] [PubMed]
23. Senapati, L.; Schrier, J.; Whaley, K.B. Electronic Transport, Structure, and Energetics of Endohedral Gd@$C_{82}$ Metallofullerenes. *Nano Lett.* **2004**, *4*, 2073–2078. [CrossRef]
24. Iwasaki, K.; Wanita, N.; Hino, S.; Yoshimura, D.; Okazaki, T.; Shinohara, H. Ultraviolet photoelectron spectra of Tb@$C_{82}$. *Chem. Phys. Lett.* **2004**, *398*, 389–392. [CrossRef]
25. Iida, S.; Kubozono, Y.; Slovokhotov, Y.; Takabayashi, Y.; Kanbara, T.; Fukunaga, T.; Fujiki, S.; Emura, S.; Kashino, S. Structure and electronic properties of Dy@$C_{82}$ studied by UV–VIS absorption, X-ray powder diffraction and XAFS. *Chem. Phys. Lett.* **2001**, *338*, 21–28. [CrossRef]
26. Huang, H.J.; Yang, S.H.; Zhang, X.X. Magnetic Behavior of Pure Endohedral Metallofullerene Ho@$C_{82}$: A Comparison with Gd@$C_{82}$. *J. Phys. Chem. B* **1999**, *103*, 5928–5932. [CrossRef]
27. Sanakis, Y.; Tagmatarchis, N.; Aslanis, E.; Ioannidis, N.; Petrouleas, V.; Shinohara, H.; Prassides, K. Dual-Mode X-Band EPR Study of Two Isomers of the Endohedral Metallofullerene Er@$C_{82}$. *J. Am. Chem. Soc.* **2001**, *123*, 9924–9925. [CrossRef]
28. Suzuki, M.; Slanina, Z.; Mizorogi, N.; Lu, X.; Nagase, S.; Olmstead, M.M.; Balch, A.L.; Akasaka, T. Single-Crystal X-ray Diffraction Study of Three Yb@$C_{82}$ Isomers Cocrystallized with $Ni^{II}$(octaethylporphyrin). *J. Am. Chem. Soc.* **2012**, *134*, 18772–18778. [CrossRef]
29. Iwamoto, M.; Ogawa, D.; Yasutake, Y.; Azuma, Y.; Umemoto, H.; Ohashi, K.; Izumi, N.; Shinohara, H.; Majima, Y. Molecular Orientation of Individual Lu@$C_{82}$ Molecules Demonstrated by Scanning Tunneling Microscopy. *J. Phys. Chem. C* **2010**, *114*, 14704–14709. [CrossRef]
30. Wang, Y.F.; Morales-Martinez, R.; Zhang, X.X.; Yang, W.; Wang, Y.X.; Rodriguez-Fortea, A.; Poblet, J.M.; Feng, L.; Wang, S.; Chen, N. Unique Four-Electron Metal-to-Cage Charge Transfer of Th to a $C_{82}$ Fullerene Cage: Complete Structural Characterization of Th@$C_{3v}$(8)-$C_{82}$. *J. Am. Chem. Soc.* **2017**, *139*, 5110–5116. [CrossRef]
31. Jin, M.; Zhuang, J.; Wang, Y.; Yang, W.; Liu, X.; Chen, N. Th@$T_d$(19151)-$C_{76}$: A Highly Symmetric Fullerene Cage Stabilized by a Tetravalent Actinide Metal Ion. *Inorg. Chem.* **2019**, *58*, 16722–16726. [CrossRef] [PubMed]
32. Yan, Y.; Morales-Martinez, R.; Zhuang, J.; Yao, Y.R.; Li, X.; Poblet, J.M.; Rodriguez-Fortea, A.; Chen, N. Th@$D_{5h}$(6)-$C_{80}$: A highly symmetric fullerene cage stabilized by a single metal ion. *Chem. Commun.* **2021**, *57*, 6624–6627. [CrossRef] [PubMed]
33. Meng, Q.; Morales-Martinez, R.; Zhuang, J.; Yao, Y.R.; Wang, Y.; Feng, L.; Poblet, J.M.; Rodriguez-Fortea, A.; Chen, N. Synthesis and Characterization of Two Isomers of Th@$C_{82}$: Th@$C_{2v}$(9)-$C_{82}$ and Th@$C_2$(5)-$C_{82}$. *Inorg. Chem.* **2021**, *60*, 11496–11502. [CrossRef] [PubMed]
34. Wang, Y.; Morales-Martinez, R.; Cai, W.; Zhuang, J.; Yang, W.; Echegoyen, L.; Poblet, J.M.; Rodriguez-Fortea, A.; Chen, N. Th@$C_1$(11)-$C_{86}$: An actinide encapsulated in an unexpected $C_{86}$ fullerene cage. *Chem. Commun.* **2019**, *55*, 9271–9274. [CrossRef]
35. Cai, W.; Morales-Martínez, R.; Zhang, X.; Najera, D.; Romero, E.L.; Metta-Magaña, A.; Rodríguez-Fortea, A.; Fortier, S.; Chen, N.; Poblet, J.M.; et al. Single crystal structures and theoretical calculations of uranium endohedral metallofullerenes (U@$C_{2n}$, 2n = 74, 82) show cage isomer dependent oxidation states for U. *Chem. Sci.* **2017**, *8*, 5282–5290. [CrossRef] [PubMed]
36. Cai, W.; Abella, L.; Zhuang, J.; Zhang, X.; Feng, L.; Wang, Y.; Morales-Martínez, R.; Esper, R.; Boero, M.; Metta-Magaña, A.; et al. Synthesis and Characterization of Non-Isolated-Pentagon-Rule Actinide Endohedral Metallofullerenes U@$C_1$(17418)-$C_{76}$, U@$C_1$(28324)-$C_{80}$, and Th@$C_1$(28324)-$C_{80}$: Low-Symmetry Cage Selection Directed by a Tetravalent Ion. *J. Am. Chem. Soc.* **2018**, *140*, 18039–18050. [CrossRef]
37. Jin, P.; Liu, C.; Li, Y.; Li, L.; Zhao, Y. Th@$C_{76}$. Computational characterization of larger actinide endohedral fullerenes. *Int. J. Quantum. Chem.* **2018**, *118*, e25501. [CrossRef]
38. Zhao, P.; Zhao, X.; Ehara, M. Theoretical Insights into Monometallofullerene Th@$C_{76}$: Strong Covalent Interaction between Thorium and the Carbon Cage. *Inorg. Chem.* **2018**, *57*, 2961–2964. [CrossRef]

39. Bao, L.; Li, Y.; Yu, P.; Shen, W.; Jin, P.; Lu, X. Preferential Formation of Mono-Metallofullerenes Governed by the Encapsulation Energy of the Metal Elements: A Case Study on Eu@$C_{2n}$ (2n = 74–84) Revealing a General Rule. *Angew. Chem. Int. Ed.* **2020**, *59*, 5259–5262. [CrossRef]
40. Hao, Y.; Feng, L.; Xu, W.; Gu, Z.; Hu, Z.; Shi, Z.; Slanina, Z.; Uhlik, F. Sm@$C_{2v}$(19138)-$C_{76}$: A Non-IPR Cage Stabilized by a Divalent Metal Ion. *Inorg. Chem.* **2015**, *54*, 4243–4248. [CrossRef]

inorganics

MDPI

Article

# Reactivity of Open-Shell Metallofullerene Anions: Synthesis, Crystal Structures, and Electrochemical Properties of Benzylated Gd@$C_{2v}$-$C_{82}$

Xinyi Zhou [1,†], Yang-Rong Yao [2,†], Yajing Hu [1], Le Yang [3], Shaoting Yang [1], Yilu Zhang [4], Qianyan Zhang [4], Ping Peng [1,*], Peng Jin [3,*] and Fang-Fang Li [1,*]

1 State Key Laboratory of Material Processing and Die & Mould Technology, School of Materials Science and Engineering, Huazhong University of Science and Technology, Wuhan 430074, China
2 CAS Key Laboratory of Materials for Energy Conversion, Anhui Laboratory of Advanced Photon Science and Technology, Department of Materials Science and Engineering, University of Science and Technology of China, Hefei 230026, China
3 School of Materials Science and Engineering, Hebei University of Technology, Tianjin 300130, China
4 State Key Laboratory for Physical Chemistry of Solid Surfaces, Department of Chemistry, College of Chemistry and Chemical Engineering, Xiamen University, Xiamen 361005, China
* Correspondence: ppeng@hust.edu.cn (P.P.); pengjin@hebut.edu.cn (P.J.); ffli@hust.edu.cn (F.-F.L.)
† These authors contributed equally to this work.

**Abstract:** The reactivity of the open-shell Gd@$C_{2v}$-$C_{82}$ with different charge states towards benzyl bromide was investigated. $[Gd@C_{2v}\text{-}C_{82}]^{3-}$ exhibited enhanced activity relative to Gd@$C_{2v}$-$C_{82}$ and $[Gd@C_{2v}\text{-}C_{82}]^{-}$. The structural characterizations, including MALDI-TOF MS, UV-vis-NIR, and single crystal X-ray diffraction, indicate the formation of isomeric benzyl monoadducts of Gd@$C_{2v}$-$C_{82}$. All three monoadducts contain 1:1 mirror-symmetric enantiomers. Additionally, the addition of the benzyl group and its specific position result in distinct electrochemical behavior of the products compared to the parent Gd@$C_{2v}$-$C_{82}$. Theoretical studies demonstrate that only $[Gd@C_{2v}\text{-}C_{82}]^{3-}$ has a HOMO energy level that matches well with the LUMO energy level of the $PhCH_2$ radical, providing a rationalization for the observed significantly different reactivity.

**Keywords:** endohedral metallofullerene; open shell; electrosynthesis; fullerene anion; Gd@$C_{2v}$-$C_{82}$

## 1. Introduction

Endohedral metallofullerenes (EMFs) have garnered substantial attention from scientists due to their exceptional chemical and physical properties [1–9]. Since the discovery of the first solvent-extractable EMF, La@$C_{82}$ [10], there has been a boom in interest in EMFs. According to the electronic structure, EMFs are categorized into two main groups: closed-shell EMFs and open-shell EMFs. Closed-shell EMFs possess completely filled electron shells, meaning that all the molecular orbitals are paired with electrons. Consequently, they exhibit a stable and relatively non-reactive structure. In contrast, open-shell EMFs have partially filled electron shells, resulting in the presence of unpaired electrons in their molecular orbitals. This characteristic makes them more reactive and prone to chemical reactions [11–13]. Nevertheless, compared with closed-shell EMFs, the reaction of open-shell EMFs is relatively less studied.

The physical and chemical properties of open-shell EMFs can be altered by functionalization [11,14–17]. By introducing trifluoromethyl, phenyl, or dichlorophenyl groups into the Er@$C_{82}$ cages, the derived products exhibit distinct photoluminescent properties [18]. Additionally, multi-hydroxylated Pr@$C_{82}$ [19], Gd@$C_{82}$ [20–23], and $Ho_x$@$C_{82}$ [24] show enhanced solubility, rendering them suitable for use in biomedicine as novel biological radical scavengers, radiographic agents, magnetic resonance imaging agents, and "missile" drugs [20].

$Gd@C_{2v}$-$C_{82}$, with the inner Gd atom capable of transferring three electrons to the $C_{82}$ cages, represents a typical open-shell molecule. The electronic states of this system can be described as $Gd^{3+}@C_{82}{}^{3-}$ [25–27]. $Gd@C_{2v}$-$C_{82}$ has magnetic properties stemming from the interaction between the spin of the metal atom and the electron spin of the cage [28]. $Gd@C_{2v}$-$C_{82}$ has found widespread use in various biomedical applications, including MRI contrast [20,29,30], antitumor therapeutics [31–33], adjuvant therapeutic agents [34,35], metabolic disease modulation [22,36,37], and so forth. Indeed, the diverse range of applications for $Gd@C_{2v}$-$C_{82}$ in biomedicine can be attributed to appropriate derivatization. The process of modifying its chemical structure through derivatization allows for tailoring its properties to suit specific biomedical applications, thereby enhancing its effectiveness and versatility in various medical contexts. In addition, the carbon cage encasing the Gd ion serves as a protective barrier, preventing the release of $Gd^{3+}$ and thereby eliminating potential toxicity concerns. Therefore, expanding the variety of $Gd@C_{2v}$-$C_{82}$ derivatives holds great significance as it allows for a more comprehensive exploration of their potential properties and applications.

In this work, we utilized the electrochemical method to produce $Gd@C_{2v}$-$C_{82}$ anions, which were then subjected to the benzylization reaction. Interestingly, this reaction has been demonstrated to be ineffective for the neutral form of the molecule. The reduction of $Gd@C_{2v}$-$C_{82}$ to its trianions resulted in a notable increase in reactivity towards the benzylization reaction. Three benzylated $Gd@C_{2v}$-$C_{82}$ isomers (**2a**, **2b**, and **2d**) are isolated and characterized. The structures are confirmed through X-ray single crystal diffraction, revealing that they are monoadducts with a benzyl group attached to the [5,6,6] junction of the $C_{82}$ cage. In addition, each isomer exists in a mirror-symmetric form. The electrochemical properties of the benzylated $Gd@C_{2v}$-$C_{82}$ isomers show distinct characteristics compared to the pristine $Gd@C_{2v}$-$C_{82}$. Through density functional theory (DFT) calculations, the reactivity of the $Gd@C_{2v}$-$C_{82}$ trianion towards the benzyl group was unveiled.

## 2. Results and Discussion

Despite $Gd@C_{2v}$-$C_{82}$ having an unpaired electron, it does not exhibit reactivity towards benzyl bromide ($PhCH_2Br$) as shown in Scheme 1. As previously known, $Gd@C_{82}$ demonstrates its capability as an excellent electron acceptor, with the capacity to accept up to seven electrons [38,39]. Electrochemical reduction can be used to selectively synthesize fullerene anions by controlling the applied potential. To enhance the reactivity, $Gd@C_{2v}$-$C_{82}$ was subjected to electrochemical reduction, specifically to its monoanion, using controlled potential electrolysis (CPE). The resulting $Gd@C_{2v}$-$C_{82}$ monoanion was then subjected to the benzylation reaction. However, it was observed that the monoanion also shows no reactivity toward $PhCH_2Br$ (Scheme 1), indicating that the nucleophilicity of the $Gd@C_{2v}$-$C_{82}$ monoanion is not sufficiently strong to attack the carbon cation ($C^+$) of $PhCH_2Br$. Therefore, additional electrons were added to the $Gd@C_{2v}$-$C_{82}$ cage through the CPE process. According to the cyclic voltammogram of $Gd@C_{2v}$-$C_{82}$, the second reduction is a simultaneous two-electron transfer process [27,39]; therefore, it is rational to consider the anions obtained via CPE to be trianions. Once the electrogeneration of $[Gd@C_{2v}$-$C_{82}]^{3-}$ was completed, $PhCH_2Br$ was added to the solution in one portion and stirred for 3h. High performance liquid chromatography (HPLC) observation confirms the occurrence of the reaction between $[Gd@C_{2v}$-$C_{82}]^{3-}$ and $PhCH_2Br$ (Scheme 1). It is evident that $[Gd@C_{2v}$-$C_{82}]^{3-}$ shows enhanced reactivity toward $PhCH_2Br$.

As observed in Figure 1a, the HPLC results show the synthesis of multiple products through the reaction of $[Gd@C_{2v}$-$C_{82}]^{3-}$ with $PhCH_2Br$. To separate and purify the products from the crude mixture, a three-stage HPLC procedure was performed. Initially, the reaction mixtures in toluene were separated using a Buckyprep column to isolate the main fractions 1, 2, and 3 (Figure 1a). Subsequent chromatography employing a Buckyprep column (Figure 1b) and a 5PBB column (Figure 1c) revealed that fraction 2 comprises four compounds: **2a**, **2b**, **2c**, and **2d**. (Figure 1c). Notably, as we only succeeded in getting single crystal structures of **2a**, **2b**, and **2d**, the subsequent discussion in this study was only focused on these three products. Electrosynthesis has been commonly recognized as an effective strategy for the highly selective synthesis of fullerene derivations. The production of multiple adducts is likely ascribed to the delocalization of charges caused by the low symmetry of the $C_{82}$ cage and the high reactivity of trianionic Gd@$C_{2v}$-$C_{82}$.

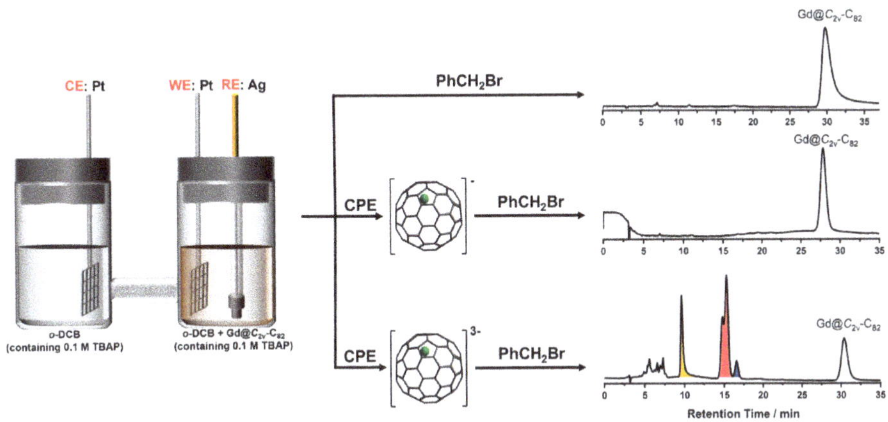


**Scheme 1.** Benzylation reactions of Gd@$C_{2v}-C_{82}$, [Gd@$C_{2v}-C_{82}$]$^{-}$, and [Gd@$C_{2v}-C_{82}$]$^{3-}$, and the corresponding HPLC profiles of the reaction crude. Conditions: Buckyprep (φ 4.6 × 250 mm), toluene, 1 mL/min. (colors indicate the areas of each fraction).

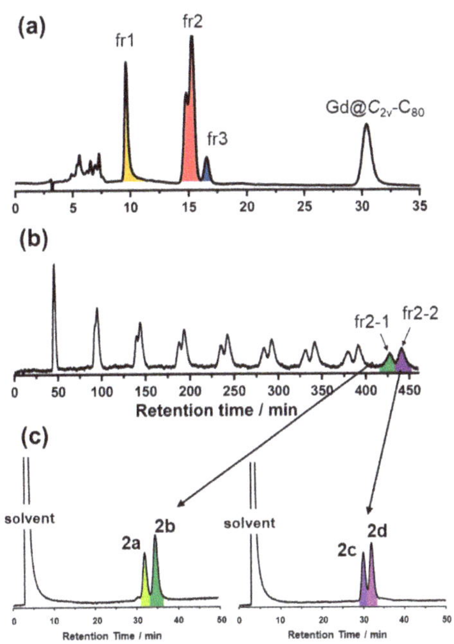


**Figure 1.** (**a**) HPLC profile of the reaction mixture (colors indicate the areas of each fraction). Conditions: Buckyprep (φ 4.6 × 250 mm), toluene, 1 mL/min. (**b**) Recycling HPLC profile of fraction 2. Conditions: Buckyprep (φ 20 × 250 mm), toluene, 7 mL/min. (**c**) HPLC profiles of fractions 2-1 and 2-2. Conditions: 5PBB (φ 4.6 × 250 mm), toluene, 1 mL/min.

The matrix-assisted laser desorption/ionization time-of-flight (MALDI–TOF) mass spectrometry gave preliminary speculation on the structure of the three products (Figure 2). The mass spectra of **2a**, **2b**, and **2d** reveal very similar molecular ion peaks, positioned around $m/z$ = 1233. This observation indicates that they are isomers, each featuring a single benzyl group ($CH_2C_6H_5$) attached to the Gd@$C_{2v}$-$C_{82}$ cage, i.e., Gd@$C_{2v}$-$C_{82}$($CH_2C_6H_5$) (theoretical $m/z$ = 1232.979). The addition pattern is consistent with that of the reported open-shell EMFs derivatives [12,40–44]. This can be attributed to the paramagnetic behavior of Gd@$C_{2v}$-$C_{82}$, which leads to specific addition patterns of functional groups in its derivatives.

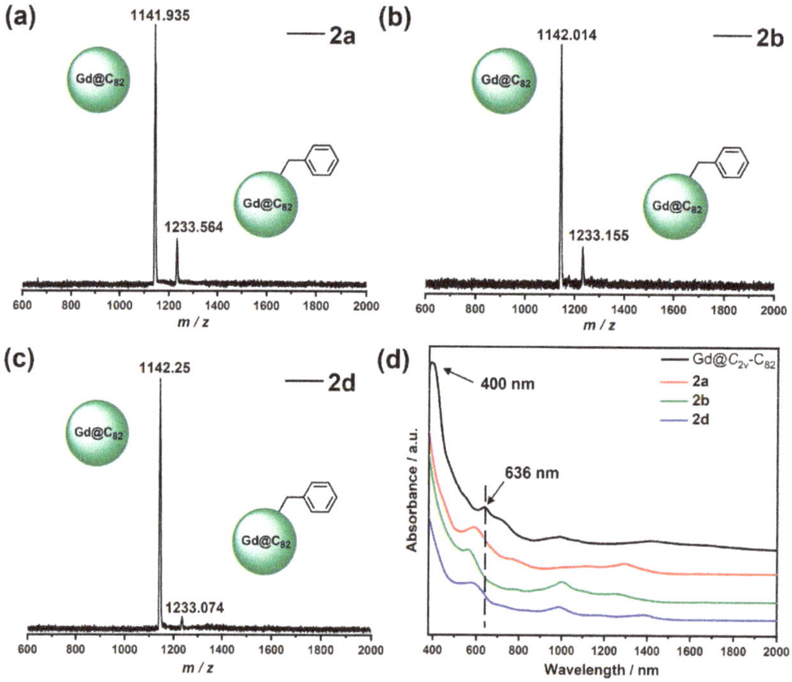


**Figure 2.** MALDI-TOF mass spectra of **2a** (**a**), **2b** (**b**), and **2d** (**c**) with TPB as the matrix. (**d**) UV-vis-NIR spectra of **2a**, **2b**, **2d**, and Gd@$C_{2v}$-$C_{82}$ in $CS_2$ solution.

The optical properties that can provide crucial insights into the electronic structures were studied through the analysis of the UV-vis-NIR absorption spectra. As shown in Figure 2, the UV-vis-NIR spectra of **2a**, **2b**, and **2d** show remarkably different behavior compared to that of the parent Gd@$C_{2v}$-$C_{82}$, which is attributed to the changes in the electronic structure of the cage caused by the addition of the benzyl groups [45]. The characteristic absorption of Gd@$C_{2v}$-$C_{82}$ at $\lambda$ = 636 nm undergoes a blue shift upon the addition of the benzyl groups, which can be explained by the reduction of the $\pi$-conjugated region in the molecule [44]. The characteristic band of Gd@$C_{82}$ at 400 nm disappears in all benzyl adducts, which is ascribed to the quenching of unpaired electrons by the odd number of benzyl groups on the Gd@$C_{82}$ cage. A similar phenomenon was observed in the absorption of Gd@$C_{82}$ morpholine derivatives [44].

Single crystal X-ray diffraction measurement is widely accepted as the most reliable technique for analyzing the structure of isomers [46–49]. It provides solid evidence for the structure confirmation of **2a**, **2b**, and **2d**. We made efforts to obtain single crystals for all the products, unfortunately, only products **2a**, **2b**, and **2d** were successfully crystallized. Figure 3 displays the single crystal structures of the three isomers, namely, **2a**, **2b**, and **2d**.

Notably, **2a** and **2b** were co-crystallized with decapyrrylcorannulene (DPC) molecules [50]. The DPC molecules were found to form a V-shaped structure when combined with **2a** and **2b**, with the angle between the DPC molecules being 64.9° and 64.8°, respectively. The arrangement of these two molecules is very similar to other reported structures [51–54]. In **2a**·2DPC, the distances between the centroid of the $C_{82}$ cage and the central five-membered rings of DPC molecules are 7.361 Å and 7.394 Å, while they are 7.439 Å and 7.474 Å in **2b**·2DPC. From there, it can be deduced that the distance between the carbon cage and the DPC is approximately 2.8–3.2 Å, indicating the presence of $\pi$-$\pi$ interactions between the carbon cage and DPC molecules. These $\pi$-$\pi$ interactions play a significant role in stabilizing the co-crystallized structure of **2a** and **2b** with DPC molecules. All three crystals feature a monoclinic $P2_1/c$ space group. The carbon cages and embedded Gd atoms (in cyan color) display varying degrees of disorder. An interesting observation is that, irrespective of the location of the benzyl addition, the Gd atoms with major occupancy are consistently found underneath a hexagonal ring along the $C_2$ axis of the $C_{2v}$-$C_{82}$ cage (marked in orange in Figure 3). And, the distances between Gd and the adjacent carbon atoms, ranging from 1.9 Å to 2.4 Å, indicate a strong interaction between the endohedral Gd and the outer $C_{2v}$-$C_{82}$ cage.

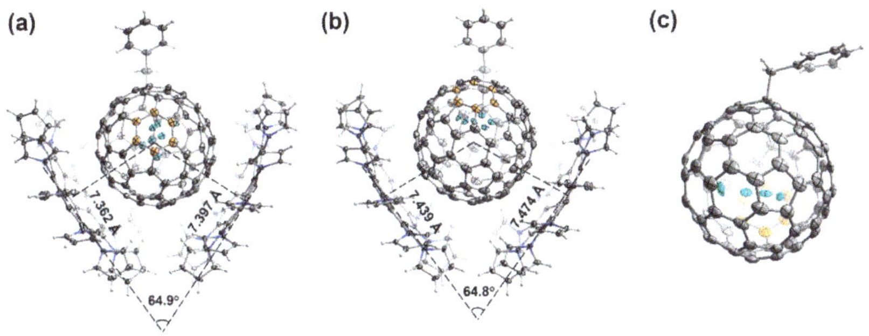


**Figure 3.** Single crystal structures of **2a**·2DPC (**a**), **2b**·2DPC (**b**), and **2d** (**c**) with thermal ellipsoids set at the 20% probability level. The solvent molecules are omitted and only one cage orientation is shown.

As depicted in Figure 3, the benzyl group was added in the form of a single bond at the [5,6,6]-ring junction of the $C_{2v}$-$C_{82}$ cage. Despite the addition, the $C_{2v}$ symmetry of the cage is preserved, indicating that the attachment of the benzyl group did not disrupt the overall symmetry of the molecule (Figure 4). Interestingly, all three crystal structures of **2a**, **2b**, and **2d** contain mirror-symmetric enantiomers, suggesting that each compound exists in a pair of structures that are mirror images of each other. To gain a better understanding of the addition sites and the arrangement of the benzyl groups in each enantiomer, a Schlegel diagram of the $C_{2v}$-$C_{82}$ cage was mapped. In this diagram, the addition sites of the benzyl groups in each enantiomer are marked (1–2 for **2a**, 3–4 for **2b**, and 5–6 for **2d**) (Figure 4a), providing valuable insights into the spatial orientation and symmetry of the derivatives. It can be observed that the addition sites in structures that are mutually enantiomeric are mirror-symmetric. The denoted sites 1 and 2 refer to cage 1 and cage 2 of **2a**, sites 3 and 4 refer to cage 1 and cage 2 of **2b**, and sites 5 and 6 refer to cage 1 and cage 2 of **2d**. Figure 4b displays the mirror-symmetric enantiomers for **2a**, **2b**, and **2d**. For every isomer, there is an identical structure that is a mirror image of the original isomer. Owing to the existence of the $C_2$ axis, the addition sites labeled as $1'$ to $6'$ on the cage are equivalent to the corresponding sites of 1 to 6.

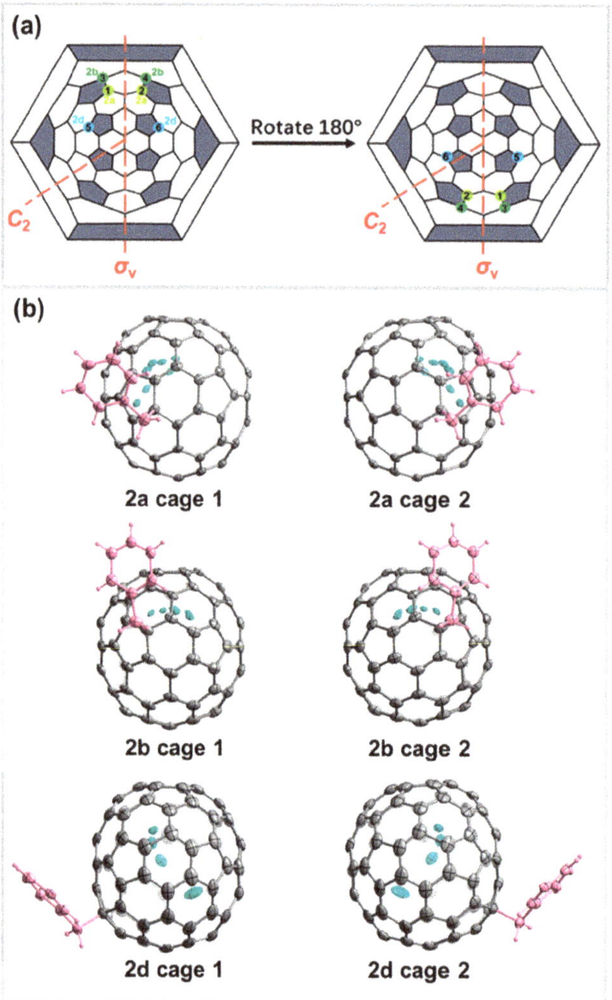


**Figure 4.** (**a**) Schlegel diagram showing the addition sites of **2a**, **2b**, and **2d**. (**b**) Structures of enantiomers of **2a**, **2b**, and **2d** with 20% thermal probability. Internal disordered sites of gadolinium atoms were also shown. Benzyl groups are labeled in pink. The co-crystallization and solvent molecules were omitted for clarity.

The crystal packing and molecular organization for each isomer were studied. As shown in Figure 5, the smallest asymmetric unit for the three products consists of four fullerene cages. Furthermore, **2a** and **2b** are arranged in a head-to-tail mode with DPC molecules along the *b*-axis. Additionally, the DPC molecules adopt an S-shape configuration along the *c*-axis. The distances between the two back-to-back DPC molecules are in the range of 3.4–3.8 Å, indicating the existence of $\pi$-$\pi$ interactions. These interactions lead to favorable $\pi$-electron overlap and stabilize the crystal packing. The crystal structure of **2d** without DPC co-crystallization reveals the formation of a one-dimensional zigzag supramolecular chain along the *b*-axis. The detailed crystallographic data for these compounds can be found in the experimental section.

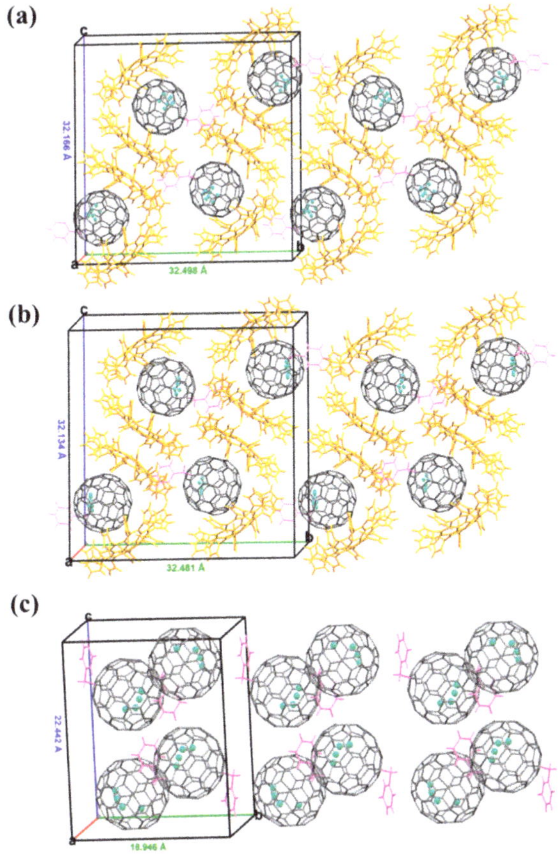


**Figure 5.** The packing structures of (**a**) **2a**·**2DPC**, (**b**) **2b**·**2DPC**, and (**c**) **2d**. The solvent molecules are omitted and only one cage orientation is shown.

The addition of functional groups on the fullerene cages can greatly alter the electrochemical properties of the compounds. Cyclic voltammetry (CV) measurements were conducted for each isomer and compared with the CV curves of the pristine Gd@$C_{2v}$-$C_{82}$ [27]. As reported in the literature, the pristine Gd@$C_{2v}$-$C_{82}$ shows three pairs of reversible redox peaks and two irreversible redox processes in its CV curves, and the second and the fourth reduction are two-electron processes [27]. However, the addition of the benzyl group to form isomers **2a**, **2b**, and **2d** has led to changes in the electrochemical characteristics of the compounds. Moreover, the addition position of the benzyl group was found to have effects on the electrochemical behavior [46,55]. As illustrated in Figure 6, the CV curves of **2a**, **2b**, and **2d** exhibit more redox signals compared with the pristine Gd@$C_{2v}$-$C_{82}$, suggesting the presence of distinct electron transfer processes, which is attributed to the addition of the benzyl group and its influence on the electronic structure of Gd@$C_{2v}$-$C_{82}$. In the case of **2a**, the CV curves display irreversible cathodic electrochemical behavior, with peak potential ($E_{pc}$) values observed at −0.97, −1.51, −1.66, −1.93, and −2.39 V vs. Fc/$Fc^+$ (Table 1). On the other hand, **2a** also shows one quasi-reversible oxidation peak with an $E_{pa}$ value of +0.14 V vs. Fc/$Fc^+$ and two irreversible oxidations with $E_{pa}$ values of +0.74 and +1.03 V vs. Fc/$Fc^+$. For compound **2b**, there are three quasi-reversible reductions with $E_{pc}$ values of −1.06, −1.89, and −2.11 V vs. Fc/$Fc^+$, while the second reduction is found to be irreversible, with an $E_{pc}$ value of −1.58 V vs. Fc/$Fc^+$. The electrochemically anodic behavior of **2b** exhibits irreversible characteristics. CV measurements for **2d** show two irreversible reductions with $E_{pc}$ values of −0.86 and −1.52 V vs. $Fc/Fc^+$ and two quasi-reversible reductions with $E_{pc}$ values of −1.77 and −2.04 V vs. $Fc/Fc^+$. Moreover, compound **2d** exhibits electrochemically irreversible anodic behavior, which is similar to the behavior observed for **2d**. The distinct electrochemical characteristics of **2a**, **2b**, and **2d** may be attributed to the specific addition position of the benzyl group and its influence on the electronic structure of the Gd@$C_{2v}$-$C_{82}$ cage. In addition, compared to the redox behavior of Gd@$C_{2v}$-$C_{82}$, the first reduction potentials of **2a**, **2b**, and **2d** show negative shifts, and the first oxidation potentials show positive shifts, which indicates a decrease in electron-donating/accepting properties after modification with the benzyl group. The enlarged electrochemical gaps in the derivatives **2a**, **2b**, and **2d** demonstrate the enhanced stability of the derivatives relative to the parent Gd@$C_{2v}$-$C_{82}$. The redox potentials for Gd@$C_{2v}$-$C_{82}$ and its benzylated products are summarized in Table 1.

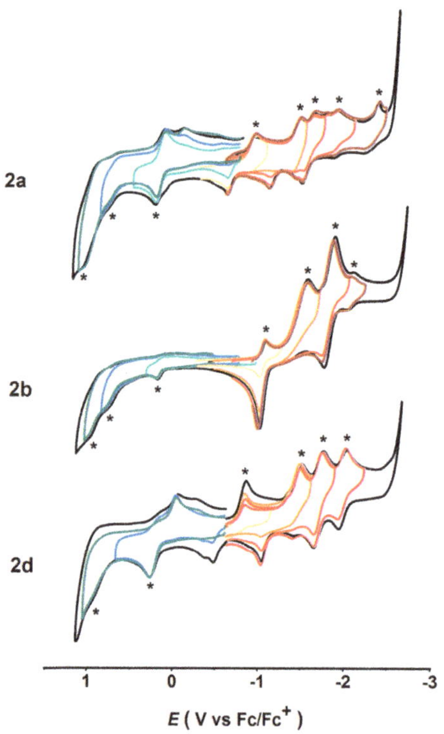


**Figure 6.** Cyclic voltammograms of **2a**, **2b**, and **2d** recorded in a 0.05 M solution of $TBAPF_6$ in *o*-DCB at a scan rate of 100 mV $s^{-1}$. (The redox processes are labeled with a star and each process is indicated in different colors).

**Table 1.** Redox potentials (V vs. $Fc/Fc^+$) of Gd@$C_{2v}$-$C_{82}$ and **2a**, **2b**, and **2d**.

| | $E^{2+/3+}$ | $E^{+/2+}$ | $E^{0/+}$ | $E^{0/-}$ | $E^{-/2-}$ | $E^{2-/3-}$ | $E^{3-/4-}$ | $E^{4-/5-}$ | $E_{gap}$ |
|---|---|---|---|---|---|---|---|---|---|
| Gd@$C_{2v}$-$C_{82}$ [a] | | +1.08 [b] | +0.09 [c] | −0.39 [c] | −1.38 [c,d] | | −2.22 [b,d] | | 0.48 |
| **2a** | +1.03 [b] | +0.74 [b] | +0.14 [c] | −0.97 [b] | −1.51 [b] | −1.66 [b] | −1.93 [b] | −2.39 [b] | 1.11 |
| **2b** | +0.94 [b] | +0.73 [b] | +0.19 [b] | −1.06 [c] | −1.58 [b] | −1.89 [c] | −2.11 [c] | | 1.25 |
| **2d** | | +0.94 [b] | +0.26 [b] | −0.86 [b] | −1.52 [b] | −1.77 [c] | −2.04 [c] | | 1.12 |

[a] From reference [21]; [b] peak potentials; [c] quasi-reversible process; and [d] two-electron process.

Density functional theory calculations at the M06-2X/6-31G*~SDD level were then carried out to disclose the underlying reactivity of Gd@$C_{2v}$-$C_{82}$ and its anions. Figure 7 illustrates the optimized structures of **2a**, **2b**, and **2d**, which are consistent with the corresponding single crystal structures. For pristine Gd@$C_{2v}$-$C_{82}$, it is well known that it has an unpaired electron distributed on the outer cage due to the $3e^-$ transfer from the internal metal to the cage. However, our calculations with different spin multiplicities reveal that all three monoadducts have the octet ground state completely stemming from the Gd-$4f^7$ electron configuration of their internal $Gd^{3+}$ cation (Table 2). Clearly, the unpaired electron (spin) on the outer cage of Gd@$C_{2v}$-$C_{82}$ has been effectively quenched as a result of the formation of a single bond with the benzyl group, which thus leads to the excellent stability of the adducts.

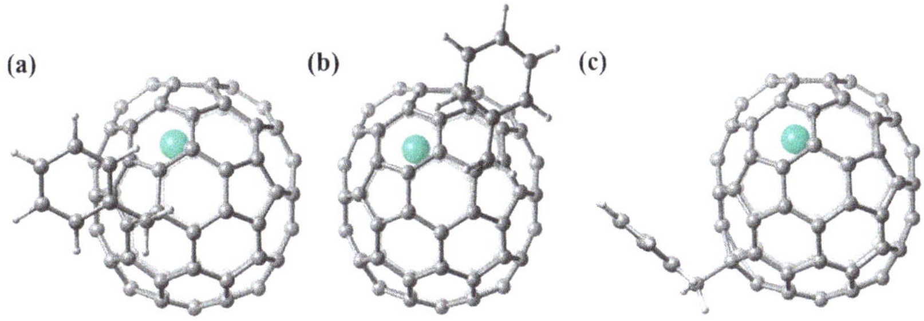


**Figure 7.** Optimized structures of (**a**) **2a**, (**b**) **2b**, and (**c**) **2d**.

**Table 2.** Relative energies (kcal/mol) of **2a**, **2b**, and **2d** as well as [Gd@$C_{2v}$-$C_{82}$]$^-$ and [Gd@$C_{2v}$-$C_{82}$]$^{3-}$ with different spin multiplicities (*M*).

| *M* | 6 | 8 | 10 |
|---|---|---|---|
| **2a** | 21.0 | 0.0 | 20.9 |
| **2b** | 17.0 | 6.8 | 14.9 |
| **2d** | 22.1 | 1.8 | 22.2 |
| [Gd@$C_{2v}$-$C_{82}$]$^-$ | 58.3 | 0.0 | 17.8 |
| [Gd@$C_{2v}$-$C_{82}$]$^{3-}$ | 0.0 | 18.3 | 26.7 |

In addition, the reactivity of different charged forms of Gd@$C_{2v}$-$C_{82}$ was elaborated through DFT simulations. Figure 8 shows the calculated energy levels of frontier molecular orbitals of Gd@$C_{2v}$-$C_{82}$ with various charge states (0, −1, −3). Consistent with our experimental findings, only [Gd@$C_{2v}$-$C_{82}$]$^{3-}$ demonstrates HOMO (highest occupied molecular orbital) energy levels that are comparable to the LUMO (lowest unoccupied molecular orbital) of $PhCH_2$ radical, thus enhancing the orbital interactions and charge transfer during the addition reaction. Through the process of electrochemical reduction, the orbital energy levels of Gd@$C_{2v}$-$C_{82}$ are greatly elevated, resulting in a significant improvement in its reactivity.

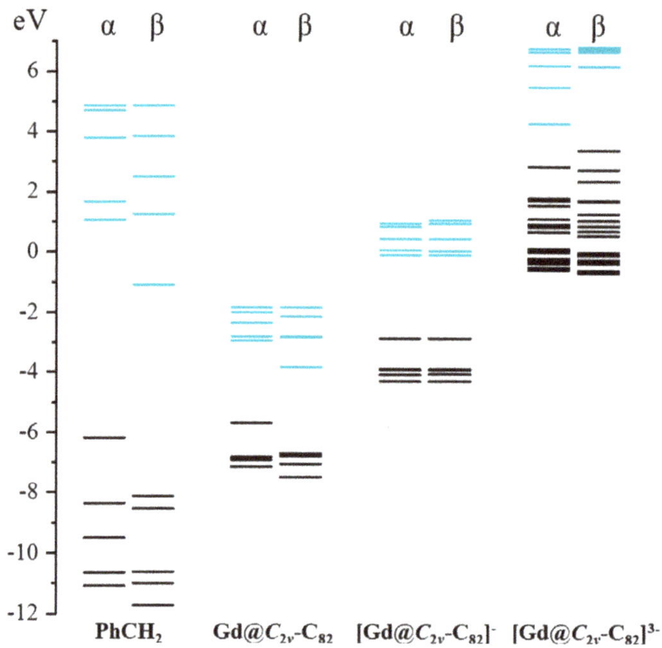


**Figure 8.** Frontier molecular orbital energy levels of $PhCH_2$ radical and Gd@$C_{2v}$-$C_{82}$ in different charge states (occupied: black; unoccupied: cyan).

## 3. Materials and Methods

Gd@$C_{2v}$-$C_{82}$ was purchased from Xiamen Funano New Material Technology Company LTD, Xiamen, China. $PhCH_2Br$, anhydrous *o*-DCB, 1,1,4,4-Tetraphenyl-1,3-butadiene (TPB), toluene, $CDCl_3$, and $CS_2$ were purchased from Aladdin and used as received. Tetrabutylammonium perchlorate (TBAP) and Tetra-n-butylammonium hexauorophosphate ($TBAPF_6$) were recrystallized from absolute ethanol and dried under vacuum before use. The purity of the products was verified by HPLC (LaboACE LC-5060, Japan Analytical Industry Co., Ltd., Tokyo, Japan) equipped with Buckyprep and 5PBB columns, with toluene used as the eluent. UV-vis-NIR spectra were recorded using a SHIMADZU UV-3600 spectrophotometer (Shimadzu Corporation, Kyoto, Japan). MALDI-TOF MS measurements were conducted using a Bruker autoex speed mass spectrometer. CV studies were conducted in a solution of *o*-DCB containing 0.05 M $TBAPF_6$ in a one-compartment cell connected to a CHI 760E workstation, Shanghai, China. A 2 mm diameter glassy carbon disk, a Pt wire, and a Ag wire were used as the working, counter, and pseudoreference electrodes, respectively. Ferrocene (Fc) was introduced into the solution at the end of each experiment as an internal standard. All reactions were carried out inside a glove box under Ar protection.

### *3.1. Synthesis of* **2a**, **2b**, *and* **2d**

A 2 mg (1.75 mmol) sample of Gd@$C_{2v}$-$C_{82}$ was electrochemically reduced at −1.36 V versus a silver wire reference electrode in 8 mL of *o*-DCB containing 0.1 M TBAP as the supporting electrolyte under an argon atmosphere. The potentiostat was turned off once the electrogeneration of $[Gd@C_{2v}\text{-}C_{82}]^{3-}$ was completed. Subsequently, 100 equivalents of $PhCH_2Br$ (21.5 μL) were added into the solution in one portion under stirring. The reaction was allowed to proceed for 2 h with stirring, and, finally, it was electrochemically oxidized at a potential of 0 V. The solvent was evaporated under a vacuum, and the crude was washed with plenty of methanol to remove the unreacted $PhCH_2Br$ and the supporting electrolyte. The slurry was separated by filtration. The residue was dried at 70 °C under a vacuum and dissolved in toluene. The soluble fraction was then purified by HPLC using Buckyprep and 5PBB columns.

### 3.2. Single Crystal X-ray Crystallography

The single crystals of **2a** and **2b** were obtained by co-crystallization with DPC molecules. In a 5.0 mL centrifuge tube, 0.5 mg of **2a** was dissolved in 0.5 mL of carbon disulfide. Then, a toluene solution containing DPC was added to the tube, with a molar ratio of DPC to fullerene in the mixed solution being 2:1. After around three weeks, black sheet-like crystals were formed. A crystal of 0.4 mm × 0.25 mm × 0.03 mm dimension was mounted in the 100 K nitrogen cold stream on an XtaLAB PRO MM007HF diffractometer with Cu-Kα radiation (λ = 1.5406 Å). The CrystalClear software package (Rigaku) was used for data collection, cell refinement, and data reduction. The crystal structures were solved by direct methods and refined by the full-matrix method based on $F^2$ using the SHELXLTL software package. All the non-hydrogen atoms were refined anisotropically, and the positions of the hydrogen atoms were generated geometrically.

Crystal data of **2a·2DPC**: $C_{216}H_{95}Gd_{0.88}N_{20}$, $M_w$ = 3108.89 amu, monoclinic, $P2_{1/C}$, $a$ = 14.6998 (1) Å, $b$ = 32.4984 (3) Å, $c$ = 32.1664 (3) Å, $\alpha$ = 90°, $\beta$ = 102.071 (1)°, $\gamma$ = 90°, $V$ = 15,026.8 (2) Å$^3$, $T$ = 100 K, $Z$ = 4, $R$ indices (all data) $R_1$ = 0.0896 (23,666), w$R_2$ = 0.2647 (29,671), GOF = 1.054.

Black sheet-like crystals of **2b** were obtained through the same procedures as for **2a**. Crystallographic characterization of a piece of co-crystal (0.12 mm × 0.1 mm × 0.08 mm) was performed at 100 K by using synchrotron radiation (0.65250 Å) with a MarCCD detector at the beamline BL17B station of Shanghai Synchrotron Radiation Facility.

Crystal data of **2b·2DPC**: $C_{216}H_{95}Gd_{0.87}N_{20}$, $M_w$ = 3106.92 amu, monoclinic, $P2_{1/C}$, $a$ = 14.6857 (4) Å, $b$ = 32.4812 (11) Å, $c$ = 32.1342 (9) Å, $\alpha$ = 90°, $\beta$ = 101.791 (1)°, $\gamma$ = 90°, $V$ = 15,004.9 (8) Å$^3$, $T$ = 100 K, $Z$ = 4, $R$ indices (all data) $R_1$ = 0.1054 (20,309), w$R_2$ = 0.3207 (29,611), GOF = 1.052.

The crystal of **2d** was grown through solvent diffusion in which a near-saturated carbon disulfide solution was injected into the glass tube followed by slow injections of hexane, and the fullerene solution was layered at a height of about 5 cm. At this point, a clear delamination between hexane and carbon disulfide solution appears, creating a diffusion interface that provides favorable conditions for crystal growth. The tube is closed with a cap and left at a low temperature for several weeks until the solvent diffusion is complete and several crystals are attached to the tube walls. The crystal of **2d** (0.1 mm × 0.08 mm × 0.06 mm) was tested using the same instruments and methods as **2b**.

Crystal data of **2d**: $C_{87}H_7Gd_{0.91}$, $M_w$ = 1233.20 amu, monoclinic, $P2_{1/C}$, $a$ = 10.891 (2) Å, $b$ = 18.946 (4) Å, $c$ = 22.442 (5) Å, $\alpha$ = 90°, $\beta$ = 96.45 (3)°, $\gamma$ = 90°, $V$ = 4601.4 (16) Å$^3$, $T$ = 100 K, $Z$ = 4, $R$ indices (all data) $R_1$ = 0.1409 (6211), w$R_2$ = 0.3370 (8277), GOF = 1.069.

The supplementary crystallographic data for this paper can be obtained free of charge from the Cambridge Crystallographic Data Centre with CCDC numbers 2281483, 2283118, and 2283119 via www.ccdc.cam.ac.uk/data_request/cif (accessed on 1 July 2023).

### 3.3. Computational Details

The calculations were carried out at the DFT level with the M06-2X [56] function, the 6-31G* all-electron basis set of C and H atoms, the SDD basis set, and the corresponding effective core potentials of Gd [57–59]. All the above DFT calculations were performed using the Gaussian 09 software package [60]. The results were visualized with CYLview [61] and VMD [62] software.

## 4. Conclusions

In summary, this study presents the first successful synthesis of open-shell Gd@$C_{2v}$-$C_{82}$ benzyl monoadducts using the electrochemical synthesis method. Three Gd@$C_{2v}$-

$C_{82}(PhCH_2)$ isomers (**2a**, **2b**, and **2d**) were synthesized, isolated, and characterized. The crystal structures revealed that the benzyl group is attached to the [5,6,6] junction of the cage through a single bond and each isomer contains a mirror-symmetric enantiomer. The addition of the benzyl group and its specific position have notable effects on the electrochemical behavior. DFT calculations reveal that $[Gd@C_{2v}\text{-}C_{82}]^{3-}$ has a HOMO energy level that aligns effectively with the LUMO energy levels of the $PhCH_2$ radical, rationalizing the higher reactivity towards benzyl. The electrochemical method shows great potential in regulating the chemical reactivity of fullerenes, opening up possibilities for creating a wide range of new fullerene products.

**Author Contributions:** Conceptualization, F.-F.L.; data curation, X.Z. and Y.-R.Y.; formal analysis, X.Z., Y.-R.Y. and Y.H.; funding acquisition, Q.Z., P.P., P.J. and F.-F.L.; investigation, X.Z., Y.H., S.Y. and P.P.; methodology, Y.H., L.Y., Y.Z., Q.Z. and P.J.; project administration, P.P. and F.-F.L.; software, Y.-R.Y. and P.J.; supervision, P.P. and F.-F.L.; validation, Y.-R.Y.; writing—original draft, X.Z. and L.Y.; Writing—review and editing, P.J. and F.-F.L. All authors have read and agreed to the published version of the manuscript.

**Funding:** The financial support from the National Natural Science Foundation of China (Grant No. 22071070, 21925104, 22171068, and 22271238) and the Fundamental Research Funds for the Central Universities (WK2060000051) is gratefully acknowledged.

**Data Availability Statement:** Data is contained within the article.

**Conflicts of Interest:** The authors declare no conflict of interest.

## References

1. Popov, A.A.; Yang, S.; Dunsch, L. Endohedral fullerenes. *Chem. Rev.* **2013**, *113*, 5989–6113. [CrossRef] [PubMed]
2. Dunsch, L.; Yang, S. Metal nitride cluster fullerenes: Their current state and future prospects. *Small* **2007**, *3*, 1298–1320. [CrossRef] [PubMed]
3. Yang, S.; Wei, T.; Jin, F. When metal clusters meet carbon cages: Endohedral clusterfullerenes. *Chem. Soc. Rev.* **2017**, *46*, 5005–5058. [CrossRef]
4. Li, F.-F.; Rodríguez-Fortea, A.; Poblet, J.M.; Echegoyen, L. Reactivity of Metallic Nitride Endohedral Metallofullerene Anions: Electrochemical Synthesis of a $Lu_3N@I_h\text{-}C_{80}$ derivative. *J. Am. Chem. Soc.* **2011**, *133*, 2760–2765. [CrossRef]
5. Li, F.-F.; Rodríguez-Fortea, A.; Peng, P.; Campos Chavez, G.A.; Poblet, J.M.; Echegoyen, L. Electrosynthesis of $Sc_3N@I_h\text{-}C_{80}$ methano derivative from trianionic $Sc_3N@I_h\text{-}C_{80}$. *J. Am. Chem. Soc.* **2012**, *134*, 7480–7487. [CrossRef] [PubMed]
6. Duong, H.M.; Myint, S.M.; Tran, T.Q.; Le, D.K. Chapter 6—Post-spinning treatments to carbon nanotube fibers. In *Carbon Nanotube Fibers and Yarns*; Miao, M., Ed.; Woodhead Publishing: Soston, UK, 2020; pp. 103–134. [CrossRef]
7. Paukov, M.; Kramberger, C.; Begichev, I.; Kharlamova, M.; Burdanova, M. Functionalized Fullerenes and Their Applications in Electrochemistry, Solar Cells, and Nanoelectronics. *Materials* **2023**, *16*, 1276.
8. Serda, M.; Korzuch, J.; Dreszer, D.; Krzykawska-Serda, M.; Musioł, R. Interactions between modified fullerenes and proteins in cancer nanotechnology. *Drug Discov. Today* **2023**, *28*, 103704. [CrossRef]
9. Grebowski, J.; Litwinienko, G. Metallofullerenols in biomedical applications. *Eur. J. Med. Chem.* **2022**, *238*, 114481. [CrossRef]
10. Chai, Y.; Guo, T.; Jin, C.; Haufler, R.E.; Chibante, L.P.; Fure, J.; Wang, L.V.; Alford, J.M.; Smalley, R.E. Fullerenes with Metals Inside. *J. Phys. Chem.* **1991**, *95*, 7564–7568.
11. Maeda, Y.; Tsuchiya, T.; Lu, X.; Takano, Y.; Akasaka, T.; Nagase, S. Current progress on the chemical functionalization and supramolecular chemistry of $M@C_{82}$. *Nanoscale* **2011**, *3*, 2421–2429.
12. Fang, H.; Cong, H.; Suzuki, M.; Bao, L.; Yu, B.; Xie, Y.; Mizorogi, N.; Olmstead, M.M.; Balch, A.L.; Nagase, S.; et al. Regioselective Benzyl Radical Addition to an Open-Shell Cluster Metallofullerene. Crystallographic Studies of Cocrystallized $Sc_3C_2@I_h\text{-}C_{80}$ and Its Singly Bonded Derivative. *J. Am. Chem. Soc.* **2014**, *136*, 10534–10540. [CrossRef]
13. Bao, L.; Chen, M.; Pan, C.; Yamaguchi, T.; Kato, T.; Olmstead, M.M.; Balch, A.L.; Akasaka, T.; Lu, X. Crystallographic Evidence for Direct Metal–Metal Bonding in a Stable Open-Shell $La_2@I_h\text{-}C_{80}$ Derivative. *Angew. Chem. Int. Ed.* **2016**, *55*, 4242–4246. [CrossRef]
14. Huang, H.; Zhang, L.; Gao, X.J.; Guo, X.; Cui, R.; Xu, B.; Dong, J.; Li, Y.; Gan, L.; Chang, F. Regioselective Polyamination of $Gd@C_{2v}(9)\text{-}C_{82}$ and non-high performance liquid chromatography rapid separation of $Gd@C_{82}(morpholine)_7$. *Chem. Mater.* **2018**, *30*, 64–68. [CrossRef]
15. Liu, Z.; Huang, H.; Wang, Y.-X.; Dong, B.-W.; Sun, B.-Y.; Jiang, S.-D.; Gao, S. Amination of the $Gd@C_{82}$ endohedral fullerene: Tunable substitution effect on quantum coherence behaviors. *Chem. Sci.* **2020**, *11*, 10737–10743. [CrossRef]
16. Kandrashkin, Y.E.; Zaripov, R.B.; Liu, F.; Büchner, B.; Kataev, V.; Popov, A.A. Temperature-dependent dynamics of endohedral fullerene $Sc_2@C_{80}(CH_2Ph)$ studied by EPR spectroscopy. *Phys. Chem. Chem. Phys.* **2021**, *23*, 18206–18220. [CrossRef]
17. Kandrashkin, Y.E.; Zaripov, R.B. Scandium dimetallofullerene with a single-electron metal–metal bond as a spectroscopic ruler for EPR measurements. *Phys. Chem. Chem. Phys.* **2022**, *24*, 19743–19752. [CrossRef] [PubMed]

18. Xu, D.; Jiang, Y.; Wang, Y.; Zhou, T.; Shi, Z.; Omachi, H.; Shinohara, H.; Sun, B.; Wang, Z. Turning on the Near-Infrared Photoluminescence of Erbium Metallofullerenes by Covalent Modification. *Inorg. Chem.* **2019**, *58*, 14325–14330. [CrossRef] [PubMed]
19. Sun, D.; Huang, H.; Yang, S.; Liu, Z.; Liu, S. Synthesis and characterization of a water-soluble endohedral metallofullerol. *Chem. Mater.* **1999**, *11*, 1003–1006. [CrossRef]
20. Li, T.; Dorn, H.C. Biomedical applications of metal-encapsulated fullerene nanoparticles. *Small* **2017**, *13*, 1603152. [CrossRef] [PubMed]
21. Li, L.; Zhen, M.; Wang, H.; Sun, Z.; Jia, W.; Zhao, Z.; Zhou, C.; Liu, S.; Wang, C.; Bai, C. Functional Gadofullerene Nanoparticles Trigger Robust Cancer Immunotherapy Based on Rebuilding an Immunosuppressive Tumor Microenvironment. *Nano Lett.* **2020**, *20*, 4487–4496. [CrossRef]
22. Jia, W.; Li, X.; Zhang, T.; Wang, C.; Zhen, M. Efficiently normalizing leukopoiesis by gadofullerene nanoparticles to ameliorate radiation-triggered myelosuppression. *J. Mater. Chem. B* **2023**, *11*, 7401–7409. [CrossRef] [PubMed]
23. Zhou, C.; Zhen, M.; Yu, M.; Li, X.; Yu, T.; Liu, J.; Jia, W.; Liu, S.; Li, L.; Li, J.; et al. Gadofullerene inhibits the degradation of apolipoprotein B100 and boosts triglyceride transport for reversing hepatic steatosis. *Sci. Adv.* **2020**, *6*, eabc1586. [CrossRef] [PubMed]
24. Cagle, D.W.; Kennel, S.J.; Mirzadeh, S.; Alford, J.M.; Wilson, L.J. In vivo studies of fullerene-based materials using endohedral metallofullerene radiotracers. *Proc. Natl. Acad. Sci. USA* **1999**, *96*, 5182–5187. [CrossRef] [PubMed]
25. Hino, S.; Umishita, K.; Iwasaki, K.; Miyazaki, T.; Miyamae, T.; Kikuchi, K.; Achiba, Y. Photoelectron spectra of metallofullerenes, $Gd@C_{82}$ and $La_2@C_{80}$: Electron transfer from the metal to the cage. *Chem. Phys. Lett.* **1997**, *281*, 115–122. [CrossRef]
26. Hino, S.; Iwasaki, K.; Umishita, K.; Miyazaki, T.; Kikuchi, K.; Achiba, Y. Photoelectron spectra of metal fullerenes: Are metal fullerenes molecular crystals or not? *J. Electron. Spectrosc. Relat. Phenom.* **1996**, *78*, 493–496. [CrossRef]
27. Suzuki, T.; Kikuchi, K.; Oguri, F.; Nakao, Y.; Suzuki, S.; Achiba, Y.; Yamamoto, K.; Funasaka, H.; Takahashi, T. Electrochemical properties of fullerenolanthanides. *Tetrahedron* **1996**, *52*, 4973–4982. [CrossRef]
28. Takano, Y.; Aoyagi, M.; Yamada, M.; Nikawa, H.; Slanina, Z.; Mizorogi, N.; Ishitsuka, M.O.; Tsuchiya, T.; Maeda, Y.; Akasaka, T. Anisotropic magnetic behavior of anionic $Ce@C_{82}$ carbene adducts. *J. Am. Chem. Soc.* **2009**, *131*, 9340–9346. [CrossRef]
29. Li, J.; Wang, T.; Feng, Y.; Zhang, Y.; Zhen, M.; Shu, C.; Jiang, L.; Wang, Y.; Wang, C. A water-soluble gadolinium metallofullerenol: Facile preparation, magnetic properties and magnetic resonance imaging application. *Dalton Trans.* **2016**, *45*, 8696–8699. [CrossRef]
30. Kato, H.; Kanazawa, Y.; Okumura, M.; Taninaka, A.; Yokawa, T.; Shinohara, H. Lanthanoid endohedral metallofullerenols for MRI contrast agents. *J. Am. Chem. Soc.* **2003**, *125*, 4391–4397. [CrossRef]
31. Kang, S.-G.; Araya-Secchi, R.; Wang, D.; Wang, B.; Huynh, T.; Zhou, R. Dual inhibitory pathways of metallofullerenol $Gd@C_{82}(OH)_{22}$ on matrix metalloproteinase-2: Molecular insight into drug-like nanomedicine. *Sci. Rep.* **2014**, *4*, 4775. [CrossRef]
32. Chen, S.H.; Kang, S.-g.; Luo, J.; Zhou, R. Charging nanoparticles: Increased binding of $Gd@C_{82}(OH)_{22}$ derivatives to human MMP-9. *Nanoscale* **2018**, *10*, 5667–5677. [CrossRef] [PubMed]
33. Deng, R.; Wang, Y.; Zhen, M.; Li, X.; Zou, T.; Li, J.; Yu, T.; Zhou, Y.; Lu, Z.; Xu, H.; et al. Real-time monitoring of tumor vascular disruption induced by radiofrequency assisted gadofullerene. *Sci. China Mater.* **2018**, *61*, 1101–1111. [CrossRef]
34. Zhang, Y.; Shu, C.; Zhen, M.; Li, J.; Yu, T.; Jia, W.; Li, X.; Deng, R.; Zhou, Y.; Wang, C. A novel bone marrow targeted gadofullerene agent protect against oxidative injury in chemotherapy. *Sci. China Mater.* **2017**, *60*, 866–880. [CrossRef]
35. Lucas, D.; Scheiermann, C.; Chow, A.; Kunisaki, Y.; Bruns, I.; Barrick, C.; Tessarollo, L.; Frenette, P.S. Chemotherapy-induced bone marrow nerve injury impairs hematopoietic regeneration. *Nat. Med.* **2013**, *19*, 695–703. [CrossRef] [PubMed]
36. Li, X.; Zhen, M.; Zhou, C.; Deng, R.; Yu, T.; Wu, Y.; Shu, C.; Wang, C.; Bai, C. Gadofullerene nanoparticles reverse dysfunctions of pancreas and improve hepatic insulin resistance for type 2 diabetes mellitus treatment. *ACS Nano* **2019**, *13*, 8597–8608. [CrossRef]
37. Jia, W.; Zhen, M.; Li, L.; Zhou, C.; Sun, Z.; Liu, S.; Zhao, Z.; Li, J.; Wang, C.; Bai, C. Gadofullerene nanoparticles for robust treatment of aplastic anemia induced by chemotherapy drugs. *Theranostics* **2020**, *10*, 6886–6897. [CrossRef]
38. Sun, B.; Li, M.; Luo, H.; Shi, Z.; Gu, Z. Electrochemical properties of metallofullerenes and their anions. *Electrochim. Acta* **2002**, *47*, 3545–3549. [CrossRef]
39. Sun, B.; Luo, H.; Shi, Z.; Gu, Z. Electrochemical properties of $Gd@C_{82}$ and its anions. *Electrochem. Commun.* **2002**, *4*, 47–49. [CrossRef]
40. Takano, Y.; Ishitsuka, M.O.; Tsuchiya, T.; Akasaka, T.; Kato, T.; Nagase, S. Retro-reaction of singly bonded $La@C_{82}$ derivatives. *Chem. Commun.* **2010**, *46*, 8035–8036. [CrossRef]
41. Zhao, Y.-l.; Zhou, Q.; Lian, Y.-f.; Yu, H.-t. Molecular structures of $Pr@C_{72}$ and $Pr@C_{72}(C_6H_3Cl_2)$: A combined experimental–theoretical investigation. *RSC Adv.* **2015**, *5*, 97568–97578. [CrossRef]
42. Kareev, I.E.; Bubnov, V.P.; Fedutin, D.N.; Yagubskii, E.B.; Lebedkin, S.F.; Laukhina, E.E.; Kuvychko, I.V.; Strauss, S.H.; Boltalina, O.B. Trifluoromethylation of Endohedral Metallofullerenes M@C82(M = Y, Ce): Synthesis, Isolation and Structure. In *Hydrogen Materials Science and Chemistry of Carbon Nanomaterials*; Springer: Dordrecht, The Netherlands, 2007; pp. 235–242.
43. Han, X.; Xin, J.; Yao, Y.; Liang, Z.; Qiu, Y.; Chen, M.; Yang, S. Capturing the Long-Sought $Dy@C_{2v}(5)$-$C_{80}$ via Benzyl Radical Stabilization. *Nanomaterials* **2022**, *12*, 3291. [CrossRef] [PubMed]
44. Huang, H.; Liu, Z.; Zhang, L.; Guo, X.; Cui, R.; Dong, J.; Chang, F.; Jiang, S.; Gao, S.; Sun, B. Paramagnetic properties adjustment for $Gd@C_{2v}(9)$-$C_{82}$ by regioselective multi-amination. *Carbon* **2020**, *158*, 320–326. [CrossRef]

45. Shu, C.; Slebodnick, C.; Xu, L.; Champion, H.; Fuhrer, T.; Cai, T.; Reid, J.E.; Fu, W.; Harich, K.; Dorn, H.C.; et al. Highly Regioselective Derivatization of Trimetallic Nitride Templated Endohedral Metallofullerenes via a Facile Photochemical Reaction. *J. Am. Chem. Soc.* **2008**, *130*, 17755–17760. [CrossRef] [PubMed]
46. Li, F.-F.; Pinzón, J.R.; Mercado, B.Q.; Olmstead, M.M.; Balch, A.L.; Echegoyen, L. [2 + 2] Cycloaddition Reaction to $Sc_3N@I_h$-$C_{80}$. The Formation of Very Stable [5,6]- and [6,6]-Adducts. *J. Am. Chem. Soc.* **2011**, *133*, 1563–1571. [CrossRef]
47. Cai, T.; Slebodnick, C.; Xu, L.; Harich, K.; Glass, T.E.; Chancellor, C.; Fettinger, J.C.; Olmstead, M.M.; Balch, A.L.; Gibson, H.W.; et al. A Pirouette on a Metallofullerene Sphere: Interconversion of Isomers of N-tritylpyrrolidino $I_h$ $Sc_3N@C_{80}$. *J. Am. Chem. Soc.* **2006**, *128*, 6486–6492. [CrossRef]
48. Chen, M.; Guan, R.; Li, B.; Yang, L.; Niu, C.; Jin, P.; Wang, G.-W.; Yang, S. Anomalous Cis-Conformation Regioselectivity of Heterocycle-Fused $Sc_3N@D_{3h}$-$C_{78}$ Derivatives. *Angew. Chem. Int. Ed.* **2021**, *60*, 7880–7886. [CrossRef]
49. Hu, Y.; Yao, Y.-R.; Liu, X.; Yu, A.; Xie, X.; Abella, L.; Rodríguez-Fortea, A.; Poblet, J.M.; Akasaka, T.; Peng, P.; et al. Unexpected formation of 1,2- and 1,4-bismethoxyl $Sc_3N@I_h$-$C_{80}$ derivatives via regioselective anion addition: An unambiguous structural identification and mechanism study. *Chem. Sci.* **2021**, *12*, 8123–8130. [CrossRef]
50. Xu, Y.-Y.; Tian, H.-R.; Li, S.-H.; Chen, Z.-C.; Yao, Y.-R.; Wang, S.-S.; Zhang, X.; Zhu, Z.-Z.; Deng, S.-L.; Zhang, Q.; et al. Flexible decapyrrylcorannulene hosts. *Nat. Commun.* **2019**, *10*, 485. [CrossRef]
51. Chen, M.; Zhao, Y.; Jin, F.; Li, M.; Guan, R.; Xin, J.; Yao, Y.-R.; Zhao, X.; Wang, G.-W.; Zhang, Q.; et al. Decisive role of non-rare earth metals in high-regioselectivity addition of $\mu_3$-carbido clusterfullerene. *Inorg. Chem. Front.* **2022**, *9*, 5688–5696. [CrossRef]
52. Jin, F.; Xin, J.; Guan, R.; Xie, X.-M.; Chen, M.; Zhang, Q.; Popov, A.A.; Xie, S.-Y.; Yang, S. Stabilizing a three-center single-electron metal–metal bond in a fullerene cage. *Chem. Sci.* **2021**, *12*, 6890–6895. [CrossRef]
53. Zhang, Y.; Guan, R.; Chen, M.; Shen, Y.; Pan, Q.; Lian, Y.; Yang, S. Favorite Orientation of the Carbon Cage and a Unique Two-Dimensional-Layered Packing Model in the Cocrystals of $Nd@C_{82}$(I,II) Isomers with Decapyrrylcorannulene. *Inorg. Chem.* **2021**, *60*, 1462–1471. [CrossRef]
54. Yang, S.; Zhou, X.; Hu, Y.; Abella, L.; Yao, Y.-R.; Peng, P.; Zhang, Q.; Rodríguez-Fortea, A.; Poblet, J.M.; Li, F.-F. Effects of Solvents on Reaction Products: Synthesis of Endohedral Metallofullerene Oxazoline and Epoxide. *J. Org. Chem.* **2023**, *88*, 4234–4243. [CrossRef] [PubMed]
55. Hu, Y.; Solé-Daura, A.; Yao, Y.-R.; Liu, X.; Liu, S.; Yu, A.; Peng, P.; Poblet, J.M.; Rodríguez-Fortea, A.; Echegoyen, L.; et al. Chemical Reactions of Cationic Metallofullerenes: An Alternative Route for Exohedral Functionalization. *Chem. Eur. J.* **2020**, *26*, 1748–1753. [CrossRef] [PubMed]
56. Zhao, Y.; Truhlar, D.G. The M06 suite of density functionals for main group thermochemistry, thermochemical kinetics, noncovalent interactions, excited states, and transition elements: Two new functionals and systematic testing of four M06-class functionals and 12 other functionals. *Theor. Chem. Acc.* **2008**, *120*, 215–241. [CrossRef]
57. Petersson, G.A.; Al-Laham, M.A. A complete basis set model chemistry. II. Open-shell systems and the total energies of the first-row atoms. *J. Chem. Phys.* **1991**, *94*, 6081–6090. [CrossRef]
58. Petersson, G.A.; Bennett, A.; Tensfeldt, T.G.; Al-Laham, M.A.; Shirley, W.A.; Mantzaris, J. A complete basis set model chemistry. I. The total energies of closed-shell atoms and hydrides of the first-row elements. *J. Chem. Phys.* **1988**, *89*, 2193–2218. [CrossRef]
59. Cao, X.; Dolg, M. Segmented contraction scheme for small-core lanthanide pseudopotential basis sets. *J. Mol. Struct. THEOCHEM* **2002**, *581*, 139–147. [CrossRef]
60. Frisch, M.J.; Trucks, G.W.; Schlegel, H.B.; Scuseria, G.E.; Robb, M.A.; Cheeseman, J.R.; Scalmani, G.; Barone, V.; Mennucci, B.; Petersson, G.A. *Gaussian 09, Revision A.01*; Gaussian Inc.: Wallingford, CT, USA, 2009.
61. Legault, C.Y. *CYLview, 1.0b*; Université de Sherbrooke: Sherbrooke, QC, Canada, 2009.
62. Humphrey, W.; Dalke, A.; Schulten, K. VMD: Visual molecular dynamics. *J. Mol. Graph.* **1996**, *14*, 33–38. [CrossRef]

inorganics

MDPI

Article

# $TmCN@C_{82}$: Monometallic Clusterfullerene Encapsulating a $Tm^{3+}$ Ion

**Huichao Zhang, Jinpeng Xin, Huaimin Jin, Wenhao Xiang, Muqing Chen, Yang-Rong Yao * and Shangfeng Yang ***

Key Laboratory of Precision and Intelligent Chemistry, Department of Materials Science and Engineering, Anhui Laboratory of Advanced Photon Science and Technology, University of Science and Technology of China, Hefei 230026, China

* Correspondence: yryao@ustc.edu.cn (Y.-R.Y.); sfyang@ustc.edu.cn (S.Y.)

**Abstract:** Metal cyanide clusterfullerenes (CYCFs) are formed via the encapsulation of a single metal atom and a cyanide unit inside fullerene cages, endowing them with excellent properties in various applications. In this work, we report the synthesis, isolation, and characterizations of the first cases of thulium (Tm)-based CYCFs with the popular $C_{82}$ carbon cages. The structural elucidation of the two $TmCN@C_{82}$ isomers was achieved via diverse analytical techniques, including mass spectrometry, Vis-NIR spectroscopy, single-crystal X-ray crystallography, and cyclic voltammetry. The crystallographic analyses unambiguously confirmed the molecular structures of the two $TmCN@C_{82}$ isomers as $TmCN@C_s(6)$-$C_{82}$ and $TmCN@C_{2v}(9)$-$C_{82}$. Both TmCN clusters adopt a well-established triangular configuration, with the Tm ion located on the symmetrical plane of the carbon cages. The electronic structures of both $TmCN@C_{82}$ isomers adopt a $Tm^{3+}(CN)^{-}@(C_{82})^{2-}$ configuration, exhibiting characteristic spectral and electrochemical properties reminiscent of divalent endohedral metallofullerenes (EMFs). Intriguingly, unlike the divalent $Tm^{2+}$ ion observed in the mono-metallofullerenes $Tm@C_{2n}$, a higher oxidation state of $Tm^{3+}$ is identified in the monometallic TmCN cluster due to bonding with the cyanide anion. This result provides valuable insight into the essential role of the non-metallic endo-units in governing the oxidation state of the metal ion and the electronic behaviors of EMFs.

**Keywords:** endohedral metallofullerenes; cyanide clusterfullerenes; thulium; X-ray crystallography

## 1. Introduction

Fullerene is an allotropic carbon with a great number of family members differentiated by the number of carbon atoms and structural isomerism [1]. A pristine fullerene comprises 12 pentagons and several hexagons according to the number of carbon atoms, forming a cage-like structure with a hollow cavity. Based on the unique structures of fullerenes, three strategies, including exohedral [2–4], cage skeletal [5–7], and endohedral [8–10] modifications, have been widely employed for the chemical functionalization of fullerenes. In particular, the hollow cavity can serve as a nanoscale container to encapsulate various metal atoms or metal clusters, resulting in the formation of endohedral metallofullerenes (EMFs), which have attracted significant interest due to their unique structures and excellent physical and chemical properties, which are different from those of empty fullerenes, in versatile research fields [11–13].

The exploration of EMFs began with the discovery of $La@C_{60}$ in 1985 [14], which sparked a surge of enthusiasm among researchers in understanding the structures, properties, and potential applications of EMFs. Over the past three decades, remarkable progress has been made, particularly with the identification of the first clusterfullerene, $Sc_3N@C_{80}$, in 1999, which opened doors to the encapsulation of various cluster species [15]. Subsequently, clusterfullerenes have developed dramatically as the most important branch of EMFs, with various clusters, such as metal nitride [15–17], carbide [18–21], oxide [22–25], sulfide [26–29], and cyanide [30–39]. Among them, metal cyanide clusterfullerenes (CYCFs), which feature a distinctive mono-metallic cluster with flexible configurations, provide ideal models for regulating properties in applications such as single-molecule magnets [30,32].

To date, a wide range of CYCFs have been successfully synthesized and characterized [30–39], including $MCN@C_{76}$ (M = Y [30], Tb [30], Lu [35]) and $MCN@C_{82}$ (M = Y [31,38], Tb [32,37], Dy [33], Lu [35], U [34]), and $MCN@C_{84}$ (M = Y [36,39], Tb [36], and Dy [36]). Notably, the oxidation state of uranium in the UCN cluster remains ambiguous as it can exhibit U(I) or U(III) due to the significant donation bonding from the fullerene cage and $CN^-$ cluster to uranium [34]. In contrast, cyanide clusters based on yttrium (Y), terbium (Tb), dysprosium (Dy), and lutetium (Lu) uniformly adopt a fixed oxidation state of +3. However, little is known about tunable lanthanide metals with variable oxidation states, such as thulium (Tm), which generally exhibit +2 or +3 oxidation states in fullerenes [40–44]. In addition, compared to the extensively studied rare earth metals, there have been limited reports on Tm-based EMFs, despite their promising applications as fluorescent probes. For example, Shinohara and co-workers demonstrated that the divalent $Tm^{2+}$ in $Tm@C_{88}$ isomers and trivalent $Tm^{3+}$ in $Tm_2C_2@C_{82}$ isomers exhibit characteristic photoluminescence at 1200 and 1300–2000 nm, respectively, owing to their distinct energy level transitions [45]. Hence, it is highly desirable to explore novel Tm-based EMFs to gain insights into their bonding, structures, and applications based on their variable oxidation states.

Herein, we report the synthesis, isolation, and characterizations of the first Tm-based EMFs, $TmCN@C_s(6)-C_{82}$ and $TmCN@C_{2v}(9)-C_{82}$, which, to the best of our knowledge, represent the first isolated Tm-based CYCFs. These two compounds were characterized well via mass spectrometry, single-crystal X-ray diffraction, visible-near-infrared (Vis-NIR) spectroscopy, and cyclic voltammetry. The crystallographic analysis revealed a triangular configuration of the TmCN cluster within both fullerene cages. The electronic structures of both $TmCN@C_{82}$ isomers are proposed to exhibit a $Tm^{3+}(CN)^-@(C_{82})^{2-}$ configuration, displaying characteristic spectral and electrochemical properties reminiscent of divalent EMFs. Notably, a higher oxidation state of $Tm^{3+}$ in the mono-metallic TmCN cluster, in contrast to the divalent $Tm^{2+}$ observed in mono-metallofullerenes $Tm@C_{2n}$, can be attributed to the presence of the encapsulated cyanide anion.

## 2. Results and Discussion

Soot containing Tm-based cyanide clusterfullerenes was synthesized utilizing a modified Krätschmer–Huffman direct current (DC) arc discharge method [46]. Briefly, graphite rods were densely packed with graphite powder, $Tm_2O_3$, and $TiO_2$, which acts as a catalyst to enhance the yields of the Tm-based cyanide clusterfullerenes. The molar ratio of C/Tm/Ti was maintained at 15/1/1, and the arc-discharge synthesis was performed under a low-vacuum atmosphere with 200 mbar He and 10 mbar $N_2$. The two targeted $TmCN@C_{82}$ isomers were extracted from the carbon soot samples using the Soxhlet extraction method, followed by isolation via multiple-stage high-performance liquid chromatography (HPLC). The high purity of the isolated $TmCN@C_{82}$ isomers was verified by observing a solitary peak in the HPLC chromatograms via an analytical Buckyprep column, as depicted in Figure 1. The inserted mass spectra of the two $TmCN@C_{82}$ isomers also exhibited a solitary peak at m/z = 1179, further confirming their high purity. In addition, the experimental isotopic distributions of the two $TmCN@C_{82}$ isomers show good agreement with the theoretical predictions.

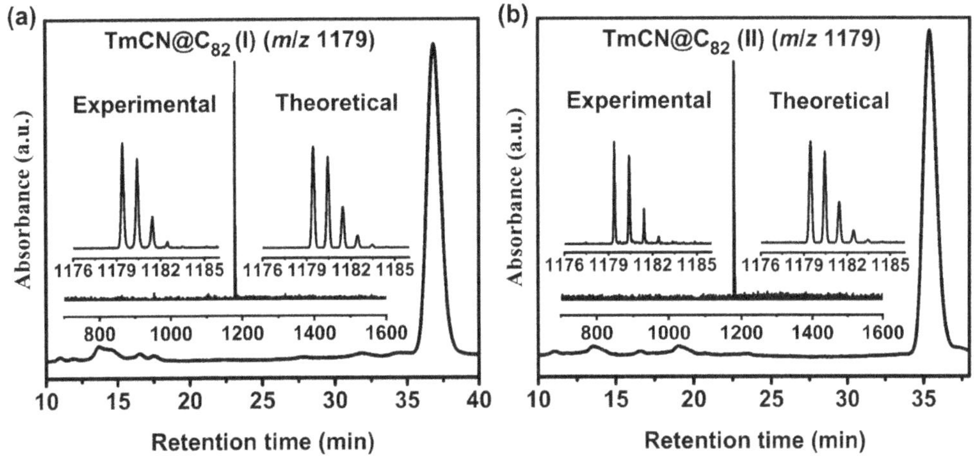


**Figure 1.** HPLC chromatograms of purified (**a**) TmCN@$C_{82}$ (I) and (**b**) TmCN@$C_{82}$ (II), obtained using an analytical Buckyprep column with toluene as the eluent and a flow rate of 1 mL/min. The inserts show the mass spectra of two TmCN@$C_{82}$ isomers with experimental versus theoretical isotopic distributions.

The molecular structures of the TmCN@$C_{82}$ (I) and TmCN@$C_{82}$ (II) were precisely characterized as TmCN@$C_s$(6)-$C_{82}$ and TmCN@$C_{2v}$(9)-$C_{82}$, respectively, via single-crystal X-ray diffraction and through their co-crystalization with decapyrrylcorannulene (DPC) molecules, which enhance the crystallinity of EMFs [47]. Both co-crystals were found to crystallize in $P2_1/c$, a space group of No. 14, which is commonly seen in the EMF-DPC co-crystallization systems, and the utilization of the low-symmetry primitive lattice for the EMF-DPC co-crystals could effectively circumvent the "crystallographic mirror dilemma" that generally appears in the EMF-$Ni^{II}$(OEP) co-crystals crystalized in the C2/m space group [48]. As shown in Figure 2a,b, the TmCN@$C_{82}$ isomers were co-crystallized with two DPC molecules, exhibiting a distinctive V-shaped configuration with a dihedral angle of about 65°. The closest C···C distances between both TmCN@$C_{82}$ isomers and the DPC molecules were measured to be ~3.37 Å; this is close to the interlayer spacing observed in graphite (3.35 Å), suggesting their $\pi$–$\pi$ stacking interactions.

The encapsulated TmCN cluster, with a CN non-metallic unit supported by a mono-metallic Tm ion, adopts a triangular configuration inside both $C_{82}$ isomers, similar to their analogs MCN@$C_s$(6)-$C_{82}$ and MCN@$C_{2v}$(9)-$C_{82}$ (M = Y, Tb, Dy, Lu). Due to the larger electronegativity of nitrogen versus carbon, the distance of the Tm-N bond is significantly shorter than that of the Tm-C bond, and the C-N bond lengths fall in the range of typical C-N triple bonds, as shown in Figure 2c,d. The encapsulated Tm ion shows some slight disorder which is probably caused by thermal vibration. The dominant Tm site, with the most significant occupancy, is located over an s-indacene-type hexagon of the carbon cages, exhibiting Tm-cage distances in the range of 2.23–2.49 Å and 2.23–2.68 Å inside the $C_s$(6)-$C_{82}$ and $C_{2v}$(9)-$C_{82}$ carbon cages, respectively. Interestingly, the dominant Tm location is aligned with the symmetrical plane of both the $C_s$(6)-$C_{82}$ and $C_{2v}$(9)-$C_{82}$ cages, which is in perfect agreement with the recent finding of the "mirror observation" for mono-metallofullerenes, suggesting that the mono-metallic cyanide clusterfullerenes display some structural characteristics that are similar to those of mono-metallofullerenes [48].

The crystallographic results suggest the preference of the TmCN cluster for encapsulation inside the cages of $C_s$(6)-$C_{82}$ and $C_{2v}$(9)-$C_{82}$, which can be inferred from our previous work on TbCN@$C_{82}$ isomers [32]. TbCN@$C_{2v}$(9)-$C_{82}$ is the most stable isomer among all TbCN@$C_{82}$ isomers, followed by TbCN@$C_s$(6)-$C_{82}$, which is similar to the relative energy order found for the YCN@$C_{82}$ isomers. The results reveal that the cage cavity volumes of $C_s$(6)-$C_{82}$ and $C_{2v}$(9)-$C_{82}$ are suitable for encapsulation of a metal cyanide cluster; this is in agreement with the theoretical studies on $C_{84}$ isomers [49], which provided insights into the relationship between the stability of an encapsulated endo-unit and the volume of a specific carbon cage.

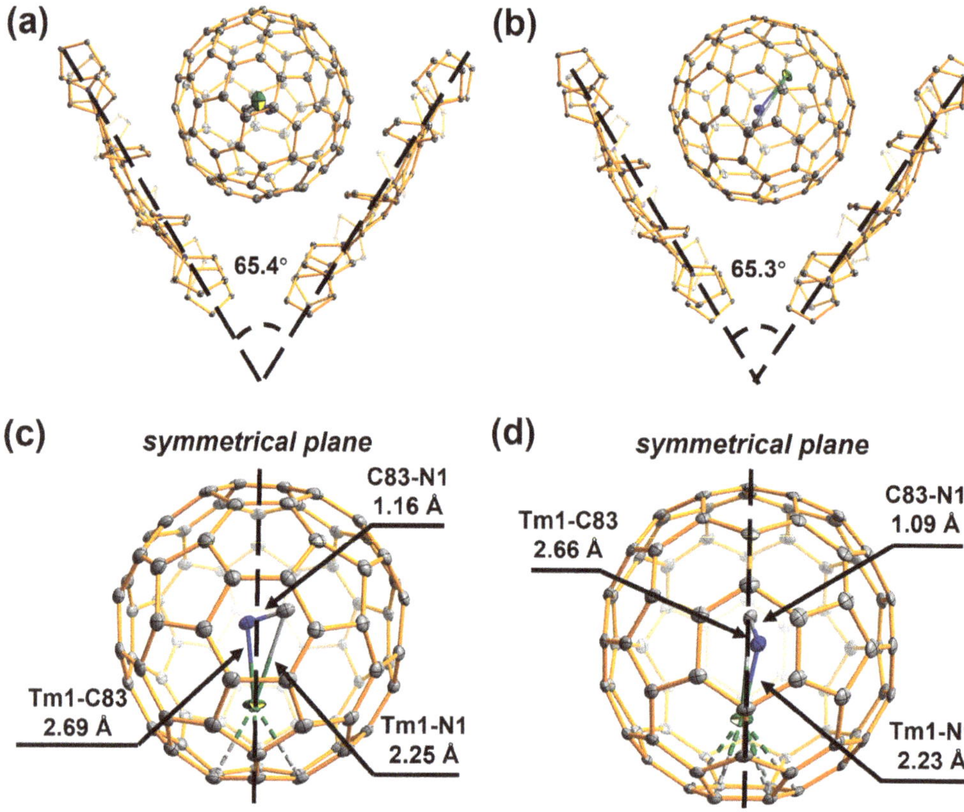


**Figure 2.** Oak Ridge thermal ellipsoid plot (ORTEP) drawings of crystallographic structures of the two co-crystals. (**a**,**b**) The relative orientations between two DPC molecules and (**a**) TmCN@$C_s$(6)-$C_{82}$ and (**b**) TmCN@$C_{2v}$(9)-$C_{82}$. (**c**,**d**) The relative orientations between the encapsulated TmCN cluster with detailed structural information and the (**c**) $C_s$(6)-$C_{82}$ and (**d**) $C_{2v}$(9)-$C_{82}$ cages. The Tm, N, and C atoms are presented as green, blue, and gray, respectively.

The electronic properties of TmCN@$C_s$(6)-$C_{82}$ and TmCN@$C_{2v}$(9)-$C_{82}$ were investigated via visible-near infrared (Vis-NIR) spectroscopy. It is well-established that the absorption spectra of EMFs are primarily governed by the $\pi$–$\pi^*$ excitations of the carbon cages; thus, Vis-NIR spectroscopy is commonly employed to study the electronic structures of EMFs [12]. As shown in Figure 3, TmCN@$C_s$(6)-$C_{82}$ exhibits a distinct spectrum with absorptions at 510, 758, 1016, 1134, and 1334 nm, while TmCN@$C_{2v}$(9)-$C_{82}$ displays a strong absorption at 1338 nm, accompanied by relatively weak absorptions at 524, 622, 706, 770, 1016, and 1168 nm. The measured optical bandgap of TmCN@$C_s$(6)-$C_{82}$ is 0.76 V, slightly larger than that of TmCN@$C_{2v}$(9)-$C_{82}$ (0.71 V). The absorption features and optical bandgaps of both TmCN@$C_s$(6)-$C_{82}$ and TmCN@$C_{2v}$(9)-$C_{82}$ resemble those of Y-, Tb-, Dy-, and Lu-CYCFs (Table 1) [31–35], which share the same cage isomerism and exhibit a divalent electronic configuration. Consequently, the formal electronic configurations of the two Tm-based CYCFs can be described as $Tm^{3+}(CN)^{-}$@$(C_s(6)\text{-}C_{82})^{2-}$ and $Tm^{3+}(CN)^{-}$@$(C_{2v}(9)\text{-}C_{82})^{2-}$, respectively. Notably, compared to TmCN@$C_s$(6)-$C_{82}$ and TmCN@$C_{2v}$(9)-$C_{82}$, Tm@$C_s$(6)-$C_{82}$ and Tm@$C_{2v}$(9)-$C_{82}$ display distinct absorption spectra [50–52] despite sharing the same cage isomerism. In addition, previous X-ray absorption spectroscopy (XAS) and ultraviolet photoelectron spectroscopy (UPS) studies [53,54] demonstrated that the Tm atoms prefer an oxidation state of +2 in monometallofullerenes. Therefore, the obtained results indicate that on one hand, the Tm ion or the TmCN cluster significantly contributes to the molecular orbitals. On the other hand, the higher oxidation state (+3) of the Tm atom in the TmCN cluster is identified due to bonding with the non-metallic CN unit.

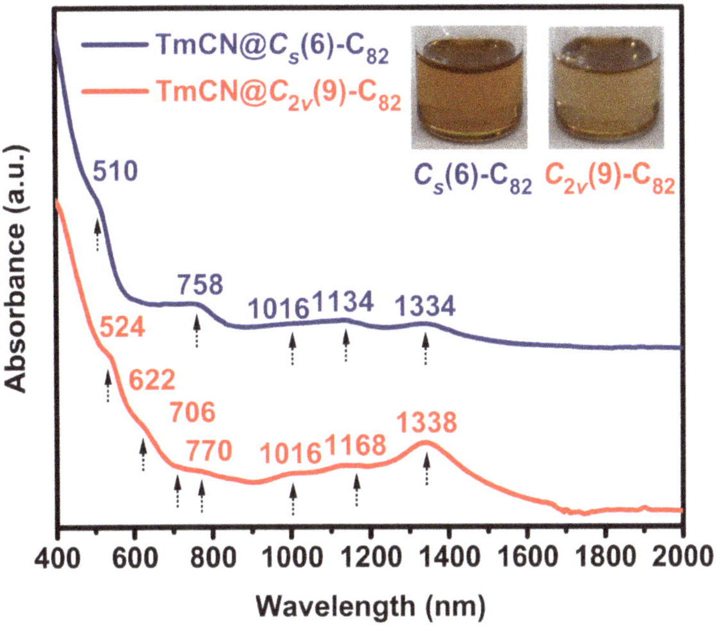


**Figure 3.** The Vis-NIR absorption spectra of TmCN@$C_s$(6)-$C_{82}$ (in blue) and TmCN@$C_{2v}$(9)-$C_{82}$ (in red). The inserts are photographs of the two CYCFs dissolved in $CS_2$. The spectra of the two TmCN@$C_{82}$ isomers are shown together by shifting one along the ordinate axis, and the absorption peaks are indicated by black arrows.

**Table 1.** Vis-NIR absorption spectral data of two TmCN@$C_{82}$ isomers as well as Y-, Tb-, Dy-, and Lu-CYCFs for comparison.

| Compounds | Absorption Peaks (nm) | $\lambda_{onset}$ (nm) | $\Delta E_{gap}$ (eV) [1] | Note |
|---|---|---|---|---|
| TmCN@$C_s$(6)-$C_{82}$ | 510, 758, 1016, 1134, 1334 | 1624 | 0.76 | this work |
| TbCN@$C_s$(6)-$C_{82}$ | 491, 744, 1072, 1132, 1318 | 1620 | 0.77 | Ref [32] |
| DyCN@$C_s$(6)-$C_{82}$ | 497, 744, 1024, 1140, 1339 | 1656 | 0.75 | Ref [33] |
| LuCN@$C_s$(6)-$C_{82}$ | 494,596, 744, 1069, 1131, 1345 | 1638 | 0.76 | Ref [35] |
| YCN@$C_s$(6)-$C_{82}$ | 491, 741,1064, 1123, 1318 | 1620 | 0.77 | Ref [31] |
| TmCN@$C_{2v}$(9)-$C_{82}$ | 524, 622, 706, 770, 1016, 1168, 1338 | 1740 | 0.71 | This work |
| TbCN@$C_{2v}$(9)-$C_{82}$ | 518, 618, 710, 772, 1144, 1320 | 1770 | 0.70 | Ref [32] |
| DyCN@$C_{2v}$(9)-$C_{82}$ | 536, 627, 1144, 1342 | 1785 | 0.69 | Ref [33] |
| LuCN@$C_{2v}$(9)-$C_{82}$ | 512, 621, 697, 782, 1096, 1296 | 1680 | 0.74 | Ref [35] |

[1] $\Delta E_{gap} = 1240/\lambda_{onset}$.

Figure 4 shows the electrochemical behaviors of TmCN@$C_s$(6)-$C_{82}$ and TmCN@$C_{2v}$(9)-$C_{82}$, determined using cyclic voltammetry (CV) with tetrabutylammonium hexafluorophosphate ($TBAPF_6$) and ferrocene (Fc) as the supporting electrolyte and internal standard, respectively. Both show four reversible reduction steps and one reversible oxidation step, recorded as +0.55, −0.67, −0.90, −1.72, and −2.16 V for TmCN@$C_s$(6) $C_{82}$ and +0.64, −0.45, −0.79, −1.60, and −1.94 for Tm@$C_{2v}$(9)-$C_{82}$. The full reduction potentials of TmCN@$C_s$(6)-$C_{82}$ are also more negatively shifted to about 0.1–0.2 V than those of Tm@$C_{2v}$(9)-$C_{82}$; thus, TmCN@$C_s$(6)-$C_{82}$ is more accessible for oxidation than Tm@$C_{2v}$(9)-$C_{82}$. Notably, the electrochemical gap of TmCN@$C_s$(6)-$C_{82}$ (1.22 V) is larger than that of Tm@$C_{2v}$(9)-$C_{82}$ (1.09 V), which results from the relatively more significant difference in their first reduction potentials.

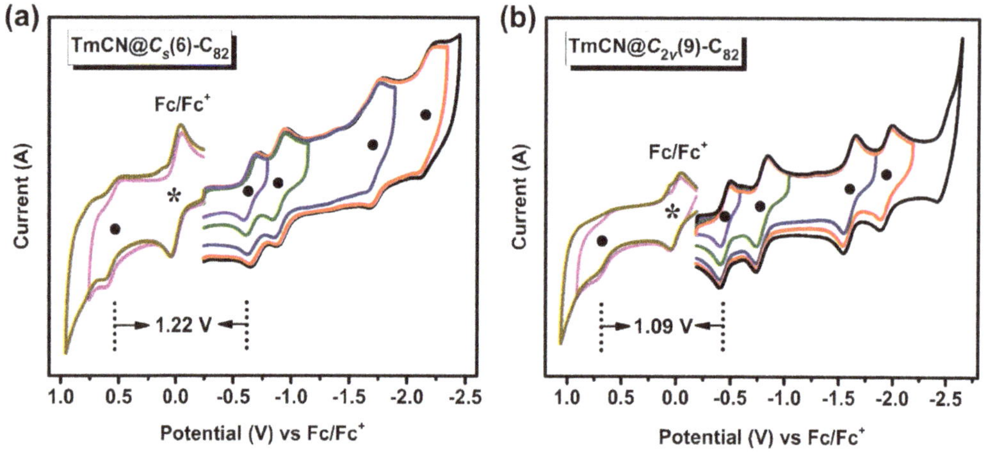


**Figure 4.** Cyclic voltammograms of (**a**) TmCN@$C_s$(6)-$C_{82}$ and (**b**) TmCN@$C_{2v}$(9)-$C_{82}$ in a *o*-DCB solution with a scan rate of 100 mV·s$^{-1}$. $TBAPF_6$ and Fc were employed as the internal standard and supporting electrolyte, respectively. Black circles and asterisks are presented to show the redox and the $Fc/Fc^+$ peaks.

The comparison of the two TmCN@$C_{82}$ isomers and other mono-metallic EMFs based on $C_s$(6)-$C_{82}$ and $C_{2v}$(9)-$C_{82}$ was also studied, as shown in Table 2. The redox behaviors of TmCN@$C_s$(6)-$C_{82}$ and TmCN@$C_{2v}$(9)-$C_{82}$ are fairly similar to those of the corresponding Y-, Tb-, Dy-, and Lu-CYCF counterparts [31–35], suggesting that both TmCN@$C_{82}$ isomers bear a divalent electronic configuration of $Tm^{3+}(CN)^-$@$(C_s(6)\text{-}C_{82})^{2-}$, which is in agreement with the spectroscopic result. Similar redox behaviors between divalent M@$C_{82}$ isomers, such as Yb@$C_s$(6)-$C_{82}$ and Yb@$C_{2v}$(9)-$C_{82}$ [55] and TmCN@$C_{82}$ isomers further confirm the electronic configurations of $Tm^{3+}(CN)^-$@$(C_{82})^{2-}$ for both $C_s$(6) and $C_{2v}$(9) isomerisms. In comparison, trivalent mono-metallofullerenes $(M^{3+})$@$(C_{82})^{3-}$ (M = Y [56], La [57,58], Pr [59], Er [60], and U [61], etc.) based on $C_s$(6) and $C_{2v}$(9) cage isomerism exhibit sharply different redox potentials with much more negatively shifted first oxidation potentials and much smaller electrochemical gaps. These phenomena originate from the three-electron charge transfer of $(M^{3+})$@$(C_{82})^{3-}$, which results in an open-shell electronic structure, while TmCN@$C_s$(6)-$C_{82}$ and TmCN@$C_{2v}$(9)-$C_{82}$ have a closed-shell electronic structures due to the two-electron transfer from the encapsulated TmCN cluster to the carbon cages.

**Table 2.** Redox potentials (V *vs.* Fc/Fc$^+$) and electrochemical gaps of TmCN@$C_s$(6)-$C_{82}$ and TmCN@$C_{2v}$(9)-$C_{82}$ and the corresponding cyanide metallofullerenes MCN@$C_{82}$ and mono-metallofullerene M@$C_{82}$ isomers.

| Compounds | ${}^{ox}E_1$ | ${}^{red}E_1$ | ${}^{red}E_2$ | ${}^{red}E_3$ | ${}^{red}E_4$ | $\Delta E_{gap}$ (eV) | Note |
|---|---|---|---|---|---|---|---|
| TmCN@$C_s$(6)-$C_{82}$ | 0.55 | −0.67 | −0.90 | −1.72 | −2.16 | 1.22 | This work |
| TbCN@$C_s$(6)-$C_{82}$ | 0.55 | −0.59 | −0.84 | −1.77 | −1.92 | 1.14 | Ref [32] |
| DyCN@$C_s$(6)-$C_{82}$ | 0.56 | −0.58 | −0.84 | 1.77 | −1.92 | 1.14 | Ref [33] |
| LuCN@$C_s$(6)-$C_{82}$ | 0.52 | −0.58 | −0.90 | −1.69 | −1.86 | 1.10 | Ref [35] |
| YCN@$C_s$(6)-$C_{82}$ | 0.56 | −0.59 | −0.84 | −1.76 | −1.92 | 1.15 | Ref [31] |
| La@$C_s$(6)-$C_{82}$ | −0.07 | −0.47 | −1.40 | −2.01 | −2.40 | 0.40 | Ref [57] |
| Pr@$C_s$(6)-$C_{82}$ | −0.07 | −0.48 | −1.39 | −1.99 | / | 0.41 | Ref [59] |
| Er@$C_s$(6)-$C_{82}$ | −0.09 | −0.44 | −1.46 | −2.04 | −2.46 | 0.35 | Ref [60] |
| Y@$C_s$(6)-$C_{82}$ | −0.07 | −0.43 | −1.43 | −2.05 | / | 0.36 | Ref [56] |
| Yb@$C_s$(6)-$C_{82}$ | 0.34 | −0.62 | −0.92 | −1.81 | −2.01 | 0.94 | Ref [55] |
| TmCN@$C_{2v}$(9)-$C_{82}$ | 0.64 | −0.45 | −0.79 | −1.60 | −1.94 | 1.09 | This work |
| TbCN@$C_{2v}$(9)-$C_{82}$ | 0.55 | −0.46 | −0.81 | −1.78 | −1.96 | 1.01 | Ref [32] |
| DyCN@$C_{2v}$(9)-$C_{82}$ | 0.56 | −0.45 | −0.81 | −1.78 | −1.96 | 1.01 | Ref [33] |
| LuCN@$C_{2v}$(9)-$C_{82}$ | 0.50 | −0.54 | −0.96 | −1.71 | −1.93 | 1.04 | Ref [35] |
| La@$C_{2v}$(9)-$C_{82}$ | 0.07 | −0.42 | −1.37 | −1.53 | −2.26 | 0.49 | Ref [58] |
| Pr@$C_{2v}$(9)-$C_{82}$ | 0.07 | −0.39 | −1.35 | −1.46 | −2.21 | 0.46 | Ref [59] |
| Er@$C_{2v}$(9)-$C_{82}$ | 0.08 | −0.42 | −1.40 | −2.18 | / | 0.50 | Ref [60] |
| U@$C_{2v}$(9)-$C_{82}$ | 0.10 | −0.43 | −1.42 | −1.76 | −1.77 | 0.53 | Ref [61] |
| Y@$C_{2v}$(9)-$C_{82}$ | 0.10 | −0.34 | −1.34 | −2.22 | / | 0.44 | Ref [56] |
| Yb@$C_{2v}$(9)-$C_{82}$ | 0.61 | −0.46 | −0.78 | −1.60 | −1.90 | 1.07 | Ref [55] |

## 3. Materials and Methods

Materials. The ragents used in this work were $Tm_2O_3$ (99.99%, Suzhou Rare Earth New Materials Co., Suzhou, China), $TiO_2$ (99%, Sinopharm Chemical Reagent Co., Ltd., Ningbo, China), $CS_2$ (99%, Sinopharm Chemical Reagent Co., Ltd.), toluene (99%, Sinopharm Chemical Reagent Co., Ltd.), tetrabutylammonium hexafluorophosphate (electrochemical-grade, Shanghai Aladdin Biochemical Technology Co., Ltd., Shanghai, China), 1,2-dichlorobenzene (99%, Shanghai Aladdin Biochemical Technology Co., Ltd.), and graphite rods and graphite powder (Sinosteel Shanghai New Graphite Material Co., Shanghai, China).

Synthesis of TmCN@$C_{82}$ isomers. Soot containing TmCN@$C_{82}$ isomers was synthesized using a modified Krätschmer–Huffman DC arc discharge method. Graphite rods packed with graphite powder, $Tm_2O_3$, and $TiO_2$, with a C/Tm/Ti molar ratio of 15/1/1, were preheated in a tube furnace at 1000 °C for 20 h under a nitrogen atmosphere. Subsequently, the preheated graphite rods were installed in a VDK-250 arc discharge reactor (Beijing Technol Science Co., Ltd., Beijing, China) under a low-vacuum atmosphere with 200 mbar He and 10 mbar $N_2$, followed by vaporization in a DC of 120 A. The two TmCN@$C_{82}$ isomers were extracted from the carbon soot with $CS_2$ using the Soxhlet extraction method. The resulting fullerene solution in $CS_2$ was evaporated and then dissolved again in toluene, followed by concentration for the following multi-stage HPLC procedures [62,63].

Isolation of TmCN@$C_{82}$ isomers. As illustrated in Figure 5, a four-stage HPLC process was employed to purify the two TmCN@$C_{82}$ isomers, and toluene was employed as the eluent in all stages. The first stage was performed on a preparative 5PYE column (20 × 250 mm) with a 15 mL/min flow rate, and fraction A, which contained the two TmCN@$C_{82}$ isomers and other fullerenes, was collected, concentrated, and then re-injected into the HPLC. In the second stage, a recycling mode was employed on a preparative Buckyprep-M column (20 × 250 mm) with a 12 mL/min flow rate. After three runs of recycling, fraction A-4, which contained the TmCN@$C_{82}$ isomers, was collected for the next stage. A semi-preparative 5PBB column (10 × 250 mm, 4 mL/min flow rate) was used in the third separation stage to remove the other mixed fullerenes in the A-4 fraction. Fraction A-4-3 was found to contain only two TmCN@$C_{82}$ isomers, and all the impurity fullerenes had been removed. In the last stage, a recycling HPLC process was employed to purify the two isomers, using a semi-preparative Buckyprep column (10 × 250 mm) with a 4 mL/min flow rate. After 13 recycling runs, purified TmCN@$C_s$(6)-$C_{82}$, labeled as A-4-3-1, and TmCN@$C_{2v}$(9)-$C_{82}$, labeled as A-4-3-2, were individually collected for further characterizations. The detailed fraction information for each stage is shown in Table 3.

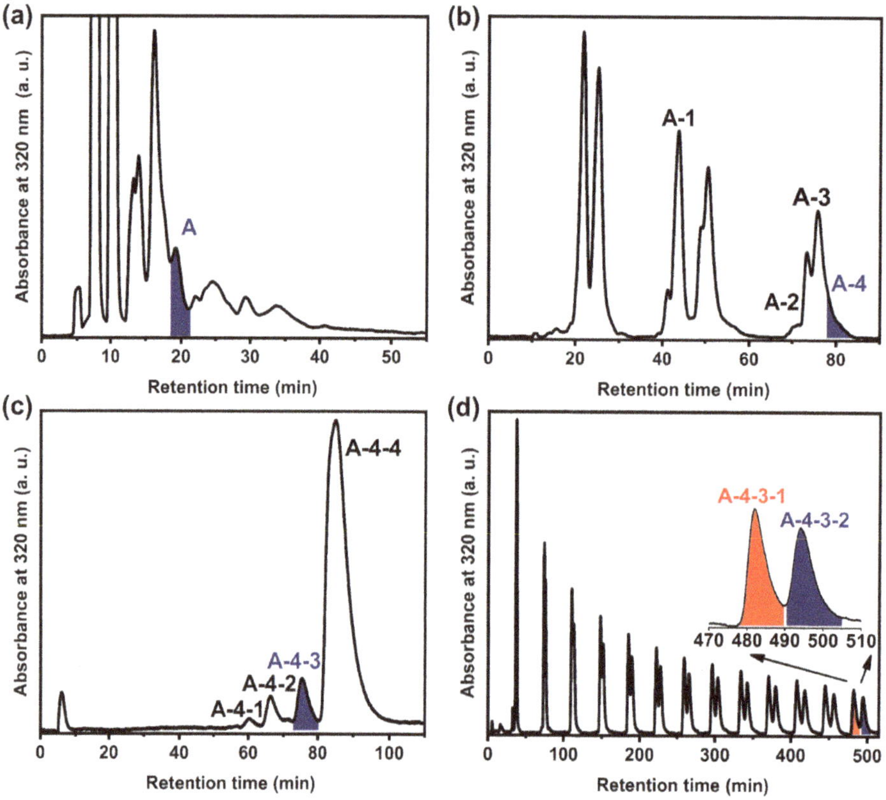


**Figure 5.** HPLC isolation procedure for TmCN@$C_s$(6)-$C_{82}$ and TmCN@$C_{2v}$(9)-$C_{82}$. (**a**) the first stage of HPLC isolation procedure on a preparative 5PYE column (20 × 250 mm). (**b**) the second stage of HPLC isolation procedure on a preparative Buckyprep-M column (20 × 250 mm). (**c**) the third stage of HPLC isolation procedure on a semi-preparative 5PBB column (10 × 250 mm). (**d**) the fourth stage of HPLC isolation procedure on a semi-preparative Buckyprep column (10 × 250 mm). The insert in (**d**) shows the recycled fractions of A-4-3-1 (red) and A-4-3-2 (blue), which represent the purified TmCN@$C_s$(6)-$C_{82}$ and TmCN@$C_{2v}$(9)-$C_{82}$, respectively.

**Table 3.** The major eluted components in each HPLC fraction and their relative abundances.

| Fraction | Subfraction | Major Component | Relative Abundance |
|---|---|---|---|
| A | A-1 | $C_{84–86}$ | 46.1% |
| | A-2 | Tm@$C_{82}$/Tm$C_{77}$N | 3.4% |
| | A-3 | Tm@$C_{80–84}$/Tm$C_{83}$N | 36.7% |
| | A-4 | Tm@$C_{80,84}$/$Ti_2C_{80}$/TmCN@$C_{82}$ | 13.8% |
| A-4 | A-4-1 | $Ti_2C_{80}$ | 2.2% |
| | A-4-2 | Tm@$C_{80}$ | 6.5% |
| | A-4-3 | TmCN@$C_{82}$ ($C_s$(6), $C_{2v}$(9)) | 8.7% |
| | A-4-1 | Tm@$C_{84}$ | 82.6% |
| A-4-3 | A-4-3-1 | TmCN@$C_s$(6)-$C_{82}$ | 52.1% |
| | A-4-3-2 | TmCN@$C_{2v}$(9)-$C_{82}$ | 47.9% |

Spectroscopic studies. The mass spectra of the two $TmCN@C_{82}$ isomers were recorded on an Autoflex III matrix-assisted laser desorption ionization time-of-flight mass spectrometer (MALDI-TOF-MS, Bruker, Germany) under positive mode. The Vis-NIR spectra of the purified samples were measured using a UV-3600 spectrometer (Shimadzu, Japan).

Electrochemical analysis. The electrochemical study was performed in a CHI 660 potentiostat (CH Instrument, USA) using a three-electrode system, including a glassy carbon working electrode, a platinum (Pt) wire counter electrode, and a silver (Ag) wire reference electrode. Tetrabutylammonium hexafluorophosphate ($TBAPF_6$) and ferrocene (Fc) were employed as the supporting electrolyte and the internal standard, respectively. $TBAPF_6$, Fc, and the samples were all dissolved in 1,2-dichlorobenzene (*o*-DCB).

X-ray crystallographic study. The co-crystals of the DPC molecules and EMF samples were obtained by slowly laying a DPC toluene solution onto an EMF $CS_2$ solution in a nuclear magnetic resonance (NMR) tube. Black block co-crystals of a suitable size for X-ray diffraction analysis were observed on the wall or in the bottom after 10–15 days. The crystallographic data were collected in the BL17B beamline station at Shanghai Synchrotron Radiation Facility. The crystallographic structures were solved using SHELXT-2014 via the intrinsic phasing method and then refined with SHELXL-2018 via the full-matrix least-squares method based on $F^2$ within the software Olex2 [64,65]. The SQUEEZE program [66] included in the PLATON program was employed to calculate the contribution of the disordered toluene solvents. The detailed crystallographic data of the two co-crystals are shown in Table 4.

**Table 4.** Crystallographic information of two $TmCN@C_{82}$ co-crystals.

| Crystal | $TmCN@C_s(6)$-$C_{82}$ | $TmCN@C_{2v}(9)$-$C_{82}$ |
|---|---|---|
| Empirical formula | $C_{210}H_{88}N_{21}Tm$ | $C_{210}H_{88}N_{21}Tm$ |
| Formula weight | 3073.94 | 3073.94 |
| Habit | Block | Block |
| Temperature, K | 100 | 100 |
| Crystal system | Monoclinic | Monoclinic |
| Space group | $P2_1/c$ | $P2_1/c$ |
| $a$, Å | 14.669 | 14.724 |
| $b$, Å | 32.146 | 32.032 |
| $c$, Å | 32.213 | 32.215 |
| $\alpha$, deg | 90 | 90 |
| $\beta$, deg | 101.74 | 101.91 |
| $\gamma$, deg | 90 | 90 |
| Volume, $Å^3$ | 14,872.3 | 14,866.8 |
| Z | 4 | 4 |
| $Dx$, $g/cm^3$ | 1.373 | 1.373 |
| $F(000)$ | 6256 | 6256 |
| Crystal Size, $mm^3$ | 0.06 × 0.03 × 0.01 | 0.12 × 0.08 × 0.03 |
| $2\theta$ max, ° | 20.499 | 24.712 |
| $R_1/wR_2$ ($I > 2\sigma(I)$) [1] | 0.1193/0.2831 | 0.0937/0.2531 |
| GOF | 1.116 | 1.052 |
| Completeness | 0.972 | 0.987 |
| Obs reflects | 5752 | 14,863 |
| Total reflects | 14,478 | 25,026 |
| Parameters | 2131 | 2162 |
| metal disorder | Tm1(0.60)/Tm2(0.08)/Tm3(0.17)/ Tm4(0.11)/Tm5(0.04) | Tm1(0.25)/Tm2(0.14)/Tm3(0.09)/Tm4(0.04)/ Tm5(0.08)/Tm6(0.13)/ Tm7(0.20)/Tm8(0.07) |

[1] $R_1 = \sum ||F_o| - |F_c|| / \sum |F_o|$, $wR_2 = [\sum w(F_o^2 - F_c^2)^2] / \sum w(F_o^2)^2]^{1/2.}$

## 4. Conclusions

In conclusion, we report the synthesis, isolation, and characterization of the first Tm-based CYCFs with the popular $C_{82}$ carbon cages via mass spectrometry, single-crystal X-ray crystallography, UV-vis-NIR spectroscopy, and cyclic voltammetry. The molecular structures of these two TmCN@$C_{82}$ were unambiguously determined to be TmCN@$C_s$(6)-$C_{82}$ and TmCN@$C_{2v}$(9)-$C_{82}$. The TmCN cluster bears a popular triangular configuration, and the Tm ion is located in one symmetrical plane of the carbon cage for both $C_s$(6)-$C_{82}$ and $C_{2v}$(9)-$C_{82}$. In addition, the two TmCN@$C_{82}$ isomers feature the characteristic spectral and electrochemical properties of divalent EMFs, suggesting a closed-shell electronic structure described as $Tm^{3+}(CN)^{-}@(C_{82})^{2-}$. Compared to the Tm@$C_{82}$ isomers, which contain a divalent $Tm^{2+}$ ion, the 3+ oxidation state of the Tm atom in the mono-metallic TmCN cluster is identified due to bonding with the encapsulated cyanide anion.

**Author Contributions:** Conceptualization, S.Y.; funding acquisition, Y.-R.Y. and S.Y.; investigation, H.Z., J.X., H.J., and W.X.; project administration, S.Y.; supervision, Y.-R.Y. and S.Y.; writing—original draft, H.Z. and Y.-R.Y.; writing—review and editing, M.C., Y.-R.Y. and S.Y. All authors have read and agreed to the published version of the manuscript.

**Funding:** This research was funded by the National Natural Science Foundation of China (51925206 and U1932214) and the Fundamental Research Funds for the Central Universities (WK2060000051).

**Data Availability Statement:** The crystallographic data can be obtained free of charge from the Cambridge Crystallographic Data Centre with the CCDC numbers 2253745 and 2253746 via www.ccdc.cam.ac.uk/data_request/cif (accessed on 31 August 2022). The other data are available upon reasonable request from the corresponding authors.

**Acknowledgments:** We thank the staff in the BL17B beamline of the National Facility for Protein Science Shanghai (NFPS) at the Shanghai Synchrotron Radiation Facility for their assistance during the crystallographic data collection.

**Conflicts of Interest:** The authors declare no conflict of interest.

## References

1. Kroto, H.W.; Heath, J.R.; O'Brien, S.C.; Curl, R.F.; Smalley, R.E. $C_{60}$: Buckminsterfullerene. *Nature* **1985**, *318*, 162–163. [CrossRef]
2. Xie, S.-Y.; Gao, F.; Lu, X.; Huang, R.-B.; Wang, C.-R.; Zhang, X.; Liu, M.-L.; Deng, S.-L.; Zheng, L.-S. Capturing the Labile Fullerene[50] as $C_{50}Cl_{10}$. *Science* **2004**, *304*, 699. [CrossRef] [PubMed]
3. Tan, Y.-Z.; Xie, S.-Y.; Huang, R.-B.; Zheng, L.-S. The stabilization of fused-pentagon fullerene molecules. *Nat. Chem.* **2009**, *1*, 450–460. [CrossRef] [PubMed]
4. Tian, H.-R.; Chen, M.-M.; Wang, K.; Chen, Z.-C.; Fu, C.-Y.; Zhang, Q.; Li, S.-H.; Deng, S.-L.; Yao, Y.-R.; Xie, S.-Y.; et al. An Unconventional Hydrofullerene $C_{66}H_4$ with Symmetric Heptagons Retrieved in Low-Pressure Combustion. *J. Am. Chem. Soc.* **2019**, *141*, 6651–6657. [CrossRef]
5. Su, Y.; Chen, Z.-C.; Tian, H.-R.; Xu, Y.-Y.; Zhang, Q.; Xie, S.-Y.; Zheng, L.-S. Implications of Nitrogen Doping on Geometrical and Electronic Structure of the Fullerene Dimers. *Chin. J. Chem.* **2021**, *39*, 93–98. [CrossRef]
6. Xiang, W.; Jiang, X.; Yao, Y.-R.; Xin, J.; Jin, H.; Guan, R.; Zhang, Q.; Chen, M.; Xie, S.-Y.; Popov, A.A.; et al. Monometallic Endohedral Azafullerene. *J. Am. Chem. Soc.* **2022**, *144*, 21587–21595. [CrossRef]
7. Hu, Z.; Dong, B.-W.; Liu, Z.; Liu, J.-J.; Su, J.; Yu, C.; Xiong, J.; Shi, D.-E.; Wang, Y.; Wang, B.-W.; et al. Endohedral Metallofullerene as Molecular High Spin Qubit: Diverse Rabi Cycles in $Gd_2@C_{79}N$. *J. Am. Chem. Soc.* **2018**, *140*, 1123–1130. [CrossRef]
8. Yao, Y.-R.; Chen, Z.-C.; Chen, L.; Zheng, S.-Y.; Yang, S.; Deng, S.-L.; Echegoyen, L.; Tan, Y.-Z.; Xie, S.-Y.; Zheng, L.-S. Two Metastable Endohedral Metallofullerenes $Sc_2C_2@C_1$(39656)-$C_{82}$ and $Sc_2C_2@C_1$(51383)-$C_{84}$: Direct-$C_2$-Insertion Products from Their Most Stable Precursors. *J. Am. Chem. Soc.* **2023**. [CrossRef]
9. Yao, Y.-R.; Shi, X.-M.; Zheng, S.-Y.; Chen, Z.-C.; Xie, S.-Y.; Huang, R.-B.; Zheng, L.-S. Atomically Precise Insights into Metal-Metal Bonds Using Comparable Endo-Units of $Sc_2$ and $Sc_2C_2$. *CCS Chem.* **2021**, *3*, 294–302. [CrossRef]
10. Jiang, H.; Yu, X.; Guo, M.; Yao, Y.-R.; Meng, Q.; Echegoyen, L.; Autschbach, J.; Chen, N. $USc_2C_2$ and $USc_2NC$ Clusters with U-C Triple Bond Character Stabilized Inside Fullerene Cages. *J. Am. Chem. Soc.* **2023**, *145*, 5645–5654. [CrossRef]
11. Rodríguez-Fortea, A.; Balch, A.L.; Poblet, J.M. Endohedral metallofullerenes: A unique host–guest association. *Chem. Soc. Rev.* **2011**, *40*, 3551–3563. [CrossRef] [PubMed]
12. Popov, A.A.; Yang, S.; Dunsch, L. Endohedral Fullerenes. *Chem. Rev.* **2013**, *113*, 5989–6113. [CrossRef] [PubMed]
13. Yang, S.; Wei, T.; Jin, F. When metal clusters meet carbon cages: Endohedral clusterfullerenes. *Chem. Soc. Rev.* **2017**, *46*, 5005–5058. [CrossRef] [PubMed]

14. Heath, J.R.; O'Brien, S.C.; Zhang, Q.; Liu, Y.; Curl, R.F.; Tittel, F.K.; Smalley, R.E. Lanthanum complexes of spheroidal carbon shells. *J. Am. Chem. Soc.* **1985**, *107*, 7779–7780. [CrossRef]
15. Stevenson, S.; Rice, G.; Glass, T.; Harich, K.; Cromer, F.; Jordan, M.R.; Craft, J.; Hadju, E.; Bible, R.; Olmstead, M.M.; et al. Small-bandgap endohedral metallofullerenes in high yield and purity. *Nature* **1999**, *401*, 55–57. [CrossRef]
16. Li, X.; Roselló, Y.; Yao, Y.-R.; Zhuang, J.; Zhang, X.; Rodríguez-Fortea, A.; de Graaf, C.; Echegoyen, L.; Poblet, J.M.; Chen, N. $U_2N@I_h(7)-C_{80}$: Fullerene cage encapsulating an unsymmetrical U(IV)=N=U(V) cluster. *Chem. Sci.* **2021**, *12*, 282–292. [CrossRef]
17. Meng, Q.; Abella, L.; Yao, Y.-R.; Sergentu, D.-C.; Yang, W.; Liu, X.; Zhuang, J.; Echegoyen, L.; Autschbach, J.; Chen, N. A charged diatomic triple-bonded U≡N species trapped in $C_{82}$ fullerene cages. *Nat. Commun.* **2022**, *13*, 7192. [CrossRef]
18. Wang, C.-R.; Kai, T.; Tomiyama, T.; Yoshida, T.; Kobayashi, Y.; Nishibori, E.; Takata, M.; Sakata, M.; Shinohara, H. A Scandium Carbide Endohedral Metallofullerene: $(Sc_2C_2)@C_{84}$. *Angew. Chem. Int. Ed.* **2001**, *40*, 397–399. [CrossRef]
19. Wang, T.-S.; Chen, N.; Xiang, J.-F.; Li, B.; Wu, J.-Y.; Xu, W.; Jiang, L.; Tan, K.; Shu, C.-Y.; Lu, X.; et al. Russian-Doll-Type Metal Carbide Endofullerene: Synthesis, Isolation, and Characterization of $Sc_4C_2@C_{80}$. *J. Am. Chem. Soc.* **2009**, *131*, 16646–16647. [CrossRef]
20. Zhang, X.; Li, W.; Feng, L.; Chen, X.; Hansen, A.; Grimme, S.; Fortier, S.; Sergentu, D.-C.; Duignan, T.J.; Autschbach, J.; et al. A diuranium carbide cluster stabilized inside a $C_{80}$ fullerene cage. *Nat. Commun.* **2018**, *9*, 2753. [CrossRef]
21. Zhuang, J.; Abella, L.; Sergentu, D.-C.; Yao, Y.-R.; Jin, M.; Yang, W.; Zhang, X.; Li, X.; Zhang, D.; Zhao, Y.; et al. Diuranium(IV) Carbide Cluster $U_2C_2$ Stabilized Inside Fullerene Cages. *J. Am. Chem. Soc.* **2019**, *141*, 20249–20260. [CrossRef] [PubMed]
22. Stevenson, S.; Mackey, M.A.; Stuart, M.A.; Phillips, J.P.; Easterling, M.L.; Chancellor, C.J.; Olmstead, M.M.; Balch, A.L. A Distorted Tetrahedral Metal Oxide Cluster inside an Icosahedral Carbon Cage. Synthesis, Isolation, and Structural Characterization of $Sc_4(\mu_3\text{-}O)_2@I_h\text{-}C_{80}$. *J. Am. Chem. Soc.* **2008**, *130*, 11844–11845. [CrossRef] [PubMed]
23. Mercado, B.Q.; Olmstead, M.M.; Beavers, C.M.; Easterling, M.L.; Stevenson, S.; Mackey, M.A.; Coumbe, C.E.; Phillips, J.D.; Phillips, J.P.; Poblet, J.M.; et al. A seven atom cluster in a carbon cage, the crystallographically determined structure of $Sc_4(\mu_3\text{-}O)_3@I_h\text{-}C_{80}$. *Chem. Commun.* **2010**, *46*, 279–281. [CrossRef] [PubMed]
24. Velkos, G.; Yang, W.; Yao, Y.-R.; Sudarkova, S.M.; Liu, X.; Büchner, B.; Avdoshenko, S.M.; Chen, N.; Popov, A.A. Shape-adaptive single-molecule magnetism and hysteresis up to 14 K in oxide clusterfullerenes $Dy_2O@C_{72}$ and $Dy_2O@C_{74}$ with fused pentagon pairs and flexible Dy–($\mu_2$-O)–Dy angle. *Chem. Sci.* **2020**, *11*, 4766–4772. [CrossRef]
25. Mercado, B.Q.; Stuart, M.A.; Mackey, M.A.; Pickens, J.E.; Confait, B.S.; Stevenson, S.; Easterling, M.L.; Valencia, R.; Rodríguez-Fortea, A.; Poblet, J.M.; et al. $Sc_2(\mu_2\text{-}O)$ Trapped in a Fullerene Cage: The Isolation and Structural Characterization of $Sc_2(\mu_2\text{-}O)@C_s(6)\text{-}C_{82}$ and the Relevance of the Thermal and Entropic Effects in Fullerene Isomer Selection. *J. Am. Chem. Soc.* **2010**, *132*, 12098–12105. [CrossRef]
26. Dunsch, L.; Yang, S.; Zhang, L.; Svitova, A.; Oswald, S.; Popov, A.A. Metal Sulfide in a $C_{82}$ Fullerene Cage: A New Form of Endohedral Clusterfullerenes. *J. Am. Chem. Soc.* **2010**, *132*, 5413–5421. [CrossRef]
27. Chen, N.; Beavers, C.M.; Mulet-Gas, M.; Rodríguez-Fortea, A.; Munoz, E.J.; Li, Y.-Y.; Olmstead, M.M.; Balch, A.L.; Poblet, J.M.; Echegoyen, L. $Sc_2S@C_s(10528)\text{-}C_{72}$: A Dimetallic Sulfide Endohedral Fullerene with a Non Isolated Pentagon Rule Cage. *J. Am. Chem. Soc.* **2012**, *134*, 7851–7860. [CrossRef]
28. Chen, N.; Mulet-Gas, M.; Li, Y.-Y.; Stene, R.E.; Atherton, C.W.; Rodríguez-Fortea, A.; Poblet, J.M.; Echegoyen, L. $Sc_2S@C_2(7892)\text{–}C_{70}$: A metallic sulfide cluster inside a non-IPR $C_{70}$ cage. *Chem. Sci.* **2013**, *4*, 180–186. [CrossRef]
29. Mercado, B.Q.; Chen, N.; Rodríguez-Fortea, A.; Mackey, M.A.; Stevenson, S.; Echegoyen, L.; Poblet, J.M.; Olmstead, M.M.; Balch, A.L. The Shape of the $Sc_2(\mu_2\text{-}S)$ Unit Trapped in $C_{82}$: Crystallographic, Computational, and Electrochemical Studies of the Isomers, $Sc_2(\mu_2\text{-}S)@C_s(6)\text{-}C_{82}$ and $Sc_2(\mu_2\text{-}S)@C_{3v}(8)\text{-}C_{82}$. *J. Am. Chem. Soc.* **2011**, *133*, 6752–6760. [CrossRef]
30. Liu, F.; Wang, S.; Gao, C.-L.; Deng, Q.; Zhu, X.; Kostanyan, A.; Westerström, R.; Jin, F.; Xie, S.-Y.; Popov, A.A.; et al. Mononuclear Clusterfullerene Single-Molecule Magnet Containing Strained Fused-Pentagons Stabilized by a Nearly Linear Metal Cyanide Cluster. *Angew. Chem. Int. Ed.* **2017**, *56*, 1830–1834. [CrossRef]
31. Yang, S.; Chen, C.; Liu, F.; Xie, Y.; Li, F.; Jiao, M.; Suzuki, M.; Wei, T.; Wang, S.; Chen, Z.; et al. An Improbable Monometallic Cluster Entrapped in a Popular Fullerene Cage: $YCN@C_s(6)\text{-}C_{82}$. *Sci. Rep.* **2013**, *3*, 1487. [CrossRef] [PubMed]
32. Liu, F.; Gao, C.-L.; Deng, Q.; Zhu, X.; Kostanyan, A.; Westerström, R.; Wang, S.; Tan, Y.-Z.; Tao, J.; Xie, S.-Y.; et al. Triangular Monometallic Cyanide Cluster Entrapped in Carbon Cage with Geometry-Dependent Molecular Magnetism. *J. Am. Chem. Soc.* **2016**, *138*, 14764–14771. [CrossRef] [PubMed]
33. Xin, J.; Jin, F.; Guan, R.; Chen, M.; Xie, X.-M.; Zhang, Q.; Xie, S.-Y.; Yang, S. Ancient pigment to treasure: Prussian blue as a cheap solid cyanide/nitrogen dual-source affording the high-yield syntheses of pricey endohedral clusterfullerenes. *Inorg. Chem. Front.* **2021**, *8*, 1719–1726. [CrossRef]
34. Meng, Q.; Abella, L.; Yang, W.; Yao, Y.-R.; Liu, X.; Zhuang, J.; Li, X.; Echegoyen, L.; Autschbach, J.; Chen, N. $UCN@C_s(6)\text{-}C_{82}$: An Encapsulated Triangular UCN Cluster with Ambiguous U Oxidation State [U(III) versus U(I)]. *J. Am. Chem. Soc.* **2021**, *143*, 16226–16234. [CrossRef] [PubMed]
35. Shen, W.; Hu, Z.; Yu, P.; Wei, Z.; Jin, P.; Shi, Z.; Lu, X. An experimental and theoretical study of $LuNC@C_{76,82}$ revealing a cage-cluster selection rule. *Inorg. Chem. Front.* **2020**, *7*, 4563–4571. [CrossRef]
36. Guan, R.; Chen, M.; Xin, J.; Xie, X.-M.; Jin, F.; Zhang, Q.; Xie, S.-Y.; Yang, S. Capturing the Missing Carbon Cage Isomer of $C_{84}$ via Mutual Stabilization of a Triangular Monometallic Cyanide Cluster. *J. Am. Chem. Soc.* **2021**, *143*, 8078–8085. [CrossRef]

37. Liu, F.; Wang, S.; Guan, J.; Wei, T.; Zeng, M.; Yang, S. Putting a Terbium-Monometallic Cyanide Cluster into the $C_{82}$ Fullerene Cage: TbCN@$C_2$(5)-$C_{82}$. *Inorg. Chem.* **2014**, *53*, 5201–5205. [CrossRef]
38. Jin, F.; Wang, S.; Yang, S.; Tamm, N.B.; Ioffe, I.N.; Troyanov, S.I. Trifluoromethyl Derivatives of a Monometallic Cyanide Cluster Fullerene, YCN@$C_{82}$(6)(CF3)$_{16/18}$. *Inorg. Chem.* **2016**, *55*, 12523–12526. [CrossRef]
39. Jin, F.; Wang, S.; Tamm, N.B.; Yang, S.; Troyanov, S.I. Synthesis, Isolation, and Trifluoromethylation of Two Isomers of $C_{84}$-Based Monometallic Cyanide Clusterfullerenes: Interplay between the Endohedral Cluster and the Exohedral Addends. *Angew. Chem. Int. Ed.* **2017**, *56*, 11990–11994. [CrossRef]
40. Zuo, T.; Walker, K.; Olmstead, M.M.; Melin, F.; Holloway, B.C.; Echegoyen, L.; Dorn, H.C.; Chaur, M.N.; Chancellor, C.J.; Beavers, C.M.; et al. New egg-shaped fullerenes: Non-isolated pentagon structures of $Tm_3N@C_s$(51 365)-$C_{84}$ and $Gd_3N@C_s$(51 365)-$C_{84}$. *Chem. Commun.* **2008**, *9*, 1067–1069. [CrossRef]
41. Krause, M.; Wong, J.; Dunsch, L. Expanding the World of Endohedral Fullerenes—The $Tm_3N@C_{2n}$ ($39 \leq n \leq 43$) Clusterfullerene Family. *Chem. Eur. J.* **2005**, *11*, 706–711. [CrossRef] [PubMed]
42. Krause, M.; Liu, X.; Wong, J.; Pichler, T.; Knupfer, M.; Dunsch, L. The Electronic and Vibrational Structure of Endohedral $Tm_3N@C_{80}$ (I) Fullerene—Proof of an Encaged $Tm^{3+}$. *J. Phys. Chem. A* **2005**, *109*, 7088–7093. [CrossRef] [PubMed]
43. Zuo, T.; Olmstead, M.M.; Beavers, C.M.; Balch, A.L.; Wang, G.; Yee, G.T.; Shu, C.; Xu, L.; Elliott, B.; Echegoyen, L.; et al. Preparation and Structural Characterization of the $I_h$ and the $D_{5h}$ Isomers of the Endohedral Fullerenes $Tm_3N@C_{80}$: Icosahedral $C_{80}$ Cage Encapsulation of a Trimetallic Nitride Magnetic Cluster with Three Uncoupled $Tm^{3+}$ Ions. *Inorg. Chem.* **2008**, *47*, 5234–5244. [CrossRef]
44. Che, Y.; Yang, H.; Wang, Z.; Jin, H.; Liu, Z.; Lu, C.; Zuo, T.; Dorn, H.C.; Beavers, C.M.; Olmstead, M.M.; et al. Isolation and Structural Characterization of Two Very Large, and Largely Empty, Endohedral Fullerenes: Tm@$C_{3v}$-$C_{94}$ and Ca@$C_{3v}$-$C_{94}$. *Inorg. Chem.* **2009**, *48*, 6004–6010. [CrossRef] [PubMed]
45. Wang, Z.; Izumi, N.; Nakanishi, Y.; Koyama, T.; Sugai, T.; Tange, M.; Okazaki, T.; Shinohara, H. Near-Infrared Photoluminescence Properties of Endohedral Mono- and Dithulium Metallofullerenes. *ACS Nano* **2016**, *10*, 4282–4287. [CrossRef] [PubMed]
46. Krätschmer, W.; Lamb, L.D.; Fostiropoulos, K.; Huffman, D.R. Solid $C_{60}$: A new form of carbon. *Nature* **1990**, *347*, 354–358. [CrossRef]
47. Xu, Y.-Y.; Tian, H.-R.; Li, S.-H.; Chen, Z.-C.; Yao, Y.-R.; Wang, S.-S.; Zhang, X.; Zhu, Z.-Z.; Deng, S.-L.; Zhang, Q.; et al. Flexible decapyrrylcorannulene hosts. *Nat. Commun.* **2019**, *10*, 485. [CrossRef]
48. Yao, Y.-R.; Roselló, Y.; Ma, L.; Puente Santiago, A.R.; Metta-Magaña, A.; Chen, N.; Rodríguez-Fortea, A.; Poblet, J.M.; Echegoyen, L. Crystallographic Characterization of U@$C_{2n}$ ($2n = 82$–$86$): Insights about Metal-Cage Interactions for Mono-metallofullerenes. *J. Am. Chem. Soc.* **2021**, *143*, 15309–15318. [CrossRef]
49. Sabirov, D.S.; Ori, O.; László, I. Isomers of the $C_{84}$ fullerene: A theoretical consideration within energetic, structural, and topological approaches. *Fuller. Nanotub. Car. N.* **2018**, *26*, 100–110. [CrossRef]
50. Kirbach, U.; Dunsch, L. The Existence of Stable Tm@$C_{82}$ Isomers. *Angew. Chem. Int. Ed.* **1996**, *35*, 2380–2383. [CrossRef]
51. Sado, Y.; Aoyagi, S.; Kitaura, R.; Miyata, Y.; Nishibori, E.; Sawa, H.; Sugimoto, K.; Shinohara, H. Structure of Tm@$C_{82}$(I) Metallofullerene by Single-Crystal X-ray Diffraction Using the 1:2 Co-Crystal with Octaethylporphyrin Nickel (Ni(OEP)). *J. Phys. Chem. C* **2013**, *117*, 6437–6442. [CrossRef]
52. Kodama, T.; Ozawa, N.; Miyake, Y.; Sakaguchi, K.; Nishikawa, H.; Ikemoto, I.; Kikuchi, K.; Achiba, Y. Structural Study of Three Isomers of Tm@$C_{82}$ by $^{13}$C NMR Spectroscopy. *J. Am. Chem. Soc.* **2002**, *124*, 1452–1455. [CrossRef] [PubMed]
53. Pichler, T.; Golden, M.S.; Knupfer, M.; Fink, J.; Kirbach, U.; Kuran, P.; Dunsch, L. Monometallofullerene Tm@$C_{82}$: Proof of an Encapsulated Divalent Tm Ion by High-Energy Spectroscopy. *Phys. Rev. Lett.* **1997**, *79*, 3026–3029. [CrossRef]
54. Pichler, T.; Knupfer, M.; Golden, M.S.; Böske, T.; Fink, J.; Kirbach, U.; Kuran, P.; Dunsch, L.; Jung, C. The metallofullerene Tm@$C_{82}$: Isomer-selective electronic structure. *Appl. Phys. A* **1998**, *66*, 281–285. [CrossRef]
55. Lu, X.; Slanina, Z.; Akasaka, T.; Tsuchiya, T.; Mizorogi, N.; Nagase, S. Yb@$C_{2n}$ ($n$ = 40, 41, 42): New Fullerene Allotropes with Unexplored Electrochemical Properties. *J. Am. Chem. Soc.* **2010**, *132*, 5896–5905. [CrossRef]
56. Bao, L.; Pan, C.; Slanina, Z.; Uhlik, F.; Akasaka, T.; Lu, X. Isolation and Crystallographic Characterization of the Labile Isomer of Y@$C_{82}$ Cocrystallized with Ni(OEP): Unprecedented Dimerization of Pristine Metallofullerenes. *Angew. Chem. Int. Ed.* **2016**, *55*, 9234–9238. [CrossRef]
57. Kikuchi, K.; Nakao, Y.; Suzuki, S.; Achiba, Y.; Suzuki, T.; Maruyama, Y. Characterization of the Isolated Y@$C_{82}$. *J. Am. Chem. Soc.* **1994**, *116*, 9367–9368. [CrossRef]
58. Suzuki, T.; Maruyama, Y.; Kato, T.; Kikuchi, K.; Achiba, Y. Electrochemical properties of La@$C_{82}$. *J. Am. Chem. Soc.* **1993**, *115*, 11006–11007. [CrossRef]
59. Akasaka, T.; Okubo, S.; Kondo, M.; Maeda, Y.; Wakahara, T.; Kato, T.; Suzuki, T.; Yamamoto, K.; Kobayashi, K.; Nagase, S. Isolation and characterization of two Pr@$C_{82}$ isomers. *Chem. Phys. Lett.* **2000**, *319*, 153–156. [CrossRef]
60. Hu, S.; Liu, T.; Shen, W.; Slanina, Z.; Akasaka, T.; Xie, Y.; Uhlik, F.; Huang, W.; Lu, X. Isolation and Structural Characterization of Er@$C_{2v}$(9)-$C_{82}$ and Er@$C_s$(6)-$C_{82}$: Regioselective Dimerization of a Pristine Endohedral Metallofullerene Induced by Cage Symmetry. *Inorg. Chem.* **2019**, *58*, 2177–2182. [CrossRef]
61. Cai, W.; Morales-Martínez, R.; Zhang, X.; Najera, D.; Romero, E.L.; Metta-Magaña, A.; Rodríguez-Fortea, A.; Fortier, S.; Chen, N.; Poblet, J.M.; et al. Single crystal structures and theoretical calculations of uranium endohedral metallofullerenes (U@$C_{2n}$, $2n = 74$, 82) show cage isomer dependent oxidation states for U. *Chem. Sci.* **2017**, *8*, 5282–5290. [CrossRef]

62. Liu, X.; Zuo, T.; Dorn, H.C. Polarizability Effects Dominate the Chromatographic Retention Behavior of Spheroidal and Elipsoidal Metallofullerene Nanospheres. *J. Phys. Chem. C* **2017**, *121*, 4045–4049. [CrossRef]
63. Liu, X.; Dorn, H.C. DFT prediction of chromatographic retention behavior for a trimetallic nitride metallofullerene series. *Inorg. Chim. Acta* **2017**, *468*, 316–320. [CrossRef]
64. Dolomanov, O.V.; Bourhis, L.J.; Gildea, R.J.; Howard, J.A.K.; Puschmann, H. OLEX2: A complete structure solution, refinement and analysis program. *J. Appl. Cryst.* **2009**, *42*, 339–341. [CrossRef]
65. Sheldrick, G. Crystal structure refinement with SHELXL. *Acta Crystallogr. Sect. C* **2015**, *71*, 3–8. [CrossRef] [PubMed]
66. Spek, A. PLATON SQUEEZE: A tool for the calculation of the disordered solvent contribution to the calculated structure factors. *Acta Crystallogr. Sect. C* **2015**, *71*, 9–18. [CrossRef]

*inorganics*

MDPI

*Article*

# $Pr@C_{82}$ Metallofullerene: Calculated Isomeric Populations

**Zdeněk Slanina [1,2,*], Filip Uhlík [3], Takeshi Akasaka [2], Xing Lu [2] and Ludwik Adamowicz [1]**

1 Department of Chemistry and Biochemistry, University of Arizona, Tucson, AZ 85721-0041, USA
2 State Key Laboratory of Materials Processing and Die & Mould Technology, School of Material Science and Engineering, Huazhong University of Science and Technology, Wuhan 430074, China; lux@hust.edu.cn (X.L.)
3 Department of Physical and Macromolecular Chemistry, Faculty of Science, Charles University in Prague, Albertov 6, 128 43 Praha, Czech Republic
* Correspondence: zdeneks@email.arizona.edu

**Abstract:** Relative equilibrium populations of the five lowest-energy isolated-pentagon-rule (IPR) isomeric structures of $Pr@C_{82}$ under high-temperature fullerene synthesis conditions were calculated with the Gibbs energy terms based on molecular characteristics derived using density functional theory (DFT) treatments (B3LYP/6-31+G*∼SDD energetics and B3LYP/6-31G*∼SDD entropy). Two leading isomers were identified, major $Pr@C_{2v};9\text{-}C_{82}$ and minor $Pr@C_s;6\text{-}C_{82}$. The calculated isomeric relative equilibrium populations agreed with observations.

**Keywords:** metallofullerenes; evaluations of isomeric populations; theory–experiment comparisons; quantum-chemical calculations

## 1. Introduction

Endohedral metallofullerenes have been pursued as not only targets of fundamental research but also promising species for nanoelectronics, as well as for energy conversions, superconductive properties, and applications in medicine. Such studies have also dealt with praseodymium-encapsulating metallofullerenes, in particular with $Pr@C_{82}$ [1–8] (a field-effect transistor was also fabricated [7] with one of its isomers). Actually, just two $Pr@C_{82}$ isomers (Figure 1) have been observed [3–6], namely the major species with the carbon cage obeying the isolated pentagon rule (IPR), conventionally labeled $Pr@C_{2v};9\text{-}C_{82}$, and the minor isomer $Pr@C_s;6\text{-}C_{82}$ (or $Pr@C_s(c;6)\text{-}C_{82}$). Presently, there are two labeling or coding systems [9] for the IPR $C_{82}$ cages considered in this research. The older system works only with the symmetries of the cages (namely, the highest topological symmetries, regardless of possible symmetry reductions): $C_{3v}(a)$, $C_{3v}(b)$, $C_{2v}$, $C_2(a)$, $C_2(b)$, $C_2(c)$, $C_s(a)$, $C_s(b)$, and $C_s(c)$. The second labeling system employs serial numbers from the enumeration spiral algorithm [10]; these symbols are: $C_{3v}(a);7$, $C_{3v}(b);8$, $C_{2v};9$, $C_2(a);3$, $C_2(b);1$, $C_2(c);5$, $C_s(a);2$, $C_s(b);4$, and $C_s(c);6$. Numerous metals can be encapsulated in $C_{82}$ cages, frequently producing at least two isomers [4–6,9,11–16]. In the present report, the isomeric relative equilibrium populations (or the relative concentrations) were evaluated for the isomeric system of the five lowest-potential-energy $Pr@C_{82}$ IPR species, as such stability information is needed for the interpretation of some observations.

The relative stabilities of isomeric nanocarbons are mostly evaluated by simply using potential-energy terms. Nevertheless, several examples have been reported [17–26] showing that the Gibbs energy should be applied instead, as its entropy component becomes increasingly important at elevated temperatures. Consequently, an isomer that is not the lowest on the potential-energy scale can still exhibit the highest populations at the high temperatures needed for nanocarbon syntheses. Moreover, other higher-energy structures can in some cases exhibit relative stability interchanges with increasing temperature. Clearly enough, it is not possible to predict such relative stability interchanges simply from the potential-energy terms alone. Hence, computations were performed in this report for the high-temperature relative equilibrium populations of the five $Pr@C_{82}$ IPR isomers with the lowest potential energy, employing both the enthalpy and entropy parts of the Gibbs energy in simulations of the isomeric equilibrium populations under supposed fullerene synthesis conditions.

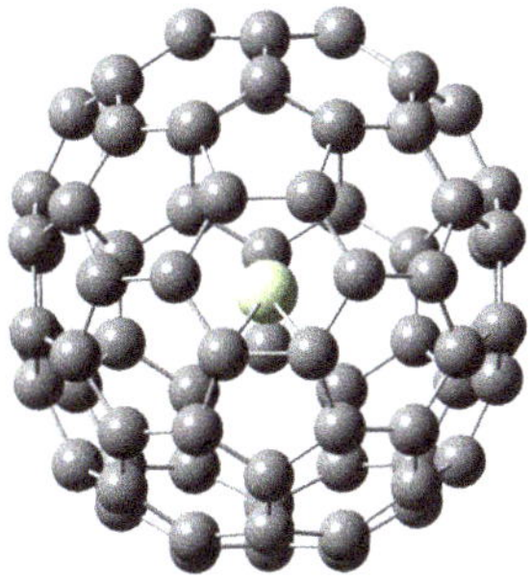
$C_{2v};9$

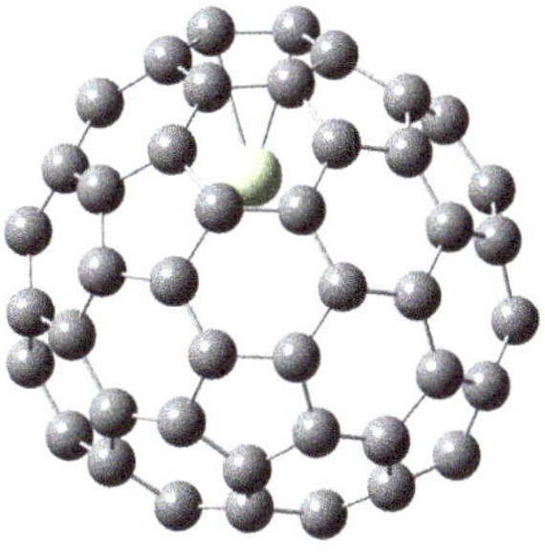
$C_s;6$

**Figure 1.** The B3LYP/6-31G*∼SDD-optimized structures of the two most populated $Pr@C_{82}$ isomers, $Pr@C_{2v};9\text{-}C_{82}$ (**left**), $Pr@C_s;6\text{-}C_{82}$ (**right**) (the shortest Pr-C contact is indicated by a link).

## 2. Calculations

The calculations started with the molecular geometry optimization of the considered isomers, which was performed in a combined atomic basis set—the standard 3-21G basis [27] for carbon atoms and the SDD basis [28] with the SDD effective core potential for praseodymium (labeled 3-21G∼SDD). We applied density functional theory (DFT) treatments employing Becke's three-parameter functional [29] combined with the Lee–Yang–Parr non-local correlation functional [30]—the unrestricted B3LYP/3-21G∼SDD treatment. Moreover, the B3LYP/3-21G∼SDD pre-optimized structures were improved further with the standard 6-31G* basis set [31] for carbon atoms—the B3LYP/6-31G*∼SDD approach. The calculations were carried out for the quartet electronic state, as the multiplicity produced significantly lower energy at the considered computational levels than the doublet and sextet electronic states. The geometry optimizations were carried out with the analytical energy gradient. All nine [9] IPR $C_{82}$ cages were subjected to geometry optimizations. The geometry optimizations with both the B3LYP/3-21G∼SDD and B3LYP/6-31G*∼SDD treatments identified five isomers (Table 1) with sufficiently low potential energy. All other $Pr@C_{82}$ IPR isomers were, according to the B3LYP/6-31G*∼SDD treatment, located 19 kcal/mol or more higher than the lowest-energy species $Pr@C_{2v};9\text{-}C_{82}$. Their energy separation (in combination with the entropy terms) produced already insignificant relative equilibrium concentrations. Further on, the inter-isomeric energetics was still improved by applying a yet more advanced B3LYP/6-31+G*∼SDD approach (still using the B3LYP/6-31G*∼SDD geometries). For the optimized B3LYP/6-31G*∼SDD geometries, the vibrational analysis was performed using the harmonic GF scheme with the analytical force–constant matrix F in order to simulate the IR vibrational spectra (Figure 2) and, in particular, to construct the partition functions of the harmonic vibrations for the stability evaluations. The SCF wavefunction stability [32,33] was systematically tested throughout in order to prevent misleading unstable SCF solutions that could produce a physically unsound picture. The calculations were performed with the Gaussian 09 suite of programs [34].

Table 1: $Pr@C_{82}$ relative potential energies $\Delta E_{pot,rel}$ for the five lowest-energy isomers calculated in the B3LYP/6-31G*~SDD-optimized structures.

| **Species** | $\Delta E_{pot,rel}$**/kcal· mol**$^{-1}$ **B3LYP/6-31G*~SDD** | **B3LYP/6-31+G*~SDD** |
|---|---|---|
| $C_2;5$ | 17.9 | 17.3 |
| $C_{3v};7$ | 13.0 | 14.2 |
| $C_{3v};8$ | 7.54 | 7.37 |
| $C_s;6$ [a] | 3.46 | 3.47 |
| $C_{2v};9$ [a] | 0.0 | 0.0 |

[a] See Figure 1.

The relative equilibrium concentrations of $m$ isomers are described as mole fractions $x_i$ that can be expressed [35,36] through their partition functions $q_i$ and the enthalpies at a temperature of absolute zero or ground-state energies $\Delta H^o_{0,i}$ (i.e., the relative potential energies corrected for the vibrational zero-point energies) by a master formula:

$$x_i = \frac{q_i exp[-\Delta H^o_{0,i}/(RT)]}{\sum_{j=1}^{m} q_j exp[-\Delta H^o_{0,j}/(RT)]}, \tag{1}$$

where $R$ is the gas constant and $T$ is the absolute temperature. Equation (1) is a rigorous formula that can be obtained [35] using statistical thermodynamics with the standard isomeric Gibbs energies under the conditions of the inter-isomeric thermodynamic equilibrium. The rotational-vibrational components of the partition functions were approximated [36] by the conventional rigid-rotor and harmonic-oscillator (RRHO) approach. Frequency scaling was not applied as it was not significant for the $x_i$ values at high temperatures [37]. Moreover, the chirality contribution [38] was considered accordingly (the partition function $q_i$ for a pair of enantiomers was doubled). The temperatures at which fullerene or metallofullerene electric-arc synthesis takes place are not precisely known; however, recent observations [39] suggest that it takes place somewhere around a temperature of 1500 K. Hence, the values discussed in this report were calculated for this particular temperature region.

Moreover, an adjusted [23,40] RRHO treatment for the description of the metal motions was applied, considering findings [41] suggesting that the encapsulated metal atoms can exercise large-amplitude vibrational motions, in particular at higher temperatures (if the motions are not restricted by derivatizations of cages [42]). It could be expected that if the metal was relatively free, then, at sufficiently elevated temperatures, its motions in different endohedral cages would provide a similar contribution to the overall partition functions. Such similar contributions would consequently be mutually canceled out in Equation (1). This approximation has been called [23,40] the floating, fluctuating, or free encapsulation model (FEM). The FEM treatment consists of two steps. The first step is the suppression of the three lowest vibrational frequencies (which belong to the metal vibrations inside the cage). The second step deals with the symmetries of the cages—they should be taken as the highest topologically possible terms, reflecting the averaging influence of the large-amplitude motions upon the symmetries. For example, with the $Pr@C_{82}$ IPR isomer related to the $C_{2v};9$ cage (Table 1), the $C_{2v}$ symmetry is indeed applied in the FEM treatment, in spite of the fact that its static [43] symmetry (i.e., derived from the geometry optimization) is only $C_s$ or even $C_1$ (Figure 1). The FEM approach is generally known to produce better agreement [23,40] with observed data than the simple RRHO treatment. The FEM approach was thus also applied in this report.

Table 2: The selected characteristics of the five lowest-energy Pr@$C_{82}$ isomers—the closest Pr-C contact [a] $r_{Pr-C}$, the Mulliken charge [b] on Pr $q_{Pr}$, and the lowest vibrational frequency [a] $\omega_{low}$.

| Species | $r_{Pr-C}$/Å | $q_{Pr}$ | $\omega_{low}$ / $cm^{-1}$ |
|---|---|---|---|
| $C_2$;5 | 2.512 | 2.449 | 39.3 |
| $C_{3v}$;7 | 2.516 | 2.293 | 17.6 |
| $C_{3v}$;8 | 2.476 | 2.462 | 22.3 |
| $C_s$;6 [c] | 2.500 | 2.337 | 20.5 |
| $C_{2v}$;9 [c] | 2.549 | 2.364 | 15.8 |

[a] B3LYP/6-31G*∼SDD terms. [b] B3LYP/3-21G∼SDD terms. [c] See Figure 1.

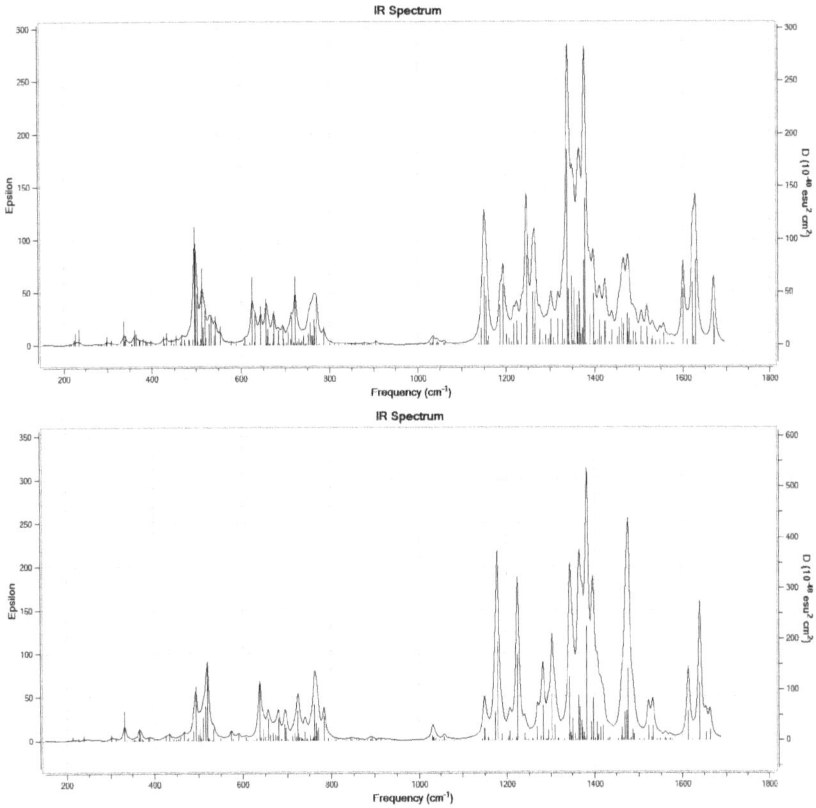


**Figure 2.** B3LYP/6-31G*∼SDD-calculated IR spectrum of the Pr@$C_{2v}$;9-$C_{82}$ (**bottom**) and Pr@$C_s$;6-$C_{82}$ (**top**) isomers.

## 3. Results and Discussion

The Pr@$C_{82}$ relative isomeric energetics calculated in the two selected approaches (namely, the relative potential energy without the inclusion of the zero-point vibrational energies) for the five lowest-energy isomers selected from the nine [9,44] IPR-satisfying $C_{82}$ cages are presented in Table 1. The lowest-energy Pr@$C_{82}$ isomer was the $C_{2v}$;9 species, followed at 3 kcal/mol higher by the $C_s$;6 endohedral species (Figure 1). Then, $C_{3v}$;8 and $C_{3v}$;7 were located about 7 and 14 kcal/mol above the $C_{2v}$;9 stabilomer, respectively. It should be noted that the B3LYP/6-31G*∼SDD and B3LYP/6-31+G*∼SDD relative energies were similar. The five species from Table 1 were subjected to equilibrium thermodynamic analysis.

Table 2 presents selected calculated molecular parameters for the five lowest-potential-energy Pr@$C_{82}$ species. The B3LYP/6-31G*∼SDD calculated shortest distances $r_{Pr-C}$ between Pr and the carbon cages at around 2.5 Å, which were thus similar to those previously reported for other $C_{82}$-based endohedrals [9,15,44]. Interestingly, the calculated position of the Pr atom in Pr@$C_{2v}$;9-$C_{82}$ (i.e., under a hexagonal ring) is also known for other systems [9]. The lowest harmonic vibrational frequencies $\omega_{low}$ in Table 1 reflect the relatively free motions of the encapsulated metals in metallofullerenes. B3LYP/3–21G∼SDD produced Mulliken atomic charges $q_{Pr}$ on Pr are close to 2.4 (electron elementary charge). It should be stressed that the Mulliken atomic charges from the 3–21G∼SDD basis are known to produce [25] for metallofullerenes good agreement with experimental charges [45] (which is not the case for, e.g., the 6-31G*∼SDD level). In addition, there are deeper physical reasons [46–49] why extended bases should not be used for Mulliken charges (as the latter can sometimes produce rather unphysical charges [47]). As for yet another basis-set effect, let us also mention that, as we were only dealing with isomers, there was no need to consider the basis set superposition error (BSSE) [49]. The BSSE correction is indeed important for association processes in order to assure that all the components are treated in the same basis set. In a set of isomers, however, the basis set used is the same for all the species, and therefore the total energies do not require a correction of the BSSE type. If such a treatment was still performed, there would be a relatively constant additive term that would then cancel out in Equation (1).

Figure 3 gives the key results of this report—the temperature dependencies of the relative equilibrium concentrations for the five lowest-potential-energy Pr@$C_{82}$ isomers at moderate and high temperatures. The presented relative isomeric populations were obtained with the FEM approach. In any case, at very low temperatures, the species with the lowest $\Delta H^o_{0,i}$ had to be the most populated, i.e., the Pr@$C_{2v}$;9-$C_{82}$ isomer. However, the second (Pr@$C_s$;6-$C_{82}$) and, somewhat, the third (Pr@$C_{3v}$;8-$C_{82}$) lowest-potential-energy isomers showed a rapid increase in their relative populations, while the remaining isomers were negligible. At the suggested [39] representative temperature of 1500 K for fullerene synthesis, the relative equilibrium populations were 59.3, 34.9, 4.8, 0.3, and 0.8% for the $C_{2v}$;9, $C_s$;6, $C_{3v}$;8, $C_{3v}$;7, and $C_2$;5 isomers, respectively. Interestingly, in a stability evaluation without the entropic part (i.e., when just the simple Boltzmann factors [35,36] were considered), the relative isomeric populations were rather different: 71.0, 22.2, 6.0, 0.6, and 0.2% for the $C_{2v}$;9, $C_s$;6, $C_{3v}$;8, $C_{3v}$;7, and $C_2$;5 species, respectively.

Overall, the calculations agreed with the observations [3–6] for two isomers, major Pr@$C_{2v}$;9-$C_{82}$ and minor Pr@$C_s$;6-$C_{82}$. Their observed [6] population ratio has been reported as Pr@$C_{2v}$;9-$C_{82}$ : Pr@$C_s$;6-$C_{82}$ = 2.5. In our calculations, the population ratio was achieved at a quite reasonable synthetic temperature of 1130 K. Still, it is not known how close the experiments were to the inter-isomeric equilibrium predicted by the thermodynamic treatment. For the interpretation of the observations, it is also of interest that the calculations predicted just two significant isomers, while the populations of the other species were rather negligible. The calculated IR spectra (Figure 2) should also be of use for the identification of the two isomers.

The reported results presented similarities with the previous calculations [12,23,50–54] for other $C_{82}$-based metallofullerenes with a similar metal-to-cage charge transfer—in particular [12,23], La@$C_{82}$. The similarities originated from the fact that metallofullerenes are not stabilized by some new covalent bond but instead formed by an ionic bond [55–59]. It should be mentioned that experimental populations can depend on the metal sources used [60]. This aspect can be related to catalytic and kinetic factors [61–63] and indicates the variable levels to which the expected inter-isomeric thermodynamic equilibrium could actually be obtained in the synthesis. Yet another factor is the solubility [64–66] of individual isomers in the selected extraction solvents. In conclusion, the Pr@$C_{82}$ results provide more evidence of the good applicability of the Gibbs energy approach to isomeric populations, thus supporting further studies of this type into yet more complex nanocarbon isomeric sets [67–71].

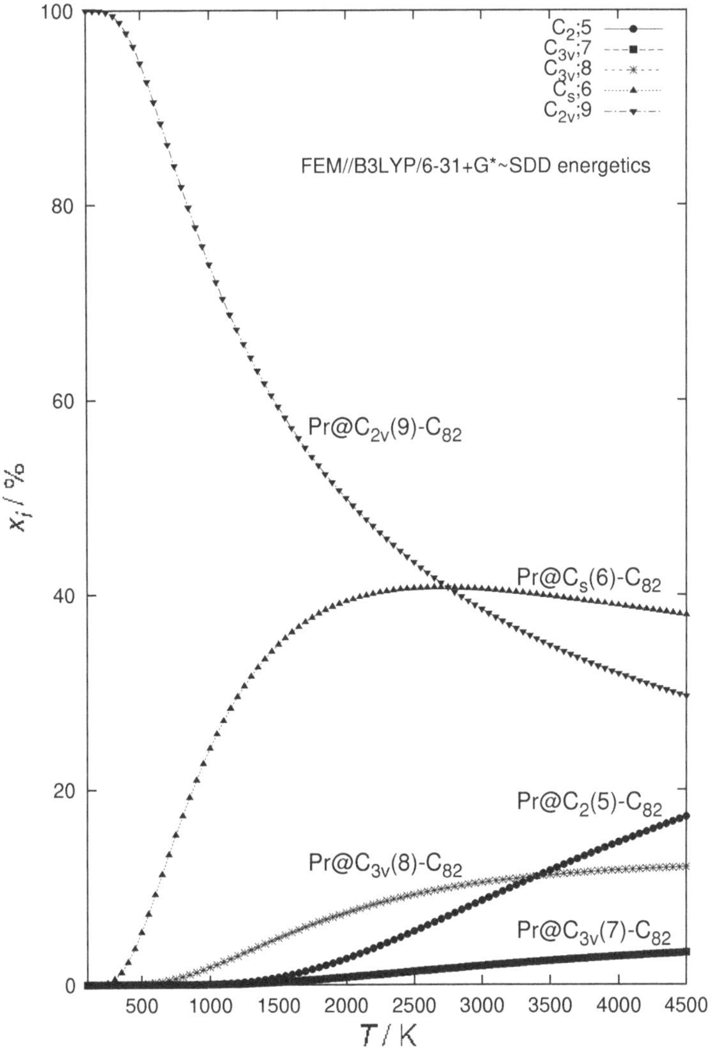


**Figure 3.** The relative populations of the $Pr@C_{82}$ isomers based on the FEM treatment with the B3LYP/6-31+G*∼SDD energetics and B3LYP/6-31G*∼SDD entropy.

**Author Contributions:** The authors contributed equally to this article. All authors have read and agreed to the published version of the manuscript.

**Funding:** The reported research was supported by the National Natural Science Foundation of China (21925104 and 92261204); the Hubei Provincial Natural Science Foundation of China (No. 2021CFA020); the International Cooperation Key Project of the Science and Technology Department of Shaanxi; and the Charles University Centre of Advanced Materials/CUCAM (CZ.02.1.01/0.0/0.0/15 003/0000417) and the MetaCentrum ((LM2010005) and CERIT-SC (CZ.1.05/3.2.00/08.0144) computing facilities. The initial phase of the research was also supported by the Alexander von Humboldt-Stiftung and Max Planck Institut für Chemie (Otto Hahn Institut).

**Data Availability Statement:** Not applicable.

**Conflicts of Interest:** The authors declare no conflict of interest.

## References


1. Wang, W.; Ding, J.; Yang, S.; Li, X.-Y. Electrochemical Properties of 4f-block metallofullerenes. In *Fullerenes. Recent Advances in the Chemistry and Physics of Fullerenes and Related Materials*; Kadish, K.M., Ruoff, R.S., Eds.; Electrochemical Society: Pennington, Australia, 1997; Volume 4, pp. 417–428.
2. Sun, D.Y.; Huang, H.J.; Yang, S.H.; Liu, Z.Y.; Liu, S.Y. Synthesis and characterization of a water-soluble endohedral metallofullerol. *Chem. Mater.* **1999**, *11*, 1003–1006. [CrossRef]
3. Akasaka, T.; Okubo, S.; Wakahara, T.; Yamamoto, K.; Kato, T.; Suzuki, T.; Nagase, S.; Kobayashi, K. Isolation and characterization of a Pr@$C_{82}$ isomer. In *Recent Advances in the Chemistry and Physics of Fullerenes and Related Materials*; Kamat, P.V., Guldi, D.M., Kadish, K.M., Eds.; Electrochemical Society: Pennington, Australia, 1999; Volume 7, pp. 771–778.
4. Akasaka, T.; Okubo, S.; Kondo, M.; Maeda, Y.; Wakahara, T.; Kato, T.; Suzuki, T.; Yamamoto, K.; Kobayashi, K.; Nagase, S. Isolation and characterization of two Pr@$C_{82}$ isomers. *Chem. Phys. Lett.* **2000**, *319*, 153–156. [CrossRef]
5. Wakahara, T.; Okubo, S.; Kondo, M.; Maeda, Y.; Akasaka, T.; Waelchli, M.; Kako, M.; Kobayashi, K.; Nagase, S.; Kato, T.; et al. Ionization and structural determination of the major isomer of Pr@$C_{82}$. *Chem. Phys. Lett.* **2002**, *360*, 235–239. [CrossRef]
6. Hosokawa, T.; Fujiki, S.; Kuwahara, E.; Kubozono, Y.; Kitagawa, H.; Fujiwara, A.; Takenobu, T.; Iwasa, Y. Electronic properties for the $C_{2v}$ and $C_s$ isomers of Pr@$C_{82}$ studied by Raman, resistivity and scanning tunneling microscopy/spectroscopy. *Chem. Phys. Lett.* **2004**, *395*, 78–81. [CrossRef]
7. Nagano, T.; Kuwahara, E.; Takayanagi, T.; Kubozono, Y.; Fujiwara, A. Fabrication and characterization of field-effect transistor device with $C_{2v}$ isomer of Pr@$C_{82}$. *Chem. Phys. Lett.* **2005**, *409*, 187–191. [CrossRef]
8. Katayanagi, H.; Kafle, B.P.; Kou, J.; Mori, T.; Mitsuke, K.; Takabayashi, Y.; Kuwahara, E.; Kubozono, Y.J. The 4d-4f dipole resonance of the Pr atom in an endohedral metallofullerene, Pr@$C_{82}$. *J. Quant. Spectrosc. Rad. Trans.* **2008**, *109*, 1590–1598. [CrossRef]
9. Slanina, Z.; Uhlík, F.; Feng, L.; Adamowicz, L. Calculated relative populations of Sm@$C_{82}$ isomers. *Fullerenes Nanotub. Carbon Nanostruct.* **2018**, *26*, 233–238. [CrossRef]
10. Fowler, P.W.; Manolopoulos, D.E. *An Atlas of Fullerenes*; Clarendon Press: Oxford, UK, 1995.
11. Slanina, Z.; Lee, S.-L.; Kobayashi, K.; Nagase, S. AM1 computed thermal effects within the nine isolated-pentagon-rule isomers of $C_{82}$. *J. Mol. Struct.* **1995**, *339*, 89–93. [CrossRef]
12. Slanina, Z.; Kobayashi, K.; Nagase, S. Ca@$C_{82}$ isomers: Computed temperature dependency of relative concentrations. *J. Chem. Phys.* **2004**, *120*, 3397–3400. [CrossRef]
13. Slanina, Z.; Kobayashi, K.; Nagase, S. Computed temperature development of the relative stabilities of La@$C_{82}$ isomers. *Chem. Phys. Lett.* **2004**, *388*, 74–78. [CrossRef]
14. Suzuki, M.; Slanina, Z.; Mizorogi, N.; Lu, X.; Nagase, S.; Olmstead, M.M.; Balch, A.L.; Akasaka, T. Single-crystal X-ray diffraction study of three Yb@C-82 isomers cocrystallized with Ni-II(octaethylporphyrin). *J. Am. Chem. Soc.* **2012**, *134*, 18772–18778. [CrossRef]
15. Slanina, Z.; Uhlík, F.; Lee, S.-L.; Suzuki, M.; Lu, X.; Mizorogi, N.; Nagase, S.; Akasaka, T. Calculated temperature development of the relative stabilities of Yb@$C_{82}$ isomers. *Fullerenes Nanotub. Carbon Nanostruct.* **2014**, *22*, 147–154. [CrossRef]
16. Hu, Z.; Hao, Y.; Slanina, Z.; Gu, Z.; Shi, Z.; Uhlík, F.; Zhao, Y.; Feng, L. Popular $C_{82}$ fullerene cage encapsulating a divalent metal ion $Sm^{2+}$: Structure and electrochemistry. *Inorg. Chem.* **2015**, *54*, 2103–2108. [CrossRef]
17. Slanina, Z.; Lee, S.-L.; Adamowicz, L. $C_{80}$, $C_{86}$, $C_{88}$: Semiempirical and ab initio SCF calculations. *Int. J. Quantum. Chem.* **1997**, *63*, 529–535. [CrossRef]
18. Slanina, Z.; Uhlík, F. Temperature dependence of the Gibbs energy ordering of isomers of $Cl_2O_2$. *J. Phys. Chem.* **1991**, *95*, 5432–5434. [CrossRef]
19. Slanina, Z.; Zhao, X.; Lee, S.-L.; Ōsawa, E. $C_{90}$-Temperature effects on relative stabilities of the IPR isomers. *Chem. Phys.* **1997**, *219*, 193–200. [CrossRef]
20. Uhlík, F.; Slanina, Z.; Ōsawa, E. $C_{78}$ IPR fullerenes: Computed B3LYP/6-31G*//HF/3-21G temperature-dependent relative concentrations. *Eur. Phys. J. D* **2001**, *16*, 349–352. [CrossRef]
21. Slanina, Z.; Zhao, X.; Uhlík, F.; Lee, S.-L.; Adamowicz, L. Computing enthalpy-entropy interplay for isomeric fullerenes. *Int. J. Quantum Chem.* **2004**, *99*, 640–653. [CrossRef]
22. Slanina, Z.; Lee, S.-L.; Adamowicz, L.; Uhlík, F.; Nagase, S. Computed structure and energetics of La@$C_{60}$. *Int. J. Quantum Chem.* **2005**, *104*, 272–277. [CrossRef]
23. Slanina, Z.; Lee, S.-L.; Uhlík, F.; Adamowicz, L.; Nagase, S. Computing relative stabilities of metallofullerenes by Gibbs energy treatments. *Theor. Chem. Acc.* **2007**, *117*, 315–322. [CrossRef]
24. Wang, Y.; Morales-Martínez, R.; Zhang, X.; Yang, W.; Wang, Y.; Rodríguez-Fortea, A.; Poblet, J.M.; Feng, L.; Wang, S.; Chen, N. Unique four-electron metal-to-cage charge transfer of Th to a $C_{82}$ fullerene cage: Complete structural characterization of Th@$C_{3v}$(8)-$C_{82}$. *J. Am. Chem. Soc.* **2017**, *139*, 5110–5116. [CrossRef] [PubMed]
25. Slanina, Z.; Uhlík, F.; Nagase, S.; Akasaka, T.; Adamowicz, L.; Lu, X. Eu@$C_{72}$: Computed comparable populations of two non-IPR isomers. *Molecules* **2017**, *22*, 1053. [CrossRef]
26. Zhao, Y.; Li, M.; Zhao, R.; Zhao, P.; Yuan, K.; Li, Q.; Zhao, X. Unmasking the optimal isomers of $Ti_2C_{84}$: $Ti_2C_2$@$C_{82}$ Instead of $Ti_2C_{84}$. *J. Phys. Chem. C.* **2018**, *122*, 13148–13155. [CrossRef]
27. Binkley, J.S.; Pople, J.A.; Hehre, W.J. Self-consistent molecular orbital methods. 21. Small split-valence basis sets for first-row elements. *J. Am. Chem. Soc.* **1980**, *102*, 939–947. [CrossRef]

28. Cao, X.Y.; Dolg, M. Segmented contraction scheme for small-core lanthanide pseudopotential basis sets. *J. Mol. Struct.* **2002**, *581*, 139–147. [CrossRef]
29. Becke, A.D. Density-functional thermochemistry. III. The role of exact exchange. *J. Chem. Phys.* **1993**, *98*, 5648–5652. [CrossRef]
30. Lee, C.; Yang, W.; Parr, R.G. Development of the Colle-Salvetti correlation-energy formula into a functional of the electron density. *Phys. Rev. B* **1988**, *37*, 785–789. [CrossRef]
31. Hehre, W.J.; Ditchfield, R.; Pople, J.A. Self-consistent molecular orbital methods. 12. Further extensions of Gaussian-type basis sets for use in molecular-orbital studies of organic-molecules. *J. Chem. Phys.* **1972**, *56*, 2257–2261. [CrossRef]
32. Schlegel, H.B.; McDouall, J.J.W. Do you have SCF stability and convergence problems? In *Computational Advances in Organic Chemistry*; Ögretir, C., Csizmadia, I.G., Eds.; Kluwer: Dordrecht, The Netherlands, 1991; pp. 167–185.
33. Slanina, Z.; Uhlík, F.; Adamowicz, L. Computations of model narrow nanotubes closed by fragments of smaller fullerenes and quasi-fullerenes. *J. Mol. Graph. Mod.* **2003**, *21*, 517–522. [CrossRef]
34. Frisch, M.J.; Trucks, G.W.; Schlegel, H.B.; Scuseria, G.E.; Robb, M.A.; Cheeseman, J.R.; Scalmani, G.; Barone, V.; Mennucci, B.; Petersson, G.A.; et al. *Gaussian 09, Rev. C.01*; Gaussian Inc.: Wallingford, CT, USA, 2013.
35. Slanina, Z. Equilibrium isomeric mixtures: Potential energy hypersurfaces as originators of the description of the overall thermodynamics and kinetics. *Int. Rev. Phys. Chem.* **1987**, *6*, 251–267. [CrossRef]
36. Slanina, Z. A Program for determination of composition and thermodynamics of the ideal gas-phase equilibrium isomeric mixtures. *Comput. Chem.* **1989**, *13*, 305–311. [CrossRef]
37. Slanina, Z.; Uhlík, F.; Zerner, M.C. $C_5H_3^+$ isomeric structures: Relative stabilities at high temperatures. *Rev. Roum. Chim.* **1991**, *36*, 965–974.
38. Slanina, Z.; Adamowicz, L. On relative stabilities of dodecahedron-shaped and bowl-shaped structures of $C_{20}$. *Thermochim. Acta* **1992**, *205*, 299–306. [CrossRef]
39. Cross, R.J.; Saunders, M. Transmutation of fullerenes. *J. Am. Chem. Soc.* **2005**, *127*, 3044–3047. [CrossRef]
40. Slanina, Z.; Adamowicz, L.; Kobayashi, K.; Nagase, S. Gibbs energy-based treatment of metallofullerenes: Ca@$C_{72}$, Ca@$C_{74}$, Ca@$C_{82}$, and La@$C_{82}$. *Mol. Simul.* **2005**, *31*, 71–77. [CrossRef]
41. Akasaka, T.; Nagase, S.; Kobayashi, K.; Walchli, M.; Yamamoto, K.; Funasaka, H.; Kako, M.; Hoshino, T.; Erata, T. $^{13}$C and $^{139}$La NMR studies of $La_2$@$C_{80}$: First evidence for circular motion of metal atoms in endohedral dimetallofullerenes. *Angew. Chem. Int. Ed.* **1997**, *36*, 1643–1645. [CrossRef]
42. Kobayashi, K.; Nagase, S.; Maeda, Y.; Wakahara, T.; Akasaka, T. $La_2$@$C_{80}$: Is the circular motion of two La atoms controllable by exohedral addition? *Chem. Phys. Lett.* **2003**, *374*, 562–566. [CrossRef]
43. Slanina, Z. *Contemporary Theory of Chemical Isomerism*; Academia: Prague, Czech Republic; D. Reidel Publ. Comp.: Dordrecht, The Netherlands, 1986; pp. 21–23.
44. Slanina, Z.; Uhlík, F.; Bao, L.; Akasaka, T.; Lu, X.; Adamowicz, L. Calculated relative populations for the Eu@$C_{82}$ isomers. *Chem. Phys. Lett.* **2019**, *726*, 29–33. [CrossRef]
45. Takata, M.; Nishibori, E.; Sakata, M.; Shinohara, H. Charge density level structures of endohedral metallofullerenes determined by synchrotron radiation powder method. *New Diam. Front. Carb. Technol.* **2002**, *12*, 271–286.
46. Hehre, W.J. *A Guide to Molecular Mechanics and Quantum Chemical Calculations*; Wavefunction: Irvine, CA, USA, 2003; p. 435.
47. Jensen, F. *Introduction to Computational Chemistry*; Wiley: Chichester, UK, 2017; p. 319.
48. Campanera, J.M.; Bo, C.; Poblet, J.M. General rule for the stabilization of fullerene cages encapsulating trimetallic nitride templates. *Angew. Chem. Int. Ed.* **2005**, *44*, 7230–7233. [CrossRef]
49. Slanina, Z.; Uhlík, F.; Pan, C.; Akasaka, T.; Lu, X.; Adamowicz, L. Computed stabilization for a giant fullerene endohedral: $Y_2C_2$@$C_1$(1660)-$C_{108}$. *Chem. Phys. Lett.* **2018**, *710*, 147–149. [CrossRef]
50. Slanina, Z.; Uhlík, F.; Lee, S.-L.; Nagase, S. Structural and bonding features of Z@$C_{82}$ (Z = Al, Sc, Y, La) endohedrals. *J. Comput. Meth. Sci. Engn.* **2010**, *10*, 569–574. [CrossRef]
51. Slanina, Z.; Uhlík, F.; Shen, W.; Akasaka, T.; Lu, X.; Adamowicz, L. Calculations of the relative populations of Lu@$C_{82}$ isomers. *Nanotub. Carbon Nanostruct.* **2019**, *27*, 710–714. [CrossRef]
52. Slanina, Z.; Uhlík, F.; Akasaka, T.; Lu, X.; Adamowicz, L. Calculated relative thermodynamic stabilities of the Gd@$C_{82}$ isomers. *ECS J. Solid State Sci. Technol.* **2021**, *10*, 071013-1–071013-4. [CrossRef]
53. Meng, Q.Y.; Morales-Martínez, R.; Zhuang, J.X.; Yao, Y. R.; Wang, Y.F.; Feng, L.; Poblet, J.M.; Rodríguez-Fortea, A.; Chen, N. Synthesis and characterization of two isomers of Th@$C_{82}$: Th@C2*v*(9)-$C_{82}$ and Th@$C_2$(5)-$C_{82}$. *Inorg. Chem.* **2021**, *60*, 11496–11502. [CrossRef]
54. Slanina, Z.; Uhlík, F.; Feng, L.; Adamowicz, L. Ho@$C_{82}$ metallofullerene: Calculated isomeric composition. *ECS J. Solid State Sci. Technol.* **2022**, *11*, 053018-1–053018-4. [CrossRef]
55. Andreoni, W.; Curioni, A. Freedom and constraints of a metal atom encapsulated in fullerene cages. *Phys. Rev. Lett.* **1996**, *77*, 834–837. [CrossRef]
56. Popov, A.A.; Dunsch, L. Bonding in endohedral metallofullerenes as studied by quantum theory of atoms in molecules. *Chem. Eur. J.* **2009**, *15*, 9707–9729. [CrossRef]
57. Slanina, Z.; Uhlík, F.; Lee, S.-L.; Adamowicz, L.; Akasaka, T.; Nagase, S. Computed stabilities in metallofullerene series: Al@$C_{82}$, Sc@$C_{82}$, Y@$C_{82}$, and La@$C_{82}$. *Int. J. Quant. Chem.* **2011**, *111*, 2712–2718. [CrossRef]

58. Rodríguez-Fortea, A.; Balch, A.L.; Poblet, J.M. Endohedral metallofullerenes: A unique host-guest association. *Chem. Soc. Rev.* **2011**, *40*, 3551–3563. [CrossRef]
59. Popov, A.A.; Yang, S.; Dunsch, L. Endohedral fullerenes. *Chem. Rev.* **2013**, *113*, 5989–6113. [CrossRef]
60. Yang, H.; Yu, M.; Jin, H.; Liu, Z.; Yao, M.; Liu, B.; Olmstead, M.M.; Balch, A.L. Isolation of three isomers of Sm@$C_{84}$ and X-ray crystallographic characterization of Sm@$D_{3d}$(19)-$_{84}$ and Sm@$C_2$(13)-$C_{84}$. *J. Am. Chem. Soc.* **2012**, *134*, 5331 [CrossRef]
61. Slanina, Z.; Zhao, X.; Uhlík, F.; Ozawa, M.; Osawa, E. Computational modelling of the metal and other elemental catalysis in the Stone-Wales fullerene rearrangements. *J. Organomet. Chem.* **2000**, *599*, 57–61. [CrossRef]
62. Hao, Y.; Feng, L.; Xu, W.; Gu, Z.; Hu, Z.; Shi, Z.; Slanina, Z.; Uhlík, F. Sm@$C_{2v}$(19138)-$C_{76}$: A Non-IPR Cage Stabilized by a Divalent Metal Ion. *Inorg. Chem.* **2015**, *54*, 4243–4248. [CrossRef]
63. Hao, Y.; Tang, Q.; Li, X.; Zhang, M.; Wan, Y.; Feng, L.; Chen, N.; Slanina, Z.; Adamowicz, L.; Uhlík, F. Isomeric $Sc_2O$@$C_{78}$ related by a single-step Stone–Wales. transformation: Key links in an unprecedented fullerene formation pathway. *Inorg. Chem.* **2016**, *55*, 11354–11361. [CrossRef]
64. Jehlička, J.; Svatoš, A.; Frank, O.; Uhlík, F. Evidence for fullerenes in solid bitumen from pillow lavas of proterozoic age from Mítov (Bohemian Massif, Czech Republic) Geochem. *Cosmochem. Acta* **2003**, *67*, 1495–1506. [CrossRef]
65. Lian, Y.; Shi, Z.; Zhou, X.; Gu, Z. Different extraction behaviors between divalent and trivalent endohedral metallofullerenes. *Chem. Mater.* **2004**, *16*, 1704–1714. [CrossRef]
66. Maeda, Y.; Tsuchiya, T.; Kikuchi, T.; Nikawa, H.; Yang, T.; Zhao, X.; Slanina, Z.; Suzuki, M.; Yamada, M.; Lian, Y.; et al. Effective derivatization and extraction of insoluble missing lanthanum metallofullerenes La@$C_{2n}$ (n = 36–38) with iodobenzene. *Carbon* **2016**, *98*, 67–73. [CrossRef]
67. An, D.-Y.; Su, J.-G.; Li, C.-H.; Li, J.-Y. Computational studies on the interactions of nanomaterials with proteins and their impacts. *Chin. Phys. B* **2015**, *24*, 120504-1–120504-8. [CrossRef]
68. Basiuk, V.A.; Tahuilan-Anguiano, D.E. Complexation of free-base and 3d transition metal(II) phthalocyanines. with endohedral fullerene $Sc_3N$@$C_{80}$. *Chem. Phys. Lett.* **2019**, *722*, 146–152. [CrossRef]
69. Tahuilan-Anguiano, D.E.; Basiuk, V.A. Complexation of free-base and 3d transition metal(II) phthalocyanines with endohedral fullerenes H@$C_{60}$, $H_2$@$C_{60}$ and He@$C_{60}$: The effect of encapsulated species. *Diam. Rel. Mat.* **2021**, *118*, 108510-1–108510-5. [CrossRef]
70. Li, M.; Zhao, R.; Dang, J.; Zhao, X. Theoretical study on the stabilities, electronic structures, and reaction and formation mechanisms of fullerenes and endohedral metallofullerenes. *Coor. Chem. Rev.* **2022**, *471*, 214762-1–214762-12. [CrossRef]
71. Slanina, Z.; Uhlík, F.; Adamowicz, L. Theoretical predictions of fullerene stabilities. In *Handbook of Fullerene Science and Technology*; Lu, X., Akasaka, T., Slanina, Z., Eds.; Springer: Singapore, 2022; pp. 111–179.

inorganics

Article

# Lu-Lu Bond in $Lu_2@C_{60}$ Metallofullerenes

Yaoxiao Zhao [1], Wangqiang Shen [2,3,*], Weixing Chen [1] and Xing Lu [3,*]

1 Engineering Research Center of Light Stabilizers for Polymer Materials, School of Materials Science and Chemical Engineering, Xi'an Technological University, Xi'an 710021, China; zhaoyaoxiao@xatu.edu.cn (Y.Z.); chenwx@xatu.edu.cn (W.C.)
2 School of Materials Science and Engineering, Hefei University of Technology, Hefei 230009, China
3 State Key Laboratory of Materials Processing and Die & Mould Technology, School of Materials Science and Engineering, Huazhong University of Science and Technology, 1037 Luoyu Road, Wuhan 430074, China
* Correspondence: shenwq@hfut.edu.cn (W.S.); lux@hust.edu.cn (X.L.)

**Abstract:** This study on $Lu_2@C_{60}$ isomers provides insights into the metal–metal bond through the confinement effect of fullerene cages. Density functional theory calculations were used to study the nature of the Lu-Lu bond in two stable endohedral metallofullerenes (EMFs), $Lu_2@C_{2v}\_C_{60}$ and $Lu_2@I_h\_C_{60}$, both with negative endohedral energy. These two isomers are geometrically connected through a simple Stone–Wales (SW) transformation. The electronic configuration of $(Lu_2)^{4+}@C_{60}{}^{4-}$ was also confirmed, leading to the formation of a two-center two-electron (2c–2e) Lu-Lu σ single bond. By comparing the Lu-Lu bonds in $Lu_2@C_{60}$ with those in acknowledged $Lu_2@C_{2n}$, the smaller $C_{60}$ fullerene compressed the geometry of $Lu_2$ resulting in a much shorter Lu-Lu bond length. However, the Lu-Lu bond strength is slightly weaker in $Lu_2@C_{60}$ than that in large fullerenes, as the Lu-Lu bond in $C_{60}$ is likely a p-p σ bond with an above the 40% contribution of p orbital and a strong metal–cage interaction. Additionally, the vis-NIR spectra of $Lu_2@C_{2v}\_C_{60}$ and $Lu_2@I_h\_C_{60}$ were simulated, which could provide valuable information for future experimental studies on Lu-based EMFs.

**Keywords:** endohedral metallofullerenes; metal–metal bond; size effect

## 1. Introduction

Metal–metal bonds have been a subject of significant research interest in past decades, which show important roles in catalysis and biology [1,2]. Metal–metal bonds provide a large perturbation in electronic structure and the unique properties of these dinuclear fragments can be harnessed in a broad range of applications [3]. Since the synthesis and characterization of $Re_2Cl_8{}^{2-}$ [4], there is growing evidence for the formation of metal–metal bonds under specific conditions [5,6], and growing research interest in understanding the nature of the metal–metal bonds. Generally, the transition metals could form multinuclear complexes with direct metal–metal interactions, but a complex coordination environment makes it difficult to understand the nature of metal–metal interaction [7].

Luckily, endohedral dimetallofullerenes (di-EMFs), i.e., fullerenes with two metal atoms trapped inside are considered as the ideal model to study the metal–metal bond [8]. In 2014, di-EMF $Sc_2@C_1(4059)$-$C_{66}$ was determined as the geometry of $Sc_2C_{66}$ molecule by the single-crystal X-ray diffraction (XRD) [9]. $La_2@C_{94}$ are also proposed with experimental and theoretical methods [10], and later, theoretical study revealed that the La–La bond plays a key role in the stability of dimetallofullerenes $[La_2@C_{2n}]^-$ ($2n$ = 92–96) [11]. Recently, Chen et al. have studied Sc-Y $\sigma^2$ bond in $ScY@C_{3v}(8)$-$C_{82}$ [12].

Additionally, the previous reports clarified strong confinement effects of fullerene size on the metal metal bond in EMFs [7,13–15]. For example, Poblet et al. theoretically reveal a U-U triple bond in $U_2@C_{60}$ with the effective bond order of 2.52, singly occupied molecular orbitals (MOs) with metal–metal bond characters in $U_2@C_{80}$, non-negligible U···U interaction in $U_2@C_{78}$, but the metal–metal interaction almost disappears in $U_2@C_{104}$ [8]. In 2017, di-EMFs $Lu_2@C_{2n}$ ($2n$ = 82, 84, and 86) were synthesized and characterized with the single-crystal XRD method, which gives the crystallographic evidence of direct Lu-Lu bond between two divalent lutetium ions inside fullerenes [16]. Later, additional $Lu_2C_{2n}$ ($2n$ = 76–90) molecules were reported with a wide range cage size, and the successful isolation and unambiguous crystallographic assignment of a series of lutetium-containing EMFs, $Lu_2C_{2n}$ ($2n$ = 76, 78, 80, 84, 86, 88, and 90), reveal an unrecognized decisive effect of the cage size on the configuration of the encapsulated clusters [17]. Following theoretical studies reveal a stable two-center two-electron (2c–2e) Lu-Lu $\sigma$ bond in $Lu_2@C_{84}$ and $Lu_2@C_{86}$ [18,19]. However, experimental and theoretical study focused on the Lu-containing di-EMFs with medium- or large-size fullerene cages and Lu-Lu bond in small fullerene cages is still unclear.

Herein, Lu-Lu bond is studied in the smaller $C_{60}$ fullerene cage with the highest yield in the fullerene family by using density functional theory calculations and bonding analysis. The aim is to evaluate the confinement effect of fullerene size on the Lu-Lu bond. Two isomers, $C_{2v}$_$C_{60}$ and $I_h$_$C_{60}$, which have been previously verified as the stable host cage of $C_{60}$ fullerene to encage inner moieties, were selected as they have well-established thermodynamic stability and were used to encapsulate the $Lu_2$ dimer. The bonding features of the Lu-Lu in the $C_{60}$ were evaluated via frontier molecular orbital (FMO), natural bond orbital (NBO), and bond orders analyses. Additionally, the visible-near-infrared (vis-NIR) spectra of $Lu_2@C_{2v}$_$C_{60}$ and $Lu_2@I_h$_$C_{60}$ were simulated to study their electronic features and gain a better understanding of the Lu-Lu bonds both in theory and experiment.

## 2. Results and Discussion

### 2.1. Stability and Geometries of $Lu_2@C_{60}$

The good stability of fullerenes $C_{2v}$_$C_{60}$ and $I_h$_$C_{60}$, connected through a single-step Stone-Wales transformation, has been previously reported [20], and they have been characterized as suitable isomers of $C_{60}$ fullerenes to encage metal atoms or clusters [8,21]. In this work, we focus on both $Lu_2@C_{2v}$_$C_{60}$ and $Lu_2@I_h$_$C_{60}$ to generally elucidate the confinement effect of the fullerene on metal–metal bond.

Figure 1 shows the optimized geometries of $Lu_2@C_{2v}$_$C_{60}$ and $Lu_2@I_h$_$C_{60}$. The lowest frequencies for $Lu_2@C_{2v}$_$C_{60}$ and $Lu_2@I_h$_$C_{60}$ are calculated to be 36 and 45 $cm^{-1}$, respectively, suggesting their possible existence, as shown in Figures S1 and S2. The single-point energy calculations showed that the $Lu_2@C_{2v}$_$C_{60}$ possessed energy of 5.3 kcal·$mol^{-1}$ higher than the energy of $Lu_2@I_h$_$C_{60}$, of which results are confirmed by the hybrid functional PBE0 (PBE0/6-311G(d,p)~SDD//PBE0/6-31G(d)~CEP-4G) showing $Lu_2@C_{2v}$_$C_{60}$ with energy of 0.7 kcal·$mol^{-1}$ higher than the energy of $Lu_2@I_h$_$C_{60}$.

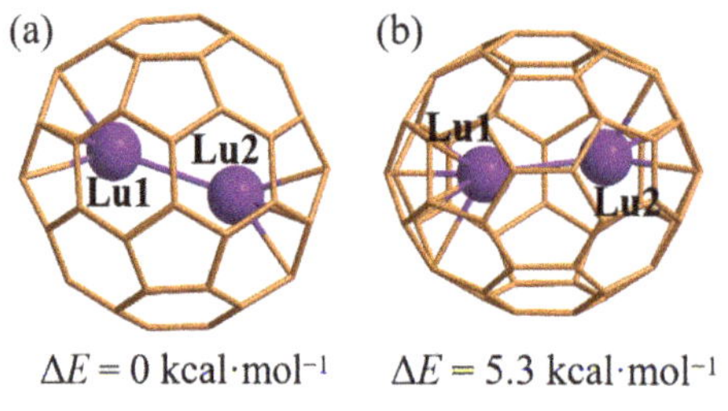


**Figure 1.** Geometries of optimized $Lu_2@I_h$_$C_{60}$ (**a**) and $Lu_2@C_{2v}$_$C_{60}$ (**b**) on the PBE/6-31G(d)~CEP-4G, including the relative energy ($\Delta E$) on the theoretical level of PBE/6-311G(d,p)~SDD with single-point calculations.

In order to evaluate the effect of temperature, the Boltzmann distribution has been calculated for $Lu_2@C_{2v}\_C_{60}$ and $Lu_2@I_h\_C_{60}$. The results indicate that the Boltzmann distribution of $Lu_2@I_h\_C_{60}$ is about 97%, 84%, 75%, 70%, 66%, and 63% at 500, 1000, 1500, 2000, 2500, and 3000 K, respectively, which is higher than $Lu_2@C_{2v}\_C_{60}$ with the Boltzmann distribution values of about 4%, 16%, 25%, 30%, 34%, and 37% at 500, 1000, 1500, 2000, 2500, and 3000 K, respectively. The energy gaps between the highest occupied molecular orbital (HOMO) and the lowest unoccupied molecular orbital (LUMO) are 0.19 and 0.35 eV for $Lu_2@I_h\_C_{60}$ and $Lu_2@C_{2v}\_C_{60}$, respectively, which are larger than the DFT-computed HOMO-LUMO gaps of $Lu_2@C_s(17,490)\_C_{76}$ and $Lu_2@C_{2v}(19,138)\_C_{76}$ [22].

The formation energy for $Lu_2@C_{2v}\_C_{60}$ and $Lu_2@I_h\_C_{60}$ has been calculated by basis set superposition error (BSSE) correction with a counterpoise calculation on the PBE/6-311G(d,p)~SDD considering the Grimme's dispersion with the original D3 damping function. The formation energy for $Lu_2@C_{2v}\_C_{60}$ and $Lu_2@I_h\_C_{60}$ are $-179.7$ and $-156.4$ kcal·mol$^{-1}$, indicating the possible synthesis and isolation in experiment. On the other hand, the much negative formation energy indicates their good stability. The similar stability for $Lu_2@I_h\_C_{60}$ and $Lu_2@C_{2v}\_C_{60}$ might be explained by the geometrical connection with only the single-step Stone–Wales transformation, and this case is similar to the hollow cage of $C_{2v}\_C_{60}$ and $I_h\_C_{60}$ isomers.

The Lu-Lu bond length in $Lu_2@I_h\_C_{60}$ and $Lu_2@C_{2v}\_C_{60}$ are 3.04 and 3.14 Å, respectively, which are much smaller than the previously reported Lu-Lu bond length (~3.50 Å) in $Lu_2@C_{2n}$ ($2n \geq 76$) [17–19,22]. This suggests that there is a confinement effect of fullerenes on the Lu-Lu bond length. The bond length of Sc-Sc and Y-Y in $C_{82}$ are 3.201 and 3.695 Å, respectively, and bond length of La-La in $C_{80}$ is 3.826 Å, indicating the presence of a metal–metal single bond, which has been verified in both theory and experiment [22]. Based on the geometry, further study into the confinement effect of fullerenes on the Lu-Lu bond is warranted.

### 2.2. Electronic Structures of $Lu_2@C_{60}$

The electronic structure of $Lu_2@C_{60}$ was further investigated through NBO calculations. The results in Table 1 show that there is charge transfer from inner $Lu_2$ dimer to the $C_{60}$ fullerenes, and in combination with the Lu-Lu $\sigma$ bond in $C_{60}$, the 6s orbitals of Lu atoms lose four electrons in $Lu_2@C_{60}$. Therefore, the electronic configurations of both $Lu_2@I_h\_C_{60}$ and $Lu_2@C_{2v}\_C_{60}$ can be expressed as $(Lu_2)^{4+}@(C_{60})^{4-}$.

**Table 1.** Natural population analysis (NPA) on $Lu_2@I_h\_C_{60}$ and $Lu_2@C_{2v}\_C_{60}$, and the NPA charges (*e*) for metal atoms on the PBE/6-311G(d,p)~SDD.

| Molecules | Atoms | Populations | NPA Charges |
|---|---|---|---|
| $Lu_2@I_h\_C_{60}$ | Lu1 | $5d^{0.37}6s^{0.29}6p^{0.48}6d^{0.88}7p^{0.29}$ | 0.78 |
| | Lu2 | $5d^{0.36}6s^{0.37}6p^{0.46}6d^{0.85}7p^{0.28}$ | 0.78 |
| $Lu_2@C_{2v}\_C_{60}$ | Lu1 | $5d^{0.36}6s^{0.30}6p^{0.41}6d^{0.97}7p^{0.39}$ | 0.65 |
| | Lu2 | $5d^{0.34}6s^{0.30}6p^{0.32}6d^{0.89}7p^{0.40}$ | 0.83 |

In addition, the location of fractional charge on 6p, 6d, and 7p orbitals of Lu atoms implies the occurrence of back-donation, similar to that observed in other EMFs [23–37]. This is further supported by the presence of electrostatic interaction between the $Lu_2$ dimer and fullerene cages, as depicted in Figure 2. These observations are consistent with the ionic model of EMFs [13,38,39].

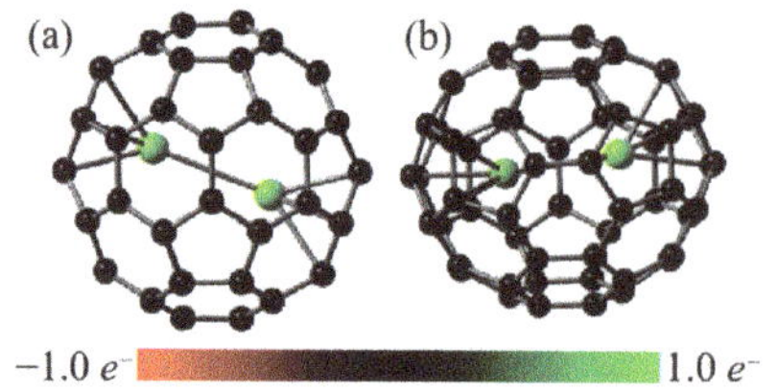


**Figure 2.** NPA charges for $Lu_2@I_h\_C_{60}$ (**a**) and $Lu_2@C_{2v}\_C_{60}$ (**b**) on the PBE/6-311G(d,p)~SDD, in which the atoms are colored by corresponding NPA charges.

Furthermore, the FMOs have been mapped as shown in Figure 3, in which the $C_{60}$ fullerenes based on single-point calculations are obtained from the optimized $Lu_2@I_h\_C_{60}$ and $Lu_2@C_{2v}\_C_{60}$. The highest occupied molecular orbitals (HOMOs) of $Lu_2@C_{60}$ are contributed by the $Lu_2$ dimer, presenting the Lu-Lu bond. It is clear that the LUMO and LUMO + 1 of the $C_{60}$ fullerenes become the HOMO-1 and HOMO-2 of the $Lu_2@C_{60}$ indicating the four-electron transfer from inner moiety to fullerene cage, and this result is in line with the natural bond order analysis again confirming the electronic configuration of $(Lu_2)^{4+}@(C_{60})^{4-}$. This electronic configuration has been verified in the previous report on $Lu_2@C_{2n}$ ($2n \geq 76$) [17–19,22]. Additionally, the $t_{1u}$ LUMO has been split after encapsulation of $Lu_2$ dimer. As shown in Figure 3, after encapsulation of $Lu_2$ dimer in $I_h$-$C_{60}$, its energy level of LUMO + 1 and LUMO is equal, but the energy level of LUMO + 2 is −4.22 eV. This is derived from the distortion of $C_{60}$ fullerene. The symmetry of $I_h$-$C_{60}$ and $C_{2v}$-$C_{60}$ have reduced to $D_{3d}$ and $C_1$, respectively, after encapsulation, and the calculated distortion energies of $I_h$-$C_{60}$ and $C_{2v}$-$C_{60}$ are 35.4 and 29.6 kcal·mol$^{-1}$, respectively, on the theoretical level of PBE/6-311G(d,p)~SDD.

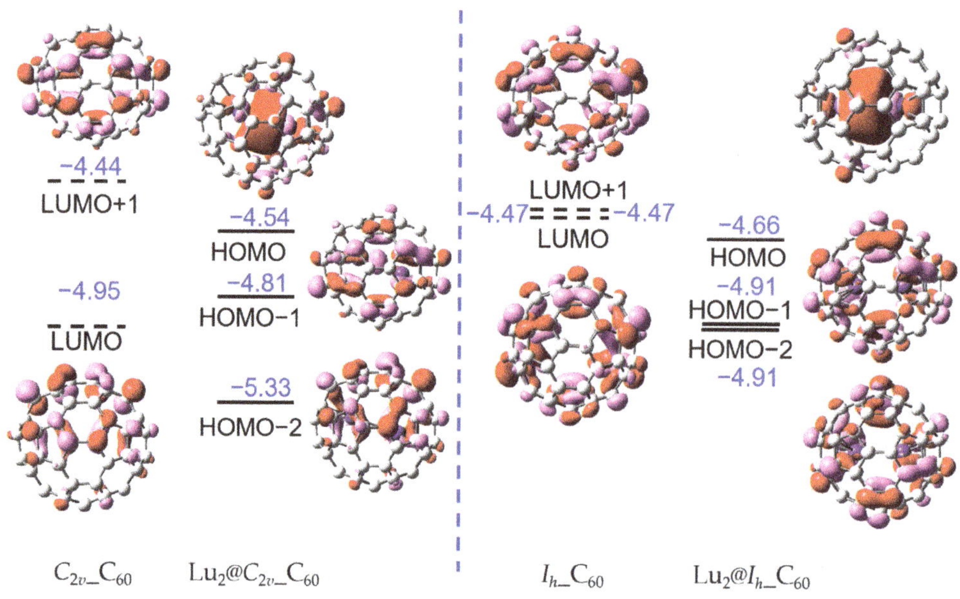


**Figure 3.** FMOs for stable $C_{2v}\_C_{60}$, $I_h\_C_{60}$, $Lu_2@I_h\_C_{60}$, and $Lu_2@C_{2v}\_C_{60}$ molecules with isovalue of 0.03 a.u. for the surface with the PBE functional, 6-311G(d,p) basis set for carbon atoms, and SDD basis set for Lu atoms. The highest occupied molecular orbitals (HOMO) represented the Lu-Lu bond in $C_{60}$. The blue numbers represent the energies of orbitals.

### 2.3. Bonding Features of $Lu_2@C_{60}$

Figure 3 illustrates the presence of a clearly localized bonding molecular orbital between two Lu atoms in $Lu_2@C_{60}$, which has been reported in previous studies [16–19]. The energy level of the bonding molecular orbitals is −4.66 and −4.54 eV for $Lu_2@I_h\_C_{60}$ and $Lu_2@C_{2v}\_C_{60}$, respectively. However, due to the confinement effect of the smaller fullerene cage size, such as $C_{60}$, further investigation is needed to determine the strength of the Lu-Lu bond. The LUMOs are shown in Figure S3, which display both a metal–metal antibonding orbital and a $\pi$ antibonding orbital.

To study the nature of the Lu-Lu bond in small fullerene sizes, we calculated the Mayer bond order (MBO) and Wiberg bond order (WBO) and the results are presented in Table 2. Despite the much shorter bond length of Lu-Lu in fullerene $C_{60}$, the MBO for Lu-Lu bond in $C_{60}$ is smaller than that in larger fullerenes. For example, in $Lu_2@C_{2n}$ ($2n$ = 84 and 86), the MBO is a little larger than 1 for Lu-Lu bond with a larger bond length [18,19]. The WBOs for Lu-Lu bond in $C_{60}$ are also smaller than 1. To gain further insights into this abnormal phenomenon, the hybrid compositions of M-M bond in $Lu_2@C_{60}$ were studied with NBO calculations. As shown in Table 2, the p orbital contributes more than 40% to the Lu-Lu bond, while the rest of the contributions come from the s and d atomic orbitals. Thus, the Lu-Lu $\sigma$ bond in $C_{60}$ is likely a p-p $\sigma$ bond. Previous studies have shown that the Lu-Lu bond is mainly contributed by s orbitals [18,19], in which the metal–metal bond is more like a s-s $\sigma$ bond. Generally, the overlap of s orbitals is more effective than that of p orbitals, meaning that the s-s $\sigma$ bond is much stronger than the p-p $\sigma$ bond. On the other hand, the interaction between $Lu_2$ dimer and fullerene cage $C_{60}$ is stronger with MBO values of 4.19 and 4.82 for $Lu_2@I_h\_C_{60}$ and $Lu_2@C_{2v}\_C_{60}$, respectively.

**Table 2.** Mayer bond order (MBO) and Wiberg bond order (WBO) of Lu-Lu bond, and hybrid compositions of M-M bond in the $Lu_2@C_{60}$ at the theoretical level of PBE/6-311G(d,p)~SDD.

| Molecules | MBO | WBO | Atoms | Hybrid Composition |
|---|---|---|---|---|
| $Lu_2@I_h\_C_{60}$ | 0.99 | 0.73 | Lu1<br>Lu2 | s(26%)p(46%)d(28%)<br>s(36%)p(43%)d(21%) |
| $Lu_2@C_{2v}\_C_{60}$ | 0.84 | 0.67 | Lu1<br>Lu2 | s(26%)p(44%)d(30%)<br>s(28%)p(45%)d(27%) |

In $Lu_2@I_h\_C_{60}$ and $Lu_2@C_{2v}\_C_{60}$, the Lu-Lu $\sigma$ bond is mainly derived from p orbitals, for which the bond strength is related to the bond length. The Lu-Lu bond in $I_h\_C_{60}$ with higher values of bond orders (0.99 for MBO and 0.73 for WBO, Table 2) show stronger bond strength than that in $C_{2v}\_C_{60}$, because the shorter bond length in $I_h\_C_{60}$ than that in $C_{2v}\_C_{60}$.

### 2.4. Simulated Spectra of $Lu_2@C_{60}$

Figure 4 illustrates that the excitation energies of the first excited state of $Lu_2@I_h\_C_{60}$ and $Lu_2@C_{2v}\_C_{60}$ are 0.49 eV (2542 nm) and 0.47 eV (2637 nm), respectively, which are smaller than the first excitation energy of approximately 0.90 eV (1384 nm) for $Lu_2@C_{86}$ in both theory and experiment [16,19]. The excitation form $S_0$ to $S_1$ is primarily derived from the transition of electrons from the highest occupied molecular orbital (HOMO) to the lowest unoccupied molecular orbital (LUMO) of $Lu_2@C_{60}$, and the energy required for this transition is 0.78 $e^-$ and 0.61 $e^-$ for $Lu_2@I_h\_C_{60}$ and $Lu_2@C_{2v}\_C_{60}$, respectively. As demonstrated in Figure 3 and Figure S3, the HOMO corresponds to the metal–metal bonding orbital and the LUMO corresponds to the metal–metal antibonding orbital and the $\pi$ antibonding orbital in $Lu_2@C_{60}$, which is potential reason why their first excitation energy is low. The much lower first excitation energy and the ease of electron transition suggest the poor photochemical stability of $Lu_2@C_{60}$.

In addition, although the absorption peaks of $Lu_2@I_h\_C_{60}$ and $Lu_2@C_{2v}\_C_{60}$ isomers have similar shapes (Figure 4), there is a clear difference in their absorption strength, which can be useful for distinguishing them in the future experiments. Specifically, $Lu_2@I_h\_C_{60}$ exhibits a much stronger absorption peak at around 750 nm compared to $Lu_2@C_{2v}\_C_{60}$. The absorption band at 200–700 nm for $Lu_2@C_{2v}\_C_{60}$ is slightly wider than that of $Lu_2@I_h\_C_{60}$. Furthermore, the IR spectra of $Lu_2@I_h\_C_{60}$ and $Lu_2@C_{2v}\_C_{60}$ were simulated (Figure S4) to evaluate their vibration models. These characterized absorption peaks in the vis-NIR and IR spectra can be beneficial for future characterization of $Lu_2@C_{60}$ in experiments.

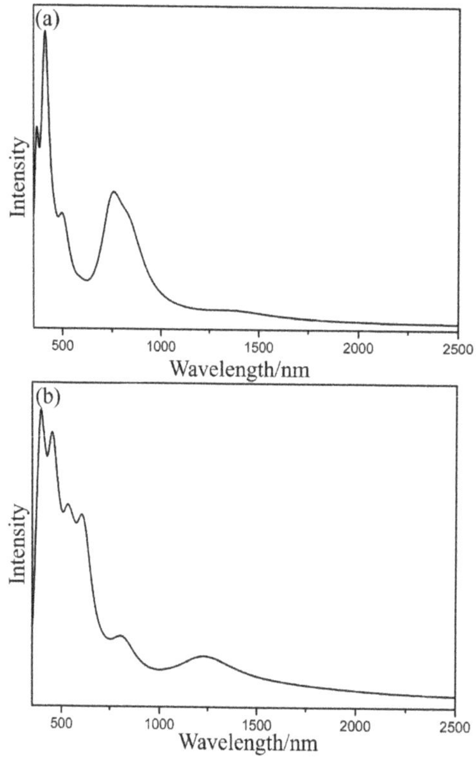


**Figure 4.** Simulated vis-NIR spectra of the $Lu_2@I_h\_C_{60}$ (**a**) and $Lu_2@C_{2v}\_C_{60}$ (**b**) isomers. The broadening function is Lorentzian and full width at half maximum is 0.30 eV.

## 3. Computational Methods

The optimizations of $Lu_2@C_{2v}\_C_{60}$ and $Lu_2@I_h\_C_{60}$ were carried out on the PBE/6-31G(d)~CEP-4G theoretical level [40,41], in which the basis function 6-31G(d) was used for the carbon atom and CEP-4G containing pseudopotential was used for the lutetium atom with valance electronic configuration of $4f^{14}5d^16s^2$. The frequencies were calculated on the same theoretical level for the optimized geometries of $Lu_2@C_{60}$, and all of them are free from imaginary frequency meaning the existence of local minima point for $Lu_2@C_{2v}\_C_{60}$ and $Lu_2@I_h\_C_{60}$. The PBE has been previously proved to be the suitable functional for the lutetium-based EMFs [18,19,42,43]. To obtain accurate energy and frontier molecular orbitals (FMOs), a single-point calculation was performed on the PBE/6-311G(d,p)~SDD theoretical level, in which a larger basis set 6-311G(d,p) [40] with polarization functions and SDD with effective core pseudopotential were used for the carbon and lutetium atoms, respectively. In order to confirm the calculated results on PBE, the hybrid functional PBE0 (PBE0/6-311G(d,p)~SDD//PBE0/6-31G(d)~CEP-4G) calculations were also carried out for both $Lu_2@C_{2v}\_C_{60}$ and $Lu_2@I_h\_C_{60}$. In order to evaluate the effect of temperature, the Boltzmann distribution has been calculated for $Lu_2@C_{2v}\_C_{60}$ and $Lu_2@I_h\_C_{60}$. The formation energy for $Lu_2@C_{2v}\_C_{60}$ and $Lu_2@I_h\_C_{60}$ was calculated based on basis set superposition errors via counterpoise corrections on the theoretical level of PBE/6-311G(d,p)~SDD. NBO [44] analyses were also conducted on the PBE/6-311G(d,p)~SDD theoretical level. The vis-NIR spectra were simulated on the PBE/6-31G(d)~CEP-4G theoretical level. All the above calculations were carried out by Gaussian16 software package [45]. In addition, Mayer bond order (MBO) [46] together with localized molecular orbitals (LMOs) [47,48] analyses for $Lu_2@C_{60}$ were performed with the Multiwfn program [49].

## 4. Conclusions

Based on the $Lu_2@I_h\_C_{60}$ and $Lu_2@C_{2v}\_C_{60}$ which are related by a single-step Stone–Wales transformation, we provide insight into the confinement effects of fullerene on the metal–metal bonding. Although, in the $Lu_2@C_{60}$, there is a much shorter Lu-Lu bond length, its bond strength is a little weaker than the Lu-Lu bond in large fullerenes, because the Lu-Lu σ bond in $C_{60}$ is likely a p-p σ bond with the contribution p orbital above 40% and a strong metal–cage interaction. Clearly, the confinement effects of fullerene play important roles in the geometry of the inner cluster, especially the bond length in the present work, and the electronic effect is more important for the bonding nature. Furthermore, the electronic configurations of $(Lu_2)^{4+}@C_{60}{}^{4-}$ were confirmed. Additionally, the vis-NIR spectra of $Lu_2@C_{2v}\_C_{60}$ and $Lu_2@I_h\_C_{60}$ were simulated, which could give some valuable information for the future experimental study on Lu-based EMFs.

**Supplementary Materials:** The following supporting information can be downloaded at: https://www.mdpi.com/2304-6740/11/7/277/s1, Figure S1: Several selected displacement vectors of $Lu_2@I_h\_C_{60}$ with vibration frequencies, including the lowest vibration one; Figure S2: Several selected displacement vectors of $Lu_2@C_{2v}\_C_{60}$ with vibration frequencies, including the lowest vibration one; Figure S3: Lowest unoccupied molecular orbitals (LUMO) with localization of $Lu_2@I_h\_C_{60}$ (a) and $Lu_2@C_{2v}\_C_{60}$ (b) with isovalue of 0.03 a.u. for the surface; Figure S4: IR vibration spectra for $Lu_2@I_h\_C_{60}$ (a) and $Lu_2@C_{2v}\_C_{60}$ (b). The broadening function is Gaussian function and full width at half maximum is 12 $cm^{-1}$; coordinates for optimized geometries: $Lu_2@I_h\_C_{60}$ and $Lu_2@C_{2v}\_C_{60}$.

**Author Contributions:** Conceptualization, X.L. and W.S.; methodology, Y.Z.; software, Y.Z.; validation, X.L., W.S., W.C. and Y.Z.; formal analysis, Y.Z. and W.S.; investigation, Y.Z.; data curation, Y.Z.; writing—original draft preparation, Y.Z. and W.S.; writing—review and editing, W.C. and X.L.; visualization, Y.Z.; supervision, X.L.; project administration, X.L.; funding acquisition, X.L. and W.S. All authors have read and agreed to the published version of the manuscript.

**Funding:** This research was funded by the National Natural Science Foundation of China (21925104, 22001084, and 92261204).

**Data Availability Statement:** Data are available on request from the authors.

**Acknowledgments:** We thank the Analytical and Testing Center at the Hefei University of Technology and Huazhong University of Science and Technology for all related measurements, and national supercomputing center in Xi'an.

**Conflicts of Interest:** The authors declare no conflict of interest.

## References

1. Rodriguez, J.A.; Goodman, D.W. The nature of the metal-metal bond in bimetallic surfaces. *Science* **1992**, *257*, 897–903. [CrossRef] [PubMed]
2. Lindahl, P.A. Metal–metal bonds in biology. *J. Inorg. Biochem.* **2012**, *106*, 172–178. [CrossRef] [PubMed]
3. Powers, I.G.; Uyeda, C. Metal–metal bonds in catalysis. *ACS Catal.* **2017**, *7*, 936–958. [CrossRef]
4. Cotton, F.A. Strong homonuclear metal-metal bonds. *Acc. Chem. Res.* **1969**, *2*, 240–247. [CrossRef]
5. Foroutan-Nejad, C.; Vicha, J.; Marek, R.; Patzschke, M.; Straka, M. Unwilling U-U bonding in $U_2@C_{80}$: Cage-driven metal-metal bonds in di-uranium fullerenes. *Phys. Chem. Chem. Phys.* **2015**, *17*, 24182–24192. [CrossRef] [PubMed]

6. Zhuang, J.; Morales-Martínez, R.; Zhang, J.; Wang, Y.; Yao, Y.R.; Pei, C.; Rodríguez-Fortea, A.; Wang, S.; Echegoyen, L.; de Graaf, C.; et al. Characterization of a strong covalent $Th^{3+}$-$Th^{3+}$ bond inside an $I_h(7)$-$C_{80}$ fullerene cage. *Nat. Commun.* **2021**, *12*, 2372. [CrossRef]
7. Li, M.; Zhao, R.; Dang, J.; Zhao, X. Theoretical study on the stabilities, electronic structures, and reaction and formation mechanisms of fullerenes and endohedral metallofullerenes. *Coordin. Chem. Rev.* **2022**, *471*, 214762. [CrossRef]
8. Moreno-Vicente, A.; Roselló, Y.; Chen, N.; Echegoyen, L.; Dunk, P.W.; Rodríguez-Fortea, A.; de Graaf, C.; Poblet, J.M. Are U–U Bonds Inside Fullerenes Really Unwilling Bonds? *J. Am. Chem. Soc.* **2023**, *145*, 6710–6718. [CrossRef]
9. Yamada, M.; Kurihara, H.; Suzuki, M.; Guo, J.D.; Waelchli, M.; Olmstead, M.M.; Balch, A.L.; Nagase, S.; Maeda, Y.; Hasegawa, T. $Sc_2@C_{66}$ revisited: An endohedral fullerene with scandium ions nestled within two unsaturated linear triquinanes. *J. Am. Chem. Soc.* **2014**, *136*, 7611–7614. [CrossRef]
10. Zhao, S.; Zhao, P.; Cai, W.; Bao, L.; Chen, M.; Xie, Y.; Zhao, X.; Lu, X. Stabilization of Giant Fullerenes $C_2(41)$-$C_{90}$, $D_3(85)$-$C_{92}$, $C_1(132)$-$C_{94}$, $C_2(157)$-$C_{96}$, and $C_1(175)$-$C_{98}$ by Encapsulation of a Large $La_2C_2$ Cluster: The Importance of Cluster–Cage Matching. *J. Am. Chem. Soc.* **2017**, *139*, 4724–4728. [CrossRef]
11. Li, Q.-Z.; Zheng, J.-J.; He, L.; Nagase, S.; Zhao, X. La–La bonded dimetallofullerenes $[La_2@C_{2n}]^-$: Species for stabilizing $C_{2n}$ ($2n$ = 92–96) besides $La_2C_2@C_{2n}$. *Phys. Chem. Chem. Phys.* **2018**, *20*, 14671–14678. [CrossRef] [PubMed]
12. Zheng, L.; Roselló, Y.; Yan, Y.; Yao, Y.R.; Fan, X.; Poblet, J.M.; Rodríguez-Fortea, A.; Chen, N. $ScY@C_{3v}(8)$-$C_{82}$: Metal-Metal $\sigma^2$ Bond in Mixed Rare-Earth Di-metallofullerenes. *Chin. J. Chem.* 2023; *early view*. [CrossRef]
13. Popov, A.A.; Yang, S.; Dunsch, L. Endohedral fullerenes. *Chem. Rev.* **2013**, *113*, 5989–6113. [CrossRef]
14. Lu, X.; Akasaka, T.; Nagase, S. Chemistry of endohedral metallofullerenes: The role of metals. *Chem. Commun.* **2011**, *47*, 5942–5957. [CrossRef]
15. Shen, W.; Bao, L.; Lu, X. Endohedral Metallofullerenes: An Ideal Platform of Sub-Nano Chemistry. *Chin. J. Chem.* **2022**, *40*, 275–284. [CrossRef]
16. Shen, W.; Bao, L.; Wu, Y.; Pan, C.; Zhao, S.; Fang, H.; Xie, Y.; Jin, P.; Peng, P.; Li, F.-F. $Lu_2@C_{2n}$ ($2n$ = 82, 84, 86): Crystallographic evidence of direct Lu–Lu bonding between two divalent lutetium ions inside fullerene cages. *J. Am. Chem. Soc.* **2017**, *139*, 9979–9984. [CrossRef] [PubMed]
17. Shen, W.; Bao, L.; Hu, S.; Yang, L.; Jin, P.; Xie, Y.; Akasaka, T.; Lu, X. Crystallographic characterization of $Lu_2C_{2n}$ ($2n$ = 76–90): Cluster selection by cage size. *Chem. Sci.* **2019**, *10*, 829–836. [CrossRef]
18. Zheng, H.; Dang, H.; Zhao, Y.; Gu, Y.-X.; Li, M.; Li, Q.-Z.; Zhao, X. Theoretical Investigations of $Lu_2C_{84}$: Unexpected Impact of Metal Electronic Configuration toward the Metal–Metal σ-Bond in Fullerene. *Inorg. Chem.* **2020**, *59*, 10113–10122. [CrossRef]
19. Zhang, K.; Zheng, H.; Li, M.; Li, Q.-Z.; Zhao, Y.; Zhao, X. Significant roles of a particularly stable two-center two-electron Lu–Lu σ bond in $Lu_2@C_{86}$: Electronic structure of Lu and radius of $Lu^{2+}$. *Inorg. Chem.* **2021**, *60*, 2425–2436. [CrossRef]
20. Stone, A.J.; Wales, D.J. Theoretical studies of icosahedral $C_{60}$ and some related species. *Chem. Phys. Lett.* **1986**, *128*, 501–503. [CrossRef]
21. Wang, Y.; Tomanek, D.; Bertsch, G.F.; Ruoff, R.S. Stability of $C_{60}$ fullerite intercalation compounds. *Phys. Rev. B* **1993**, *47*, 6711–6720. [CrossRef]
22. Popov, A.A.; Avdoshenko, S.M.; Pendás, A.M.; Dunsch, L. Bonding between strongly repulsive metal atoms: An oxymoron made real in a confined space of endohedral metallofullerenes. *Chem. Commun.* **2012**, *48*, 8031–8050. [CrossRef]
23. Fu, W.; Zhang, J.; Fuhrer, T.; Champion, H.; Furukawa, K.; Kato, T.; Mahaney, J.E.; Burke, B.G.; Williams, K.A.; Walker, K.; et al. $Gd_2@C_{79}N$: Isolation, characterization, and monoadduct formation of a very stable heterofullerene with a magnetic spin state of S = 15/2. *J. Am. Chem. Soc.* **2011**, *133*, 9741–9750. [CrossRef] [PubMed]
24. Yan, Y.; Morales-Martinez, R.; Zhuang, J.; Yao, Y.R.; Li, X.; Poblet, J.M.; Rodriguez-Fortea, A.; Chen, N. $Th@D_{5h}(6)$-$C_{80}$: A highly symmetric fullerene cage stabilized by a single metal ion. *Chem. Commun.* **2021**, *57*, 6624–6627. [CrossRef] [PubMed]
25. Meng, Q.; Abella, L.; Yang, W.; Yao, Y.R.; Liu, X.; Zhuang, J.; Li, X.; Echegoyen, L.; Autschbach, J.; Chen, N. $UCN@C_s(6)$-$C_{82}$: An Encapsulated Triangular UCN Cluster with Ambiguous U Oxidation State [U(III) versus U(I)]. *J. Am. Chem. Soc.* **2021**, *143*, 16226–16234. [CrossRef] [PubMed]
26. Dong, B.; Yu, Y.; Slanina, Z.; Wang, F.; Lian, Y.; Uhlik, F.; Feng, L. $Ho_2C_2$ Cluster with Flexible Configurations inside a Large $C_2(61)$-$C_{92}$ Cage. *Inorg. Chem.* **2022**, *61*, 605–612. [CrossRef] [PubMed]
27. Liu, X.; Li, L.; Liu, B.; Wang, D.; Zhao, Y.; Gao, X. Theoretical study on the ground state structure of uranofullerene $U@C_{82}$. *J. Phys. Chem. A* **2012**, *116*, 11651–11655. [CrossRef]
28. Mulet-Gas, M.; Abella, L.; Dunk, P.W.; Rodríguez-Fortea, A.; Kroto, H.W.; Poblet, J.M. Small endohedral metallofullerenes: Exploration of the structure and growth mechanism in the $Ti@C_{2n}$ ($2n$ = 26–50) family. *Chem. Sci.* **2015**, *6*, 675–686. [CrossRef]
29. Li, B.; Gu, X.; Jin, P. Overlooked Effects of La-4f Orbitals in Endohedral Metallofullerenes. *Inorg. Chem.* **2022**, *61*, 5891–5902. [CrossRef]
30. Cai, W.; Alvarado, J.; Metta-Magana, A.; Chen, N.; Echegoyen, L. Interconversions between Uranium Mono-metallofullerenes: Mechanistic Implications and Role of Asymmetric Cages. *J. Am. Chem. Soc.* **2020**, *142*, 13112–13119. [CrossRef]
31. Shen, W.; Hu, Z.; Yu, P.; Wei, Z.; Jin, P.; Shi, Z.; Lu, X. An experimental and theoretical study of $LuNC@C_{76,82}$ revealing a cage-cluster selection rule. *Inorg. Chem. Front.* **2020**, *7*, 4563–4571. [CrossRef]
32. Bao, L.; Li, Y.; Yu, P.; Shen, W.; Jin, P.; Lu, X. Preferential Formation of Mono-Metallofullerenes Governed by the Encapsulation Energy of the Metal Elements: A Case Study on $Eu@C_{2n}$ ($2n$ = 74–84) Revealing a General Rule. *Angew. Chem. Int. Ed.* **2020**, *59*, 5259–5262. [CrossRef] [PubMed]
33. Bao, L.; Peng, P.; Lu, X. Bonding inside and outside Fullerene Cages. *Acc. Chem. Res.* **2018**, *51*, 810–815. [CrossRef]

34. Xu, D.; Jiang, Y.; Wang, Y.; Zhou, T.; Shi, Z.; Omachi, H.; Shinohara, H.; Sun, B.; Wang, Z. Turning on the Near-Infrared Photoluminescence of Erbium Metallofullerenes by Covalent Modification. *Inorg. Chem.* **2019**, *58*, 14325–14330. [CrossRef]
35. Guan, R.; Chen, Z.-C.; Huang, J.; Tian, H.-R.; Xin, J.; Ying, S.-W.; Chen, M.; Zhang, Q.; Li, Q.; Xie, S.-Y. Self-driven carbon atom implantation into fullerene embedding metal–carbon cluster. *Proc. Natl. Acad. Sci. USA* **2022**, *119*, e22025631192022. [CrossRef] [PubMed]
36. Xiang, W.; Jiang, X.; Yao, Y.-R.; Xin, J.; Jin, H.; Guan, R.; Zhang, Q.; Chen, M.; Xie, S.-Y.; Popov, A.A. Monometallic Endohedral Azafullerene. *J. Am. Chem. Soc.* **2022**, *144*, 21587–21595. [CrossRef] [PubMed]
37. Li, M.; Zhao, Y.; Yuan, K.; Zhao, R.; Zhao, P.; Zhao, X. Insight into the thermodynamically preferred $V_3N@I_h(31924)$-$C_{80}$ and acknowledged $V_xSc_{3-x}N@I_h(31924)$-$C_{80}$ ($x$ = 0, 1 and 2). *Carbon* **2018**, *132*, 312–322. [CrossRef]
38. Zhao, Y.-X.; Yuan, K.; Li, M.-Y.; Ehara, M.; Zhao, X. In-Depth Theoretical Probe into Novel Mixed-Metal Uranium-Based Endohedral Clusterfullerenes $Sc_2UX@I_h(31924)$-$C_{80}$ (X = C, N). *Inorg. Chem.* **2019**, *58*, 10769–10777. [CrossRef] [PubMed]
39. Gómez-Torres, A.; Esper, R.; Dunk, P.W.; Morales-Martínez, R.; Rodríguez-Fortea, A.; Echegoyen, L.; Poblet, J.M. Small Cage Uranofullerenes: 27 Years after Their First Observation. *Helv. Chim. Acta* **2019**, *102*, e1900046. [CrossRef]
40. Raghavachari, K.; Binkley, J.S.; Seeger, R.; Pople, J.A. Self-Consistent Molecular Orbital Methods. 20. Basis set for correlated wave-functions. *J. Chem. Phys.* **1980**, *72*, 650–654.
41. Perdew, J.P.; Burke, K.; Ernzerhof, M. Generalized gradient approximation made simple. *Phys. Rev. Lett.* **1996**, *77*, 3865. [CrossRef]
42. Samoylova, N.A.; Avdoshenko, S.M.; Krylov, D.S.; Thompson, H.R.; Kirkhorn, A.C.; Rosenkranz, M.; Schiemenz, S.; Ziegs, F.; Wolter, A.U.; Yang, S. Confining the spin between two metal atoms within the carbon cage: Redox-active metal–metal bonds in dimetallofullerenes and their stable cation radicals. *Nanoscale* **2017**, *9*, 7977–7990. [CrossRef] [PubMed]
43. Popov, A.A. Redox-active metal–metal bonds between lanthanides in dimetallofullerenes. *Curr. Opin. Electrochem.* **2018**, *8*, 73–80. [CrossRef] [PubMed]
44. Glendening, E.D.; Landis, C.R.; Weinhold, F. Natural bond orbital methods. *WIREs Comput. Mol. Sci.* **2012**, *2*, 1–42. [CrossRef]
45. Frisch, M.; Trucks, G.; Schlegel, H.; Scuseria, G.; Robb, M.; Cheeseman, J.; Scalmani, G.; Barone, V.; Petersson, G.; Nakatsuji, H.; et al. *Gaussian 16, Revision A.03*; Gaussian, Inc.: Wallingford, CT, USA, 2016.
46. Bridgeman, A.J.; Cavigliasso, G.; Ireland, L.R.; Rothery, J. The Mayer bond order as a tool in inorganic chemistry. *J. Chem. Soc. Dalton Trans.* **2001**, *1*, 2095–2108. [CrossRef]
47. Popov, A.A.; Dunsch, L. Bonding in endohedral metallofullerenes as studied by quantum theory of atoms in molecules. *Chem.—Eur. J.* **2009**, *15*, 9707–9729. [CrossRef]
48. Foster, J.; Boys, S. Canonical configurational interaction procedure. *Rev. Modern Phys.* **1960**, *32*, 300. [CrossRef]
49. Lu, T.; Chen, F. Multiwfn: A multifunctional wavefunction analyzer. *J. Comput. Chem.* **2012**, *33*, 580–592. [CrossRef]

inorganics

MDPI

Article

# DFT Analysis of the Electronic and Structural Properties of Lanthanide Nitride Cluster Fullerenes $Ln_3N@C_{80}$

César Martínez-Flores and Vladimir A. Basiuk *

Instituto de Ciencias Nucleares, Universidad Nacional Autónoma de México, Circuito Exterior C.U., Ciudad de México 04510, Mexico; cesar.flores@correo.nucleares.unam.mx
* Correspondence: basiuk@nucleares.unam.mx

**Abstract:** We have undertaken a DFT study of the nitride cluster fullerenes (NCFs) $Ln_3N@C_{80}$ for the complete series of fourteen lanthanides plus lanthanum by using the PBE functional with the Grimme's dispersion correction (PBE-D2). We tested the DN and DND basis sets, which are equivalent to 6-31G and 6-31G(d) Pople-type basis sets, respectively. Due to the known convergence problems when treating lanthanide-containing systems, only with the DN basis set was it possible to complete the calculations (geometry optimization and analysis of selected electronic parameters) for all the fifteen NCFs. We found that the bending of the $Ln_3N$ cluster increases as the ionic radius increases, in general agreement with the available X-ray diffraction data. The $Ln_3N$ cluster becomes more planar as the Ln–N bond length is contracted, and the $C_{80}$ cavity slightly deforms. The HOMO-LUMO energies and distribution, as well as the charge and spin of the encapsulated metal ions, are analyzed.

**Keywords:** nitride cluster fullerenes; lanthanides; density functional theory; geometry; electronic parameters

## 1. Introduction

At the end of the last century, it was discovered experimentally that it was possible trap atoms or molecules (A) inside fullerene cages ($C_n$): the resulting complexes are referred to as endohedral fullerenes (EFs or endohedral metallofullerenes, EMFs, when A species contains metals) of the general formula $A@C_n$. The new structures were related to promises of application in many areas of medicine as magnetic resonance imaging contrast agents, single-molecule magnets, electron-spin quantum computing, etc. [1–3]. This was the logical next step in the chemistry of carbon nanoclusters after the experimental discovery of the $C_{60}$ molecule via mass spectrometry in 1985 [4,5], when the efforts to obtain new related systems with unique geometrical and electronic properties dramatically increased [6,7]. The $A@C_n$ structures described until present are based on the combination of different atoms/molecules and fullerenes of variable size. In 1999, Dorn and collaborators [8] reported on the synthesis of $Sc_3N@I_h(7)$-$C_{80}$, a member of a new family of EFs named nitride clusterfullerenes (NCFs). The NCFs encapsulate an $M_3N$ cluster where M usually is a rare-earth metal (Sc, Y or lanthanide), and the fullerene $C_{2n}$ cage comprises different sizes ($60 < 2n < 88$) [9]. The $M_3N$ cluster is made up of $M^{3+}$ ions located at the vertices of a triangle, with the nitride $N^{3-}$ in its center. Neither the $M_3N$ cluster nor the $C_{2n}$–$I_h$ cage can be synthesized separately due to their lability, whereas their combination leads to the stabilization of NCF via the electron transfer of six electrons from the cluster to the carbon cage as it is suggested by experimental and theoretical data [10–12].

One of the most interesting NCF types is the one where the lanthanide (Ln) species are present in the internal cluster ($Ln_3N$) due to the presence of 4*f*-electrons in their electronic configuration. The Ln-based combinations exhibit long spin relaxation lifetimes [13–15] and intriguing geometrical characteristics [16–18]. In some cases, the encapsulated $Ln_3N$

cluster suffers deformation from planar topology, depending on the Ln metal and cavity size. In addition, changes can be observed in the Ln–N bond length, the Ln–N–Ln angle between Ln and N species, as well as the $C_{2n}$ cage geometry. Such changes have been reported for a broad variety of cavity sizes and species [2,3,9].

A particularly interesting and important family is $Ln_3N@C_{80}$ NCFs. Here, one should emphasize that the previous experimental and theoretical research reports were limited to consideration of a few selected members of this family. The topology of the inner cluster has been analyzed both experimentally and theoretically. For example, this is the case for X-ray studies of $Ln_3N$ bending in $Gd_3N@C_{80}$ [19], $Tb_3N@C_{80}$ [20], $Ho_3N@C_{80}$ [21] and $Er_3N@C_{80}$ [21,22]. For $Lu_3N@C_{80}$ [23], the encapsulated species has a planar structure, explained by the fact that Lu has the smallest ionic radius of all lanthanides. Interestingly, the X-ray structural analysis turns out to be a very challenging task, since until now it has been impossible to crystallize pure $Ln_3N@C_{80}$ NCFs; the 'trick' which is used to obtain experimental geometries is to co-crystallize NCF with nickel (II) octaethylporphyrin [19–23]. The interaction between the two interacting units is very strong, so one cannot be totally sure whether it does not alter the $Ln_3N@C_{80}$ geometry. In this regard, theoretical calculations within the framework of density functional theory (DFT) on isolated NCFs have an evident advantage in terms of excluding the external influence of the above sort. Within this area, one should mention several efforts to characterize the geometrical and electronic properties of NCFs. For example, Aparicio-Anglès et al. [16] showed for $M_3N@C_{80}$ (where M = Sc, Y, La-Nd, Gd, Tb, Er, Tm and Lu), as well as for some other NCFs, that the bending of the $M_3N$ cluster increases as the atomic radius increases. Additionally, Schlesier et al. [24] used DFT calculations to analyze the geometrical features of some $M_3N@C_{80}$ members (namely, mixed dysprosium–lanthanide $DyM_2N@C_{80}$ and $Dy_2MN@C_{80}$, where M = Gd, Er, Tm and Lu), with a similar conclusion that the N atom is pushed out of the plane of Ln ions. On the other hand, to the best of our knowledge, the information on the frontier molecular orbital and spin density distribution in NCFs is rather scarce [7,25,26], with no published information on how they change during the step-by-step adding of $4f$-electrons from La to Lu.

The changes in Ln–N bond lengths were analyzed previously both experimentally [21,27,28] and theoretically [16,24,29,30] for several NCFs. In particular, Olmstead et al. [21], Yang et al. [27] and Zuo et al. [28] reported the relevant experimental data for the Ho, Er, Dy and Tm NCFs. Aparicio-Anglès et al. [16] presented the DFT-calculated values for the Ln–N bond length for selected lanthanides including La-Nd, Gd, Tb and Er-Lu, and found it shortened as the ionic radius decreased. Both the above values and the ones reported by Schlesier et al. [24] and Gan et al. [29] for Gd, Dy, Er and Lu NCFs are in good agreement with the experimental parameters. On the other hand, again, there are no published DFT-calculated values comprising the complete lanthanide series.

In the present work we made an attempt to trace how the main geometrical and electronic properties of $Ln_3N@C_{80}$ NCFs change for the complete series from lanthanum to lutetium, that is, during the step-by-step addition of $4f$-electrons. The DFT functional of choice was PBE, same as in the most systematic related study by Aparicio-Anglès et al. [16] (though, periodic plane wave calculations instead of cluster calculations were employed in their work). Nevertheless, our approach has an important difference from the latter, where for Tb, Er, Tm and Lu ions, having very deep $4f$ orbitals, the $f$ electrons were considered for simplicity to be in the core (that is, included in the pseudopotential); in our case, to fulfill our goal, all the $4f$ electrons had to be included explicitly.

## 2. Results and Discussion

Selected geometrical parameters, orbital energies and other electronic properties for $Ln_3N@C_{80}$ NCFs were calculated by using the DN and DND basis sets (Tables 1–4 and Figures 1–7). We considered the complete series of lanthanides with the purpose of tracing the effect of a step-by-step adding of $4f$-electrons on the selected geometrical and electronic parameters of NCFs. It was also important to verify to what degree the results can vary when using the basis sets of different sizes.

**Table 1.** Three Ln–N bond lengths (in Å), Ln–N–Ln angles (in °), the pyramidalization ($\theta$) angle of the $Ln_3N$ cluster (in °), Ln···Ln distances (in Å), and the N···$C_{C80}$ distance between N and the nearest $C_{80}$ aromatic ring (in Å) were calculated by using the DN basis set.

| NFC | Ln–N (Å) | Ln–N–Ln (°) | $\theta$ (°) | Ln···Ln (Å) | N···$C_{C80}$ (Å) |
|---|---|---|---|---|---|
| $La_3N@C_{80}$ | 2.247, 2.260, 2.265 | 94.6, 95.8, 96.0 | 31.3 | 3.325, 3.345, 3.352 | 2.949 |
| $Ce_3N@C_{80}$ | 2.179, 2.223, 2.242 | 90.6, 93.5, 95.2 | 33.0 | 3.174, 3.206, 3.266 | 2.962 |
| $Pr_3N@C_{80}$ | 2.068, 2.082, 2.165 | 111.1, 113.5, 117.1 | 14.5 | 3.503, 3.540, 3.540 | 3.597 |
| $Nd_3N@C_{80}$ | 2.079, 2.086, 2.090 | 115.5, 116.7, 117.9 | 10.6 | 3.533, 3.550, 3.568 | 3.758 |
| $Pm_3N@C_{80}$ | 2.076, 2.085, 2.086 | 115.7, 117.0, 118.0 | 10.2 | 3.532, 3.550, 3.568 | 3.770 |
| $Sm_3N@C_{80}$ | 2.058, 2.059, 2.084 | 117.7, 118.8, 119.9 | 6.4 | 3.543, 3.564, 3.566 | 3.925 |
| $Eu_3N@C_{80}$ | 2.064, 2.065, 2.068 | 118.3, 119.2, 119.7 | 5.6 | 3.548, 3.564, 3.569 | 3.948 |
| $Gd_3N@C_{80}$ | 2.064, 2.065, 2.066 | 118.2, 119.1, 119.5 | 5.9 | 3.546, 3.56, 3.567 | 3.957 |
| $Tb_3N@C_{80}$ | 2.090, 2.108, 2.126 | 107.8, 109.4, 112.0 | 19.2 | 3.422, 3.441, 3.481 | 3.443 |
| $Dy_3N@C_{80}$ | 2.097, 2.112, 2.116 | 109.2, 109.2, 111.2 | 18.4 | 3.431, 3.446, 3.477 | 3.461 |
| $Ho_3N@C_{80}$ | 2.081, 2.128, 2.132 | 105.4, 109.3, 112.2 | 20.0 | 3.433, 3.496, 3.889 | 3.423 |
| $Er_3N@C_{80}$ | 2.069, 2.084, 2.086 | 112.0, 112.7, 115.3 | 15.3 | 3.444, 3.472, 3.508 | 3.617 |
| $Tm_3N@C_{80}$ | 2.056, 2.065, 2.092 | 110.8, 110.9, 118.6 | 15.1 | 3.415, 3.423, 3.543 | 3.593 |
| $Yb_3N@C_{80}$ | 2.022, 2.086, 2.088 | 109.0, 114.9, 116.7 | 15.0 | 3.398, 3.464, 3.497 | 3.543 |
| $Lu_3N@C_{80}$ | 2.047, 2.049, 2.050 | 118.8, 119.8, 120.0 | 3.9 | 3.528, 3.545, 3.548 | 3.985 |

**Table 2.** Three Ln–N bond lengths (in Å), Ln–N–Ln angles (in °), the pyramidalization ($\theta$) angle of the $Ln_3N$ cluster (in °), Ln···Ln distances (in Å), and the N···$C_{C80}$ distance between N and the nearest $C_{80}$ aromatic ring (in Å) were calculated by using the DND basis set.

| NFC | Ln–N (Å) | Ln–N–Ln (°) | $\theta$ (°) | Ln···Ln (Å) | N···$C_{C80}$ (Å) |
|---|---|---|---|---|---|
| $La_3N@C_{80}$ | 2.236, 2.251, 2.252 | 95.6, 97.0, 97.2 | 30.4 | 3.360, 3.366, 3.337 | 2.983 |
| $Ce_3N@C_{80}$ | 2.177, 2.223, 2.246 | 88.9, 92.7, 94.0 | 33.9 | 3.126, 3.184, 3.235 | 2.933 |
| $Pr_3N@C_{80}$ | 2.048, 2.120, 2.129 | 111.1, 114.2, 115.7 | 14.8 | 3.503, 3.508, 3.530 | 3.563 |
| $Nd_3N@C_{80}$ | 2.054, 2.310, 2.310 | 78.3, 99.8, 99.8 | 32.5 | 2.917, 3.343, 3.343 | 2.998 |
| $Eu_3N@C_{80}$ | 2.039, 2.090, 2.223 | 105.6, 109.3, 118.4 | 17.6 | 3.446, 3.487, 3.548 | 3.462 |
| $Gd_3N@C_{80}$ | 2.078, 2.093, 2.096 | 115.1, 115.3, 115.3 | 12.8 | 3.520, 3.526, 3.539 | 3.692 |
| $Dy_3N@C_{80}$ | 1.984, 2.304, 2.477 | 63.3, 93.8, 99.4 | 35.9 | 3.275, 3.277, 2.513 | 2.935 |
| $Tm_3N@C_{80}$ | 2.060, 2.123, 2.132 | 104.8, 107.8, 110.2 | 21.3 | 3.370, 3.388, 3.429 | 3.366 |
| $Yb_3N@C_{80}$ | 2.014, 2.084, 2.086 | 117.0, 117.1, 108.0 | 14.4 | 3.373, 3.494, 3.497 | 3.551 |
| $Lu_3N@C_{80}$ | 2.054, 2.054, 2.054 | 119.1, 119.6, 119.8 | 4.2 | 3.541, 3.550, 3.554 | 3.994 |

From La to Lu, the $Ln_3N$ cluster shows geometrical changes in Ln–N bond lengths, Ln–N–Ln angles, Ln···Ln distances, N···$C_{C80}$ distances (between N and the nearest $C_{80}$ ring), and the bending degree. These changes are due to the electrostatic repulsion between the fullerene cage and the $Ln_3N$ unit, which depends on the size of the metal ion (Tables 1 and 2; Figures 1 and 2). In the case of Ln–N bond lengths, the DN results are close to the ones obtained previously experimentally [19–22,27,28,31] and theoretically [22–24,29,32,33], with the present DFT values tending to be slightly higher (except for Nd [16], Gd and Lu [16,24,33] NCFs). The average calculated values for La and Ho NCFs reach a maximum discrepancy of 4.4% and 3.7% in comparison with the computational results of Gan et al. [29] and experimental values of Olmstead et al. [21], respectively. In general, the trend in Ln–N bond length changes can be described as follows: a notable contraction is observed from $La_3N@C_{80}$ (2.247–2.265 Å; Table 1 and Figure 1) to $Gd_3N@C_{80}$ (2.064–2.066 Å), then an increase for $Tb_3N@C_{80}$ (2.090–2.126 Å), followed by a gradual decrease until the end of the series, that is, $Lu_3N@C_{80}$ (2.047–2.05 Å). The largest span of Ln–N bond lengths was found for Pr (0.097 Å). The experimental values by Olmstead et al. [21], Yang et al. [27] and Zuo et al. [28] showed a similar reduction in Ln–N bond length in the series of Dy-Tm NCFs, as with the DFT-computed data reported previously [22–24,29,32,33]. In the present study, with the DN basis set, within the entire series two general situations can be observed: in most cases, only two Ln–N bonds have the same (or almost the same) length (La, Pr-Sm and Dy-Yb), whereas for the Eu, Gd and Lu NCFs, all three bond lengths are almost exactly equal.

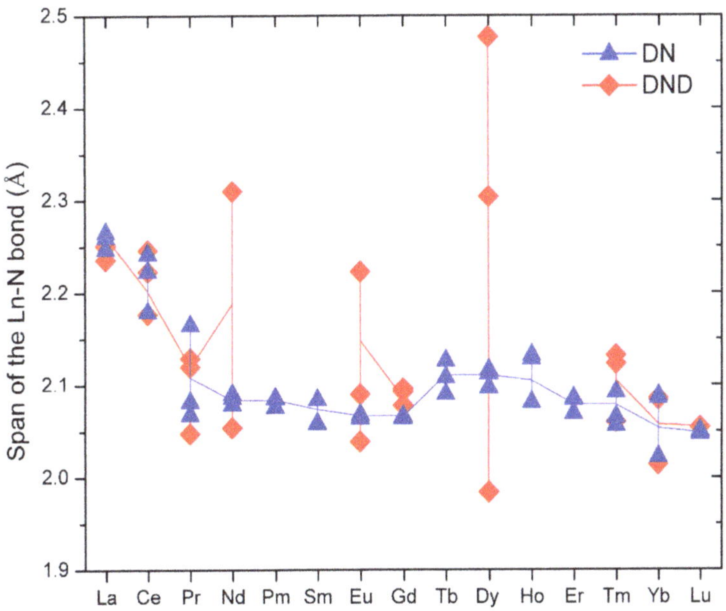


**Figure 1.** Comparison of the three Ln–N bond lengths in $Ln_3N@C_{80}$ NCFs calculated by using the DN and DND basis sets.

Regarding the geometries obtained with the DND basis sets (Table 2 and Figure 1), the general trend of Ln–N bond contraction can be seen in two (incomplete) series: La-Gd and especially Dy-Lu. However, this conclusion is not as straightforward as the one made for the DN basis set. The reason is that the use of the DND basis set results in too-rampant differences between the three Ln–N bond lengths, which are supposed to be more uniform, according to the experimental geometries ([21] and references therein). An especially 'conflictive' case is $Dy_3N@C_{80}$, for which the difference span reaches 0.493 Å, with the Dy–N bond lengths being 1.984, 2.304 and 2.477 Å; whereas the experimental ones are 2.017, 2.087 and 2.059 Å [21]; and the ones obtained with the DN basis set are 2.097, 2.112 and 2.116 Å (Table 1). The comparison of the two sets of results leads to a clear conclusion that, despite the prevailing idea that the use of a larger basis guarantees more realistic geometries (among other computed parameters), in the present case the smaller DN basis set is preferred to DND. In other words, for $Ln_3N@C_{80}$ NCFs we observe the same tendency as in the case of lanthanide 'double-decker' bisphthalocyanines that have been reported recently [34]. Taken together with the results of another study on $Ln@C_{60}$

endohedral fullerenes [35], where no substantial effect of the basis set (DN vs. DND) on the calculated parameters was found, one can strongly suggest the use of DN for further theoretical investigations of lanthanide-containing NCFs, polyazamacrocyclic and probably some other classes of Ln compounds (at least when using the $DMol^3$ module). Hereafter in this section we will place emphasis on the parameters obtained with the DN basis set for the same reasons. Coming back to the Ln–N bond lengths (Figure 1), one final observation to mention is that a reasonably good match between the DN and DND-computed values was obtained only for five lanthanides: La, Ce, Gd, Yb and Lu.

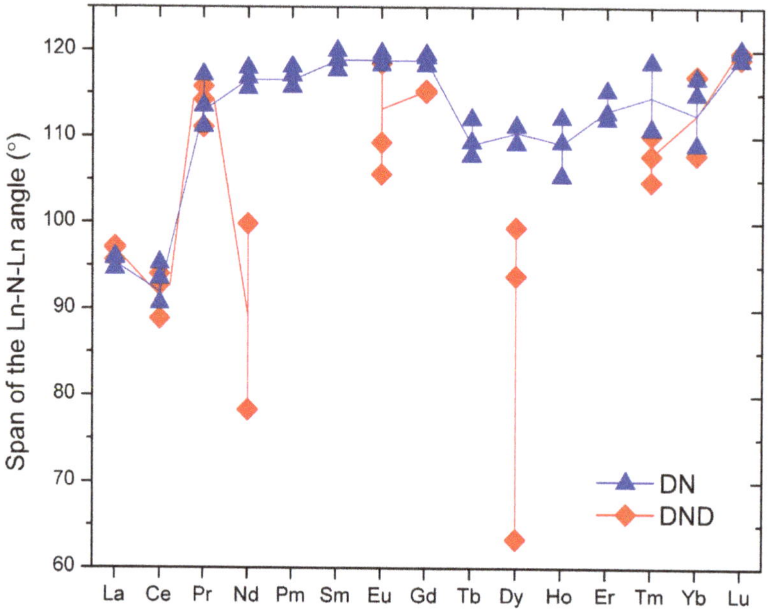


**Figure 2.** Comparison of the three Ln–N–Ln angles in $Ln_3N@C_{80}$ NCFs calculated by using the DN and DND basis sets.

Another important and closely related geometric parameter is the Ln–N–Ln angle (Tables 1 and 2; Figure 2), which directly reflects the effect of $Ln_3N$ confinement within the $C_{80}$ cage. In the case of the DN basis set, the NCFs studied can be classified into two groups (though the difference between them is rather subtle): the three Ln–N–Ln values are notably different for Ce-Sm, Tb, Ho, Er and Yb; whereas for La, Eu, Gd, Dy, Tm and Lu, one value is different from the other two. The phenomenon of lanthanide contraction manifests itself in the general trend of increasing Ln–N–Ln angles in two separate series: Pr-Gd and Tb-Lu; in other words, the effect is not continuous from Pr to Lu. In addition, with the values for La and Ce NCFs, which have the largest $Ln^{3+}$ ionic radii, the angles are much smaller than for all other $Ln_3N@C_{80}$ species, due to the strongest confinement. As a whole, this behavior correlates with the trends in the changes of Ln–N bond length; the larger Ln–N–Ln angles match the smaller Ln–N values. In the case of the DND basis set (Table 2; Figure 2), the changes in Ln–N–Ln angles from La to Lu are much more rampant, with very big differences between the three values for most NCFs, which is no surprise considering the results of the Ln–N bond length analysis presented above.

When comparing our results (obtained with the more reliable DN basis set) with that reported by other authors, for $Er_3N@C_{80}$ we found an average Er-N-Er angle of 113.3°, which differs from the experimental value [22] by 6%. In the case of $Tb_3N@C_{80}$ and $Tm_3N@C_{80}$, we obtained the average angles of 109.7° and 113.4°, respectively, which are up to 7% smaller than the values calculated by Aparicio-Anglès et al. [16], who used a very different theoretical approach (periodic calculations without inclusion of a dispersion correction and explicit consideration of the 4*f*-electrons).

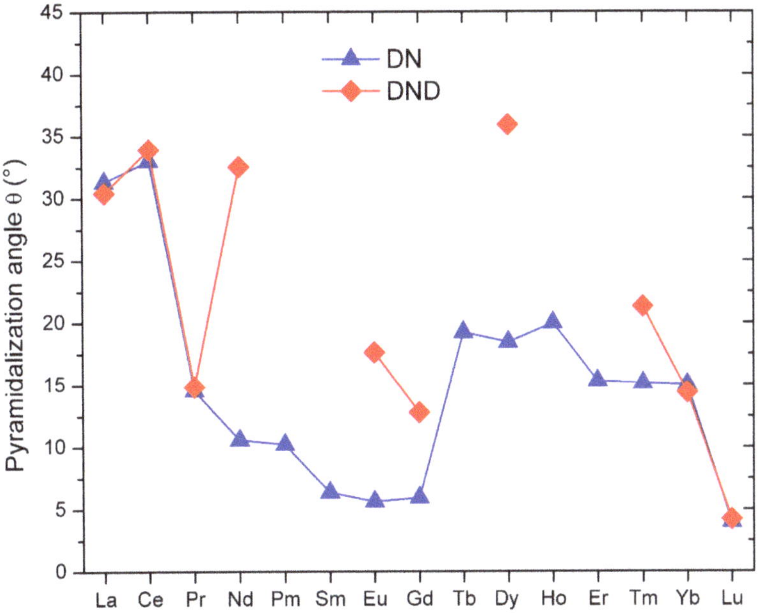


**Figure 3.** Pyramidalization ($\theta$) angle in $Ln_3N$ endohedral species calculated by using the DN and DND basis sets.

The bending of the $Ln_3N$ cluster caused by the confinement within the $C_{80}$ cage can also be characterized by the pyramidalization angle $\theta$, which is referred to as one of the most important geometric parameters for NCFs [36]. Several experimental [19–21,23,37] and theoretical [6,16,24,33] reports showed that the $Ln_3N$ cluster can acquire a planar or bent shape depending on the Ln metal and the size of the $C_{2n}$ cavity. In most cases, large fullerene cages favor planar geometry [24]. The $\theta$ values obtained are summarized in Tables 1 and 2 and Figure 3. The ones obtained with the DN basis set exhibit more uniform behavior compared to the rampant ones calculated with DND, in particular, a more uniform decrease, which is split into two series of Ce-Gd and Tb-Lu. Only for La, Ce, Yb and Lu did the DN and DND values approximately coincide, with the lowest $\theta$ values of 3.9 and 4.2°, respectively, found for the Lu3N cluster. As a whole, the $\theta$ angles presented here are larger compared to the experimental [19–21,28] and DFT-calculated ones [16–20,28] reported by other authors, and again, the DN basis set yields more realistic results. On the other hand, it is difficult to explain why the experimental $\theta$ values for the central lanthanides Gd, Tb and Dy uniformly decrease (14.6, 12.6 and 7.4°, respectively; [21] and references therein), but the DN-computed angles jump from 5.9° for Gd to 19.2° and to 18.4° for Tb and Dy, respectively. One should remember, however, that the experimental values were obtained for NCF–Ni(II) octaethylporphyrin co-crystals (where solvent molecules are present as well), and it remains unclear how the solid-state environment can influence the geometry of encapsulated $Ln_3N$ species.

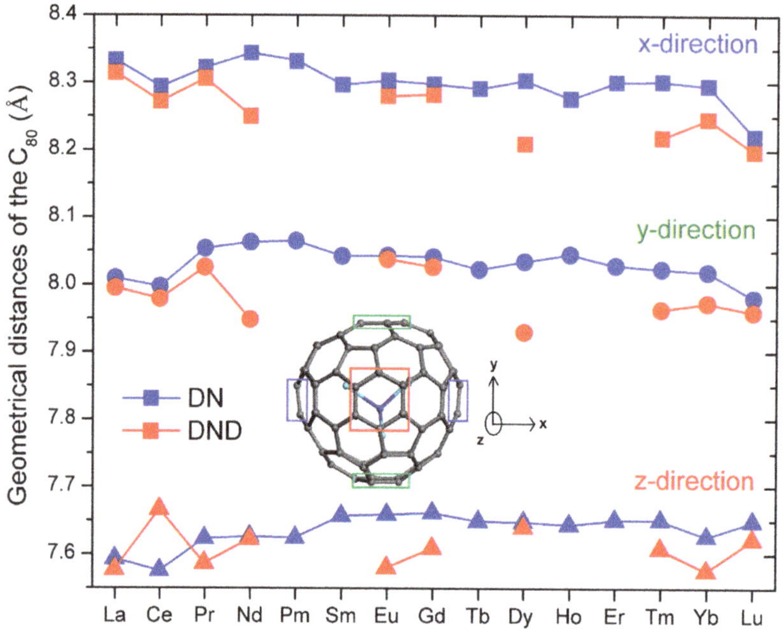


**Figure 4.** Average distances between opposite aromatic rings of fullerene cage in $Ln_3N@C_{80}$ NCFs calculated by using the DN and DND basis sets. The *x*- (horizontal), *y*- (vertical) and *z*-directions (into the plain of view) are considered, as explained by the inset.

Other geometric parameters of interest are the Ln$\cdots$Ln and N$\cdots C_{C80}$ distances (between the N atom and the nearest hexagon/pentagon ring of $C_{80}$ cage) [16,29], which directly depend on the bending of the $Ln_3N$ cluster (Tables 1 and 2; Figures S1 and S2 of the Electronic Supplementary Information). For example, the N$\cdots C_{C80}$ distances exhibit behavior opposite to the behavior of the $\theta$ values, since the increase in pyramidalization results in a closer approach of the N atom to the fullerene cage.

The presence of encapsulated bulky $Ln_3N$ clusters leads to slight geometrical changes in the geometry of the $C_{80}$ cage. It can be expressed in different ways. As Figure 4 explains, the way we did it is to measure average distances between two opposite C–C bonds in the *x*-direction, between two opposite pentagonal rings in the *y*-direction, and between two opposite hexagons in the *z*-direction. As calculated with the DN basis set, the $C_{80}$ cage exhibits relatively stable geometry with variations of 0.12 (*x*-direction), 0.08 (*y*-direction) and 0.09 Å (*z*-direction) between the maximum (La and Nd for *x*-; Pm for *y*-; Sm-Gd for *z*-direction) and minimum (Lu for *x*- and *y*-directions; Ce for *z*-direction) distances. As a general trend, the cage is compressed in the *x*- and *y*-directions, and at the same time slightly elongates in the *z*-direction, as the atomic number of lanthanides increases. Like for all the above parameters analyzed, the use of the DND basis sets yields very rampant values without clear trends.

**Table 3.** HOMO, LUMO and HOMO-LUMO gap energies (in eV), as well as the charge and spin values (in *e*; formal spin of Ln atom in parenthesis) for $Ln_3N@C_{80}$ NCFs calculated by using the DN basis set.

| NFC | $E_{HOMO}$ (eV) | $E_{LUMO}$ (eV) | $E_{gap}$ (eV) | Charge (*e*) | | Spin (*e*) [a] | |
|---|---|---|---|---|---|---|---|
| | | | | Ln | N | Ln | N |
| $La_3N@C_{80}$ | −6.012 | −4.640 | 1.371 | 0.447, 0.467, 0.486 | −0.925 | 0, 0, 0 (0) | 0 |
| $Ce_3N@C_{80}$ | −5.818 | −4.956 | 0.862 | 0.397, 0.415, 0.439 | −0.809 | 0.948, 0.992, 1.044 (1) | 0.042 |
| $Pr_3N@C_{80}$ | −5.500 | −4.798 | 0.702 | 0.303, 0.352, 0.353 | −0.453 | 2.063, 2.072, 2.103 (2) | 0.051 |
| $Nd_3N@C_{80}$ | −5.923 | −4.983 | 0.941 | 0.160, 0.178, 0.180 | −0.634 | 3.147, 3.162, 3.179 (3) | 0.179 |
| $Pm_3N@C_{80}$ | −5.976 | −5.248 | 0.728 | 0.054, 0.070, 0.080 | −0.696 | 4.223, 4.256, 4.268 (4) | 0.066 |
| $Sm_3N@C_{80}$ | −5.956 | −5.660 | 0.296 | 0.424, 0.434, 0.463 | −0.837 | 5.441, 5.446, 5.474 (5) | 0.446 |
| $Eu_3N@C_{80}$ | −5.940 | −5.868 | 0.072 | 0.426, 0.441, 0.469 | −0.857 | 6.646, 6.649, 6.661 (6) | 0.627 |
| $Gd_3N@C_{80}$ | −6.046 | −4.589 | 1.457 | 0.496, 0.512, 0.527 | −0.992 | 6.899, 6.934, 6.935 (7) | 0.028 |
| $Tb_3N@C_{80}$ | −5.979 | −5.791 | 0.189 | 0.409, 0.430, 0.454 | −0.827 | 5.519, 5.520, 5.530 (6) | 0.198 |
| $Dy_3N@C_{80}$ | −5.974 | −5.900 | 0.074 | 0.412, 0.412, 0.435 | −0.824 | 4.338, 4.387, 4.409 (5) | 0.109 |
| $Ho_3N@C_{80}$ | −5.952 | −5.882 | 0.070 | 0.457, 0.499, 0.505 | −0.862 | 3.350, 3.360, 3.384 (4) | 0.058 |
| $Er_3N@C_{80}$ | −5.961 | −5.872 | 0.089 | 0.467, 0.487, 0.542 | −0.878 | 2.377, 2.393, 2.421 (3) | 0.083 |
| $Tm_3N@C_{80}$ | −5.960 | −5.835 | 0.125 | 0.368, 0.375, 0.420 | −0.822 | 1.263, 1.268, 1.273 (2) | 0.005 |
| $Yb_3N@C_{80}$ | −5.942 | −5.859 | 0.083 | 0.417, 0.420, 0.445 | −0.855 | 0.012, 0.099, 0.110, (1) | 0.053 |
| $Lu_3N@C_{80}$ | −6.034 | −4.506 | 1.528 | 0.557, 0.584, 0.595 | −1.074 | 0, 0, 0 (0) | 0 |

[a] Absolute values.

Tables 3 and 4 summarize the calculated HOMO, LUMO and HOMO-LUMO gap ($E_{gap}$) energies, as well as the charge and spin (from Mulliken population analysis) of Ln and N atoms as calculated with the DN and DND basis sets. For the frontier orbital energies, obtained with either the DN or DND basis set, no clear trend is observed (Tables 3 and 4), with La, Gd and Lu NCFs having similar values for LUMO energies with both basis sets. The DN-calculated $E_{gap}$ energies (Table 3) tend to decrease in the entire NCF series, with the exception of the 'jumping-back' values for Nd, Gd and Lu derivatives. These species with totally empty (La), half-filled (Gd) and totally filled (Lu) 4*f* orbitals exhibit the highest gap values of 1.371, 1.457 and 1.528 eV, respectively; these results are in good agreement with the DFT-calculated values by Zhu et al. [38]. As a whole, $E_{gap}$ spans from 0.070 ($Ho_3N@C_{80}$) to 1.528 eV ($Lu_3N@C_{80}$) with DN, and from 0.084 ($Dy_3N@C_{80}$) to 1.485 eV (La and $Lu_3N@C_{80}$), with no clear trend, when the DND basis set is used (Table 4). Here, it is appropriate to mention that the GGA functionals (including PBE) tend to underestimate HOMO-LUMO gap energies, and we present the values obtained simply to illustrate their relative behavior from La to Lu.

**Table 4.** HOMO, LUMO and HOMO-LUMO gap energies (in eV), as well as the charge and spin values (in *e*; formal spin of Ln atom in parenthesis) for $Ln_3N@C_{80}$ NCFs calculated by using the DND basis set.

| NFC | $E_{HOMO}$ (eV) | $E_{LUMO}$ (eV) | $E_{gap}$ (eV) | Charge (*e*) | | Spin (*e*) [a] | |
|---|---|---|---|---|---|---|---|
| | | | | Ln | N | Ln | N |
| $La_3N@C_{80}$ | −5.769 | −4.284 | 1.485 | 0.962, 0.962, 0.962 | −1.021 | 0, 0, 0 (0) | 0 |
| $Ce_3N@C_{80}$ | −5.459 | −4.695 | 0.764 | 0.829, 0.837, 0.838 | −0.889 | 1.052, 1.079, 1.101 (1) | 0.116 |
| $Pr_3N@C_{80}$ | −5.214 | −4.537 | 0.677 | 0.050, 0.082, 0.086 | −0.758 | 2.047, 2.047, 2.065 (2) | 0.051 |
| $Nd_3N@C_{80}$ | −5.579 | −5.140 | 0.438 | 0.060, 0.164, 0.174 | −0.545 | 3.185, 3.220, 3.297 (3) | 0.127 |
| $Eu_3N@C_{80}$ | −5.649 | −5.558 | 0.090 | 0.888, 0.907, 0.907 | −0.933 | 6.567, 6.608, 6.703 (6) | 0.243 |

**Table 4.** *Cont.*

| NFC | $E_{HOMO}$ (eV) | $E_{LUMO}$ (eV) | $E_{gap}$ (eV) | Charge (*e*) | | Spin (*e*) [a] | |
|---|---|---|---|---|---|---|---|
| | | | | Ln | N | Ln | N |
| $Gd_3N@C_{80}$ | −5.739 | −4.365 | 1.373 | 0.854, 0.857, 0.857 | −1.136 | 6.866, 6.910, 6.913 (7) | 0.031 |
| $Dy_3N@C_{80}$ | −5.634 | −5.550 | 0.084 | 0.696, 0.867, 0.901 | −0.742 | 4.175, 4.159, 4.446 (5) | 0.261 |
| $Tm_3N@C_{80}$ | −5.673 | −5.573 | 0.100 | 0.819, 0.823, 0.832 | −0.964 | 1.266, 1.282, 1.288 (2) | 0.114 |
| $Yb_3N@C_{80}$ | −5.650 | −5.563 | 0.088 | 0.845, 0.862, 0.862 | −1.012 | 0.014, 0.114, 0.122 (1) | 0.085 |
| $Lu_3N@C_{80}$ | −5.758 | −4.272 | 1.485 | 0.882, 0.884, 0.884 | −1.215 | 0, 0, 0 (0) | 0 |

[a] Absolute values.

As we mentioned in Introduction, the $I_h$-$C_{80}$ fullerene is a very reactive molecule per se [1,3], but its combination with some other species ($M_3N$, $M_2$, etc.) gives stable complexes [1,2,9,10,39,40]; this stabilizing effect is due to electron transfer from the encapsulated species to the carbon cage. Formally, in the case of rare-earth derivatives $M_3N@C_{80}$, the $M_3N$ cluster is supposed to donate six electrons to $C_{80}$. Nevertheless, it turns out that the "actual" charges are very different from "formal" charges, which are integers and imply that the cluster−cage bonding is purely ionic. This aspect was addressed in several theoretical studies (on representative $Sc_3N@C_{80}$), which found that the calculated net $Sc_3N$ charges are also much smaller than the values expected for the purely ionic bonding [41,42]; for example, around +1, either from the Mulliken or Hirshfeld population analysis [42]. This effect is referred to as "back-donation" [43,44]. Additionally, in our calculations, the actual positive charge on Ln atoms (Tables 3 and 4; from the Mulliken population analysis) was much lower than the "formal" charge of +3; in particular, it spans from 0.054–0.080 *e* ($Pm_3N$) to 0.557–0.595 *e* ($Lu_3N$) with the DN basis set, and from 0.050–0.086 *e* ($Pr_3N$) to 0.962 *e* ($La_3N$) with DND. For the N atom, the values vary from −0.453 *e* ($Pr_3N$) to −1.074 *e* ($Lu_3N$) with DN, and from −0.545 *e* ($Nd_3N$) to −1.215 *e* ($Lu_3N$) with DND. The DN-calculated absolute values are generally lower than the DND ones. It is interesting and important to mention the observation that the use of a smaller basis set produces more reliable results for the Mulliken population analysis [45–47].

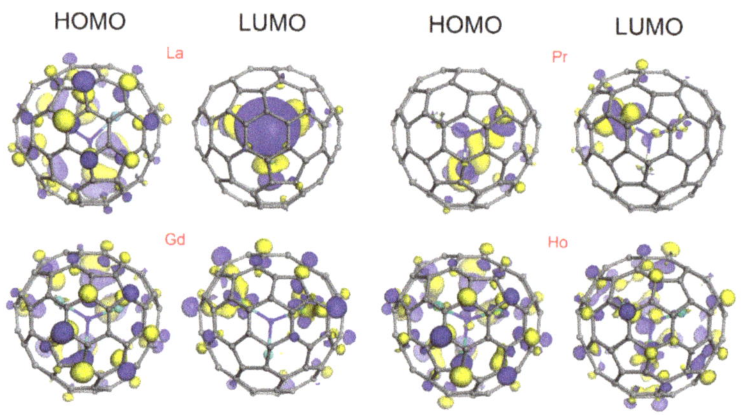


**Figure 5.** Selected patterns of HOMO-LUMO distributions (isosurfaces at 0.03 a.u.) in $Ln_3N@C_{80}$ NCFs calculated by using the DN basis set.

Figure 5 shows the selected patterns of the HOMO and LUMO distribution obtained with the DN basis set. After analyzing these plots for the complete NCF series, it was not possible to classify them in a simple way due to a large variability in the results (with both basis sets). Here, we show the four most distinctive cases (La, Pr, Gd and Ho NCFs) obtained with the DN basis set (the reader can find all the plots for the fifteen NCFs in

Electronic Supplementary Information, in Figures S3–S6; with the aim of completeness of the graphical presentation, both the front and the side views are included.) As can be seen in Figure 5, the HOMO lobes in the case of La and Gd NFCs are very similar: they are localized on the $C_{80}$ cage, with no apparent contribution from the $Ln_3N$ cluster. In the case of Ho NCF, the HOMO distribution on fullerene cage is rather similar to the one found for La and Gd complexes, but a minor contribution from the $Ho_3N$ cluster can be observed as well. For $Pr_3N@C_{80}$, the HOMO lobes are localized mostly on two Pr atoms, whereas the contribution from the $C_{80}$ cage is minimal. For the LUMO density, the lobes on Ln and N atoms are of highly variable size; of the ones presented in Figure 5, the largest ones can be seen for the $La_3N$ cluster. The LUMO lobes on the fullerene cage almost vanish for La, Ce, Nd, Pm, Sm and Tb NCFs. The results obtained with DND (Figures S5–S6) only partially match the ones calculated by using the DN basis set.

As compared to the formal spin for isolated $Ln^{3+}$ ions, the results obtained with the DN and DND basis sets (Tables 3 and 4; Figure 6) show the possibility of spin transfer both to and from lanthanide species. In the DN data set (Table 3), spin depletion from Ln can be observed for Ce (but from two cerium atoms only) and the late lanthanides Gd-Yb (by up to 0.890–0.988 $e$ for Yb, compared to to the formal spin of 1 $e$ for isolated $Yb^{3+}$); this can be also seen from Figure 6 (here we plotted the average values for simplicity). The early lanthanides Pr-Eu tend to acquire additional spin, with the strongest increase by 0.646–0.661 $e$ observed for Eu (compared to the formal spin of 6 $e$ for isolated $Eu^{3+}$). For La and Lu ions with totally empty and totally filled $4f$ orbitals, respectively, the spin remains to be zero. The limited available values calculated by using the DND basis set (Table 4 and Figure 6) follow a similar trend.

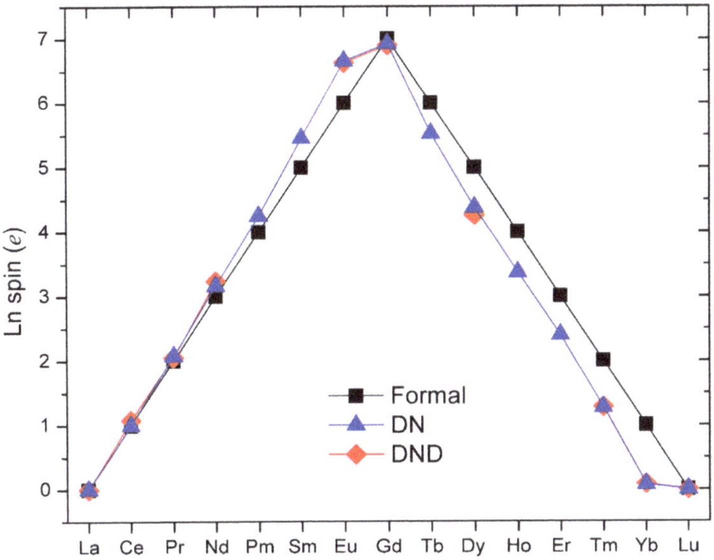


**Figure 6.** Comparison of the average spin of three Ln ions in $Ln_3N@C_{80}$ NCFs calculated by using the DN and DND basis sets with the formal spin of isolated $Ln^{3+}$ ions.

The spin density plots for $Ln_3N@C_{80}$ NCFs presented in Figure 7 (the DN data set) cannot serve as a similar quantitative measure of spin transfer, but instead are more illustrative, especially in order to show variability in spin alignment. In the most simplistic terms of spin distribution, they can be classified into three general cases. The first one includes La and Lu NCFs (with the closed-shell electronic configuration of Ln species), where no unpaired electrons can be found on either the $Ln_3N$ cluster or $C_{80}$ cage (the corresponding side views are shown in Figure S7). In the case of Ce-Nd and Gd NCFs, the main contribution comes from the $Ln_3N$ cluster, where the largest lobes are localized on metal species. For the remaining NCFs (Pm-Eu, Tb-Yb), spin density can be found on both the cluster and fullerene cage. In the calculations with the DND basis set, qualitatively similar spin density plots were obtained (Figures S8 and S9 of the Electronic Supplementary Information).

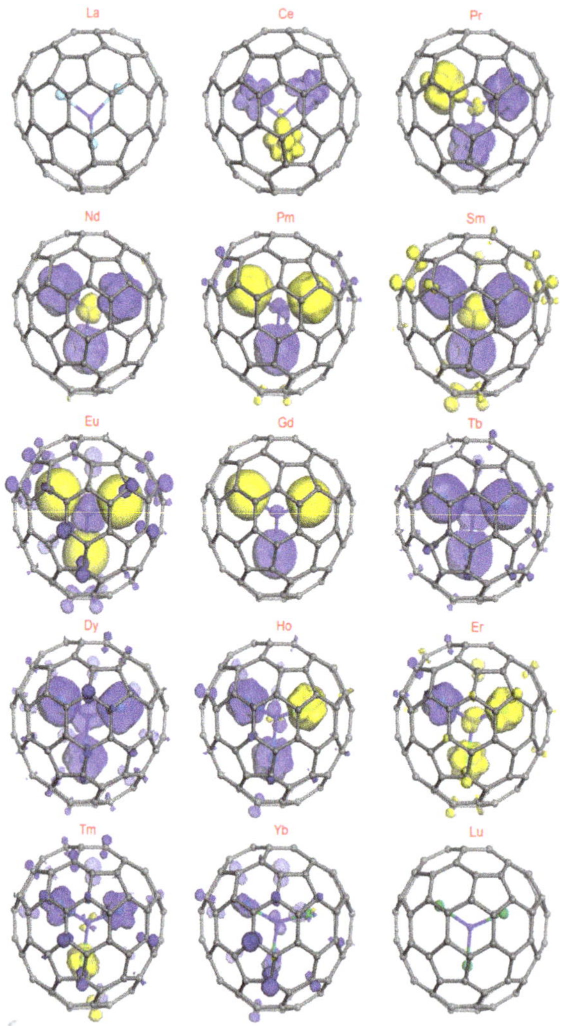


**Figure 7.** Spin density plots (isosurfaces at 0.02 a.u.) for $Ln_3N@C_{80}$ complexes as calculated by using the DN basis set. Blue and yellow lobes correspond to spin-up and spin-down electrons, respectively.

As regards the spin-up and spin-down directions, they turned out to be broadly variable as well. When considering $Ln_3N$ cluster only, all the NCFs can be classified into the following groups: (1) the spin directions of all four atoms coincide (spin-up) for Tb, Dy and Yb; (2) the spin directions of three Ln atoms coincide, but for the N atom it is the opposite, which is observed for Nd, Sm and Eu; (3) the spin directions of Ln atoms differ (2 + 1) for all the remaining NCFs (Ce, Pr, Pm, Gd, Ho, Er and Tm). One should also note that in three cases of Tb, Dy and Yb derivatives, the spin directions of $Ln_3N$ atoms (spin-up) are the same as those of the small lobes found on the carbon atoms of the $C_{80}$ cage.

## 3. Computational Methods

The theoretical analysis of the geometries and electronic properties of $Ln_3N@C_{80}$ NCFs was performed by using the numerical-based DFT module DMol$^3$, which is a part of the Materials Studio 8.0 software package [48–51]. The Perdew–Burke–Ernzerhof (PBE) general gradient approximation functional [52] was employed in combination with the empirical dispersion correction developed by Grimme [53], usually referred to as PBE-D (or PBE-D2). The previous computational method was successfully used in our previous studies involving $C_{80}$-based dyads [25,54], as well as in numerous studies by other research groups on systems that include metal species and carbon nanoclusters.

The DN (with no polarization functions added) and DND (with a polarization *d*-function added on all non-hydrogen atoms) double-numerical basis sets were employed, which are equivalent to the 6-31G and 6-31G(d) Pople-type basis sets. The full geometry optimization and the calculation of the electronic parameters were performed with the DFT semi-core pseudopotentials (DSPPs), including relativistic and spin-orbit coupling effects, whose account is important due to the presence of the lanthanide species. In the calculations with the DN basis set, the real space cutoff (defined by the presence of lanthanide species) was 4.3 Å and a DND of 5.0 Å. The convergence criteria were as follows: For the DN basis set, an energy gradient of $10^{-4}$ Ha/Å, maximum force of 0.02 Ha/Å, maximum displacement of 0.05 Å, and SCF tolerance of $10^{-4}$ Ha; for the DND basis set, an energy gradient of $2 \cdot 10^{-5}$ Ha/Å, maximum force of 0.004 Ha/Å, maximum displacement of 0.005 Å, and SCF tolerance of $10^{-5}$ Ha.

The existence of 4*f* orbitals results in a high number of degenerate states near the Fermi level, which has a serious negative impact on the self-consistent field (SCF) convergence. Thermal smearing was employed as an indispensable tool to solve the (almost always inevitable) SCF convergence problems according to the general protocol explained in the previous studies [55,56]. It was found that the use of a very low value of 0.0001 Ha (equivalent temperature of 31.6 K) yields stable and consistent results. At the same time, even with the use of thermal smearing tools, only for the DN basis set was a successful convergence afforded for all $Ln_3N@C_{80}$ NCFs. With DND, we were unable to finish the calculations for Pm-, Sm-, Tb-, Ho- and Er-containing systems.

## 4. Conclusions

The most important results can be summarized as follows.

(1) The geometry optimization and electronic parameter calculation for the series of fifteen $Ln_3N@C_{80}$ NCFs (Ln = La-Lu) was possible to complete only with the DN basis set. The use of DND basis sets leads to SCF convergence problems, due to which it was impossible to obtain the results for Pm-, Sm-, Tb-, Ho- and Er-containing systems. Furthermore, within the NCFs for which the computations seemed to be successful, too-rampant behavior was observed for most important geometric parameters (including the Ln–N bond lengths, Ln–N–Ln angles, and pyramidalization angle $\theta$), which is supposed to be more uniform and logical.

(2) For the frontier orbital energies, obtained with either the DN or DND basis set, no clear trend is observed. The DN-calculated $E_{gap}$ energies tend to decrease in the entire NCF series, with the exception of the 'jumping-back' values for Nd, Gd and Lu derivatives. These species with totally empty (La), half-filled (Gd) and totally filled (Lu) 4*f* orbitals exhibit the highest gap values of 1.371, 1.457 and 1.528 eV, respectively, which are in good agreement with the DFT results reported elsewhere. The corresponding HOMO and LUMO distribution patterns are highly variable.

(3) The calculated "actual" positive charge on Ln atoms, as estimated from the Mulliken population analysis, is much lower than the "formal" charge of +3; it spans from 0.054–0.080 *e* ($Pm_3N$) to 0.557–0.595 *e* ($Lu_3N$) with the DN basis set, and from 0.050–0.086 *e* ($Pr_3N$) to 0.962 *e* ($La_3N$) with DND. Similarly, for the N atom, the values vary from −0.453 *e* ($Pr_3N$) to −1.074 *e* ($Lu_3N$) with DN, and from −0.545 *e* ($Nd_3N$) to −1.215 *e* ($Lu_3N$) with DND. These results are in good agreement with the known effect of "back-donation".

(4) As compared to the formal spin for isolated $Ln^{3+}$ ions, the results obtained show the possibility of spin transfer both to and from lanthanide species. In the DN data set, spin depletion from Ln can be observed for Ce (but from two cerium atoms only) and the late lanthanides Gd-Yb, with it being up to 0.890–0.988 *e* for Yb. The early lanthanides Pr-Eu tend to acquire additional spin, with the strongest increase by 0.646–0.661 *e* observed for Eu. For La and Lu ions with totally empty and totally filled 4*f* orbitals, respectively, the spin remains to be zero. The spin density plots reveal interesting features in terms of localization of unpaired electrons, as well as spin alignment.

**Supplementary Materials:** Supplementary data associated with this article can be found in the online version at https://www.mdpi.com/article/10.3390/inorganics11050223/s1. Figure S1: Comparison of the Ln$\cdots$Ln distance (Å) in $Ln_3N@C_{80}$ NCFs calculated by using the DN and DND basis sets; Figure S2: Comparison of the N$\cdots C_{C80}$ distance (Å) of the $Ln_3N@C_{80}$ complexes calculated by using the DN and DND basis sets; Figure S3: Complete patterns of HOMO-LUMO distribution (front views with isosurfaces at 0.03 a.u.) in $Ln_3N@C_{80}$ NCFs calculated by using the DN basis set; Figure S4: Complete patterns of HOMO-LUMO distribution (isosurfaces at 0.03 a.u.) in $Ln_3N@C_{80}$ NCFs but rotated at an angle of 90° to the left (side view) with respect to Figure S3; Figure S5: Patterns of HOMO-LUMO distribution (front views with isosurfaces at 0.03 a.u.) in $Ln_3N@C_{80}$ NCFs calculated by using the DND basis set; Figure S6: Patterns of HOMO-LUMO distribution (isosurfaces at 0.03 a.u.) in $Ln_3N@C_{80}$ NCFs but rotated at an angle of 90° to the left (side views) with respect to Figure S5; Figure S7: Spin density plots (isosurfaces at 0.02 a.u.) for $Ln_3N@C_{80}$ NCFs calculated by using the DN basis set but rotated an angle of 90° to the left (side views) with respect to Figure 7. Blue and yellow lobes correspond to spin-up and spin-down electrons, respectively; Figure S8: Spin density plots (isosurfaces at 0.02 a.u.) for $Ln_3N@C_{80}$ NCFs were calculated by using the DND basis set. Blue and yellow lobes correspond to spin-up and spin-down electrons, respectively; Figure S9: Spin density plots (isosurfaces at 0.02 a.u.) for $Ln_3N@C_{80}$ NCFs calculated by using the DND basis set but rotated an angle of 90° to the left (side views) with respect to Figure S8.

**Author Contributions:** Conceptualization, V.A.B.; methodology, V.A.B.; validation, C.M.-F.; formal analysis, C.M.-F.; investigation, C.M.-F.; data curation, C.M.-F.; writing—original draft preparation, C.M.-F.; writing—review and editing, V.A.B.; visualization, C.M.-F.; supervision, V.A.B.; funding acquisition, V.A.B. All authors have read and agreed to the published version of the manuscript.

**Funding:** This research was funded by the National Autonomous University of Mexico (UNAM), grant DGAPA-IN103622.

**Institutional Review Board Statement:** Not applicable.

**Informed Consent Statement:** Not applicable.

**Data Availability Statement:** The raw/processed data required to reproduce these findings cannot be shared at this time due to technical limitations (their total size of a few GB).

**Acknowledgments:** C.M.-F. thanks to DGAPA-UNAM for the postdoctoral fellowship.

**Conflicts of Interest:** The authors declare no conflict of interest.

## References

1. Chaur, M.N.; Melin, F.; Ortiz, A.L.; Echegoyen, L. Chemical, Electrochemical, and Structural Properties of Endohedral Metallofullerenes. *Angew. Chem. Int. Ed.* **2009**, *48*, 7514–7538. [CrossRef] [PubMed]
2. Popov, A.A.; Yang, S.; Dunsch, L. Endohedral fullerenes. *Chem. Rev.* **2013**, *113*, 5989–6113. [CrossRef] [PubMed]
3. Yang, S.; Wei, T.; Jin, F. When metal clusters meet carbon cages: Endohedral clusterfullerenes. *Chem. Soc. Rev.* **2017**, *46*, 5005–5058. [CrossRef] [PubMed]
4. Heath, J.R.; O'Brien, S.C.; Zhang, Q.; Liu, Y.; Curl, R.F.; Tittel, F.K.; Smalley, R.E. Lanthanum complexes of spheroidal carbon shells. *J. Am. Chem. Soc.* **1985**, *107*, 7779–7780. [CrossRef]
5. Kroto, H.; Heath, J.; O'Brien, S.; Curl, R.F.; Smalley, R.E. $C_{60}$: Buckminsterfullerene. *Nature* **1985**, *318*, 162–163. [CrossRef]
6. Shen, W.; Hu, S.; Lu, X. Endohedral Metallofullerenes: New Structures and Unseen Phenomena. *Chem. Eur. J.* **2020**, *26*, 5748–5757. [CrossRef]
7. Miyazaki, T.; Hino, S. Electronic and Geometric Structures of Cluster Encapsulated Fullerenes. In *Physics and Chemistry of Carbon-Based Materials*; Kubozono, Y., Ed.; Springer: Singapore, 2019. [CrossRef]

8. Stevenson, S.; Rice, G.; Glass, T.; Harich, K.; Cromer, F.; Jordan, M.R.; Craft, J.; Hadju, E.; Bible, R.; Olmstead, M.M.; et al. Small-bandgap endohedral metallofullerenes in high yield and purity. *Nature* **1999**, *401*, 55–57. [CrossRef]
9. Dunsch, L.; Yang, S. Metal Nitride Cluster Fullerenes: Their Current State and Future Prospects. *Small* **2007**, *3*, 1298–1320. [CrossRef]
10. Popov, A.A. (Ed.) *Endohedral Fullerenes: Electron Transfer and Spin*; Springer International Publishing: Cham, Switzerland, 2017. [CrossRef]
11. Campanera, J.M.; Bo, C.; Poblet, J.M. General Rule for the Stabilization of Fullerene Cages Encapsulating Trimetallic Nitride Templates. *Angew. Chem. Int. Ed.* **2005**, *44*, 7230–7233. [CrossRef]
12. Jiang, Y.; Li, Z.; Wu, Y.; Wang, Z. $Ln_3@C_{80}^+$ (Ln = lanthanide): A new class of stable metallofullerene cations with multicenter metal-metal bonding in the sub-nanometer confined space. *Inorg. Chem. Front.* **2022**, *9*, 2173–2181. [CrossRef]
13. Bunzli, J.-C.G. Rising Stars in Science and Technology: Luminescent Lanthanide Materials. *Eur. J. Inorg. Chem.* **2017**, *2017*, 5058–5063. [CrossRef]
14. Gaita-Ariño, A.; Luis, F.; Hill, S.; Coronado, E. Molecular spins for quantum computation. *Nat. Chem.* **2019**, *11*, 301–309. [CrossRef] [PubMed]
15. Westerström, R.; Dreiser, J.; Piamonteze, C.; Muntwiler, M.; Weyeneth, S.; Brune, H.; Rusponi, S.; Nolting, F.; Popov, A.; Yang, S.; et al. An Endohedral Single-Molecule Magnet with Long Relaxation Times: $DySc_2N@C_{80}$. *J. Am. Chem. Soc.* **2012**, *134*, 9840–9843. [CrossRef] [PubMed]
16. Aparicio-Anglès, X.; Alegret, N.; Clotet, A.; Rodríguez-Fortea, A.; Poblet, J.M. Endohedral Metallofullerenes Containing Lanthanides: A Robust Yet Simple Computational Approach. *J. Phys. Chem. C* **2013**, *117*, 12916–12921. [CrossRef]
17. Nakanishi, R.; Satoh, J.; Katoh, K.; Zhang, H.; Breedlove, B.K.; Nishijima, M.; Nakanishi, Y.; Omachi, H.; Shinohara, H.; Yamashita, M. $DySc_2N@C_{80}$ Single-Molecule Magnetic Metallofullerene Encapsulated in a Single-Walled Carbon Nanotube. *J. Am. Chem. Soc.* **2018**, *140*, 10955–10959. [CrossRef] [PubMed]
18. Li, B.; Gu, X.; Jin, P. Overlooked Effects of La-4f Orbitals in Endohedral Metallofullerenes. *Inorg. Chem.* **2022**, *61*, 5891–5902. [CrossRef]
19. Stevenson, S.; Phillips, J.P.; Reid, J.E.; Olmstead, M.M.; Rath, S.P.; Balch, A.L. Pyramidalization of $Gd_3N$ inside a $C_{80}$ cage. The synthesis and structure of $Gd_3N@C_{80}$. *Chem. Commun.* **2004**, 2814–2815. [CrossRef]
20. Zuo, T.; Beavers, C.M.; Duchamp, J.C.; Campbell, A.; Dorn, H.C.; Olmstead, M.M.; Balch, A.L. Isolation and structural characterization of a family of endohedral fullerenes including the large, chiral cage fullerenes $Tb_3N@C_{88}$ and $Tb_3N@C_{86}$ as well as the $I_h$ and $D_{5h}$ isomers of $Tb_3N@C_{80}$. *J. Am. Chem. Soc.* **2007**, *129*, 2035–2043. [CrossRef]
21. Olmstead, M.M.; Zuo, T.; Dorn, H.C.; Li, T.; Balch, A.L. Metal ion size and the pyramidalization of trimetallic nitride units inside a fullerene cage: Comparisons of the crystal structures of $M_3N@I_h$-$C_{80}$ (M = Gd, Tb, Dy, Ho, Er, Tm, Lu, and Sc) and some mixed metal counterparts. *Inorg. Chim. Acta* **2017**, *1*, 321–326. [CrossRef]
22. Hu, S.; Zhao, P.; Shen, W.; Yu, P.; Huang, W.; Ehara, M.; Xie, Y.; Akasaka, T.; Lu, X. Crystallographic characterization of $Er_3N@C_{2n}$ (2n = 80, 82, 84, 88): The importance of a planar $Er_3N$ cluster. *Nanoscale* **2019**, *11*, 13415–13422. [CrossRef]
23. Shen, W.-Q.; Bao, L.-P.; Hu, S.-F.; Gao, X.-J.; Xie, Y.-P.; Gao, X.-F.; Huang, W.-H.; Lu, X. Isolation and crystallographic characterization of $Lu_3N@C_{2n}$ (2n = 80 − 88): Cage selection by cluster size. *Chem. Eur. J.* **2018**, *24*, 16692–16698. [CrossRef] [PubMed]
24. Schlesier, C.; Liu, F.; Dubrovin, V.; Spree, L.; Büchner, B.; Avdoshenko, S.M.; Popov, A.A. Mixed dysprosium-lanthanide nitride clusterfullerenes $DyM_2N@C_{80}$-$I_h$ and $Dy_2MN@C_{80}$-$I_h$ (M = Gd, Er, Tm, and Lu): Synthesis, molecular structure, and quantum motion of the endohedral nitrogen atom. *Nanoscale* **2019**, *11*, 13139–13153. [CrossRef] [PubMed]
25. Basiuk, V.A.; Tahuilan-Anguiano, D.E. Complexation of free-base an 3d transition metal(II) phthalocyanines with endohedral fullerene $Sc_3N@C_{80}$. *Chem. Phys. Lett.* **2019**, *722*, 146–152. [CrossRef]
26. Yamada, M.; Liu, M.T.H.; Nagase, S.; Akasaka, T. New Horizons in Chemical Functionalization of Endohedral Metallofullerenes. *Molecules* **2020**, *25*, 3626. [CrossRef] [PubMed]
27. Yang, S.; Troyanov, S.I.; Popov, A.A.; Krause, M.; Dunsch, L. Deviation from the Planarity-a Large $Dy_3N$ Cluster Encapsulated in an $I_h$-$C_{80}$ Cage: An X-ray Crystallographic and Vibrational Spectroscopic Study. *J. Am. Chem. Soc.* **2006**, *128*, 16733–16739. [CrossRef]
28. Zuo, T.; Olmstead, M.M.; Beavers, C.M.; Balch, A.L.; Wang, G.; Yee, G.T.; Shu, C.; Xu, L.; Elliott, B.; Echegoyen, L.; et al. Preparation and structural characterization of the $I_h$ and the $D_{5h}$ isomers of the endohedral fullerenes $Tm_3N@C_{80}$: Icosahedral $C_{80}$ cage encapsulation of a trimetallic nitride magnetic cluster with three uncoupled $Tm^{3+}$ ions. *Inorg. Chem.* **2008**, *47*, 5234–5244. [CrossRef]
29. Gan, L.-H.; Yuan, R. Influence of cluster size on the structures and stability of trimetallic nitride fullerenes $M_3N@C_{80}$. *ChemPhysChem* **2006**, *7*, 1306–1310. [CrossRef]
30. Slanina, Z.; Uhlík, F.; Feng, L.; Akasaka, T.; Lu, X.; Adamowicz, L. Calculations of the $Lu_3N@C_{80}$ two-isomer equilibrium. *Fuller. Nanotub. Carbon Nanostruct.* **2019**, *27*, 382–386. [CrossRef]
31. Krause, M.; Dunsch, L. Gadolinium nitride $Gd_3N$ in carbon cages. The influence of cluster size and bond strength. *Angew. Chem. Int. Ed.* **2005**, *44*, 1557–1560. [CrossRef]
32. Lu, J.; Sabirianov, R.F.; Mei, W.N.; Gao, Y.; Duan, C.G.; Zeng, X.C. Structural and magnetic properties of $Gd_3N@C_{80}$. *J. Phys. Chem. B* **2006**, *110*, 23637–23640. [CrossRef]

33. Zhang, Z.X.; Liu, Y.; Han, P.D.; Zhuang, S.Y.; Wang, T.S.; Luo, S.C.; Xum, B.S. Metallofullerenes encaging mixed-metal clusters: Synthesis and structural studies of $Gd_xHo_{3-x}N@C_{80}$ and $Gd_xLu_{3-x}N@C_{80}$. *ChemPhysChem* **2015**, *16*, 295–298. [CrossRef] [PubMed]
34. Martínez-Flores, C.; Bolivar-Pineda, L.M.; Basiuk, V.A. Lanthanide bisphthalocyanine single-molecule magnets: A DFT survey of their geometries and electronic properties from lanthanum to lutetium. *Mater. Chem. Phys.* **2022**, *287*, 126271. [CrossRef]
35. Martínez-Flores, C.; Basiuk, V.A. $Ln@C_{60}$ endohedral fullerenes: A DFT analysis for the complete series from lanthanum to lutetium. *Comp. Theor. Chem.* **2022**, *1217*, 113878. [CrossRef]
36. Thomsen, C.; Reich, S. Raman Scattering in Carbon Nanotubes. In *Light Scattering in Solid IX. Topics in Applied Physics*; Cardona, M., Merlin, R., Eds.; Springer: Berlin/Heidelberg, Germany, 2006; Volume 108. [CrossRef]
37. Stevenson, S.; Lee, H.M.; Olmstead, M.M.; Kozikowski, C.; Stevenson, P.; Balch, A.L. Preparation and crystallographic characterization of a new endohedral, $Lu_3N@C_{80}\cdot 5$ (o-xylene), and comparison with $Sc_3N@C_{80}\cdot 5$ (o-xylene). *Eur. J. Chem.* **2002**, *8*, 4528–4535. [CrossRef]
38. Zhu, Y.; Li, Y.; Yang, Z.Q. First-principles investigation on the electronic structures of intercalated fullerenes $M_3N@C_{80}$ (M = Sc, Y, and lanthanides). *Chem. Phys. Lett.* **2008**, *4614*, 285–289. [CrossRef]
39. Koltover, V.K.; Bubnov, V.P.; Estrin, Y.I.; Lodygina, V.P.; Davydov, R.M.; Subramoni, M.; Manoharan, P.T. Spin-transfer complexes of endohedral metallofullerenes: ENDOR and NMR evidences. *Phys. Chem. Chem. Phys.* **2003**, *5*, 2774–2777. [CrossRef]
40. Dunsch, L.; Krause, M.; Noack, J.; Georgi, P. Endohedral nitride cluster fullerenes: Formation and spectroscopic analysis of $L_{3-x}M_xN@C_{2n}$ ($0 \leq x \leq 3$; N = 39, 40). *J. Phys. Chem. Solids* **2004**, *65*, 309–315. [CrossRef]
41. Popov, A.A.; Dunsch, L. Hindered cluster rotation and $^{45}Sc$ hyperfine splitting constant in distonoid anion radical $Sc_3N@C_{80}^-$, spatial spin−charge separation as a general principle for anions of endohedral fullerenes with metal-localized lowest unoccupied molecular orbitals. *J. Am. Chem. Soc.* **2008**, *130*, 17726–17742. [CrossRef]
42. Valencia, R.; Rodríguez-Fortea, A.; Poblet, J.M. Understanding the stabilization of metal carbide endohedral fullerenes $M_2C_2@C_{82}$ and related systems. *J. Phys. Chem. A* **2008**, *112*, 4550–4555. [CrossRef]
43. Liu, D.; Hagelberg, F.; Park, S.S. Charge transfer and electron backdonation in metallofullerenes encapsulating $NSc_3$. *Chem. Phys.* **2006**, *330*, 380–386. [CrossRef]
44. Wu, J.; Hagelberg, F. Computational study on $C_{80}$ enclosing mixed trimetallic nitride clusters of the form $Gd_xM_{3-x}N$ (M = Sc, Sm, Lu). *J. Phys. Chem. C* **2008**, *112*, 5770–5777. [CrossRef]
45. Lu, J.; Zhang, X.; Zhao, X. Relativistic electronic structure calculations on endohedral $Gd@C_{60}$, $La@C_{60}$, $Gd@C_{74}$, and $La@C_{74}$. *App. Phys. A* **2000**, *70*, 461–464. [CrossRef]
46. Slanina, Z.; Uhlík, F.; Akasaka, T.; Lu, X.; Adamowicz, L. Computational modeling of the $Ce@C_{82}$ metallofullerene isomeric composition. *ECS J. Solid State Sci. Technol.* **2019**, *8*, M118–M121. [CrossRef]
47. Slanina, Z.; Uhlík, F.; Akasaka, T.; Lu, X.; Adamowicz, L. Calculated relative thermodynamic stabilities of the $Gd@C_{82}$ isomers. *ECS J. Solid State Sci. Technol.* **2021**, *10*, 071013. [CrossRef]
48. Delley, B. An all-electron numerical method for solving the local density functional for polyatomic molecules. *J. Chem. Phys.* **1990**, *92*, 508–517. [CrossRef]
49. Delley, B. Fast calculation of electrostatics in crystals and large molecules. *J. Phys. Chem.* **1996**, *100*, 6107–6110. [CrossRef]
50. Delley, B. From molecules to solids with the $DMol^3$ approach. *J. Chem. Phys.* **2000**, *113*, 7756–7764. [CrossRef]
51. Delley, B.; Ellis, D.; Freeman, A.; Baerends, E.; Post, D. Binding energy and electronic structure of small copper particles. *Phys. Rev. B* **1983**, *27*, 2132–2144. [CrossRef]
52. Perdew, J.; Burke, K.; Ernzerhof, M. Generalized gradient approximation made simple. *Phys. Rev. Lett.* **1996**, *77*, 3865–3868. [CrossRef]
53. Grimme, S. Semiempirical GGA-type density functional constructed with a long-range dispersion correction. *J. Comput. Chem.* **2006**, *27*, 1787–1799. [CrossRef]
54. Basiuk, V.A.; Basiuk, E.V. Noncovalent complexes of $I_h$-$C_{80}$ fullerene with phthalocyanines. *Fuller. Nanotub. Carbon Nanostruct.* **2018**, *26*, 69–75. [CrossRef]
55. Basiuk, V.A. Electron smearing in DFT calculations: A case study of doxorubicin interaction with single-walled carbon nanotubes. *Int. J. Quantum Chem.* **2011**, *111*, 4197–4205. [CrossRef]
56. Basiuk, V.A.; Prezhdo, O.V.; Basiuk, E.V. Thermal smearing in DFT calculations: How small is really small? A case of La and Lu atoms adsorbed on graphene. *Mater. Today Commun.* **2020**, *25*, 101595. [CrossRef]

inorganics

MDPI

Review

# Metalloborospherene Analogs to Metallofullerene

**Jordan Burkhardt [1,2], Hayden Prescott [1,2] and Wan-Lu Li [1,3,*]**

1 Aiiso Yufeng Li Family Department of Chemical and Nano Engineering, University of California San Diego, La Jolla, CA 92093, USA
2 Department of Chemistry and Biochemistry, University of California San Diego, La Jolla, CA 92093, USA
3 Program of Materials Science and Engineering, University of California San Diego, La Jolla, CA 92093, USA
* Correspondence: wal019@ucsd.edu

**Abstract:** Boron, the neighbor element to carbon in the periodic table, is characterized by unique electron deficiency that fosters multicenter delocalized bonding, contributing to its diverse chemistry. Unlike carbon cages (fullerenes), which preserve their structural integrity under endohedral or exohedral doping, larger boron cages (borospherenes) exhibit diverse structural configurations. These configurations can differ from those of pure boron cages and are stabilized by various metals through unique metal–boron bonding, resulting in a variety of metalloborospherenes. Due to boron's electron deficiency, metalloborospherenes exhibit fascinating chemical bonding patterns that vary with cluster size and the type of metal dopants. This review paper highlights recent advancements in metalloborospherene research, drawing comparisons with metallofullerenes, and focuses on the use of transition metals, lanthanides, and actinides as dopants across various cage dimensions.

**Keywords:** fullerene; borospherene; metal-containing cages; analogs; clusters

## 1. Introduction

The discovery and characterization of fullerenes, a class of carbon cage structures, have profoundly impacted various fields of chemistry and materials science. Fullerenes, composed predominantly of pentagonal and hexagonal carbon rings, exhibit remarkable stability due to their unique sp$^2$ hybridization and delocalized $\pi$-electron system [1,2]. Among these, $C_{60}$, known as buckminsterfullerene, is the most well-known and stable fullerene, characterized by a structure comprising 12 pentagons and 20 hexagons [3,4]. This stability is attributed to the Isolated Pentagon Rule (IPR), which states that no two pentagons should share an edge, minimizing strain in the carbon network [5].

Boron, a neighbor of carbon in the periodic table, forms analogous cage-like structures known as borospherenes, first discovered in the $B_{40}^-$ cage [6]. Unlike carbon, boron is electron-deficient, which gives rise to the formation of multicenter bonds to stabilize its structures [7,8]. This electron deficiency leads to diverse bonding patterns, including delocalized $\sigma$ and $\pi$ bonds, which stabilize planar or quasi-planar clusters for boron clusters up to $B_{38}^-$ [6,9]. This delocalized bonding scheme imparts extraordinary stability to planar boron clusters due to multiple aromaticities, where the number of delocalized electrons satisfies the $(4n + 2)$ Hückel rule, analogous to hydrocarbons [10]. However, larger boron cages, such as the seashell-like chiral $B_{28}^-$ cluster and the $B_{40}^-$, adopt fullerene-like structures [6,11] that are stabilized through delocalized multicenter bonding to maintain the structural integrity despite their larger size.

The unique electronic properties of boron make it an excellent candidate for forming stable metalloborospherenes through metal doping [7]. These structures can be stabilized by incorporating various metals, leading to significant changes in their geometrical and electronic configurations. Metalloborospherenes exhibit a range of fascinating bonding patterns and structural motifs that evolve with cluster size and metal composition. This review compares recent advancements in metalloborospherene research with metallofullerene studies, focusing on transition metals, lanthanides, and actinides as dopants in various cage dimensions.

Metallofullerenes are formed by encapsulating metal atoms within fullerene cages, such as $C_{60}$ or its analogs, using methods like arc discharge, laser ablation, and chemical vapor deposition [12,13]. The encapsulated metal atoms significantly influence the properties of fullerenes, leading to various applications in materials science, medicine, and catalysis. For example, in materials science, metallofullerenes are used to enhance the mechanical and electronic properties of composite materials. In medicine, they are explored for drug delivery systems and diagnostic imaging due to their ability to encapsulate therapeutic agents or contrast materials [14–16]. By comparison, metalloborospherenes, though less understood, involve substantial interactions between the metal atoms and the boron framework. These interactions result in unique structural and electronic properties that differ significantly from those of metallofullerenes. Metalloborospherenes exhibit complex bonding patterns and structural motifs that evolve with cluster size and the type of metal dopant. The incorporation of different metals can lead to a variety of electronic properties, making metalloborospherenes potentially valuable for applications in catalysis, electronics, and nanotechnology [7,17]. Understanding these properties requires further research, but the potential for discovering new materials with tailored functionalities is significant.

Recent studies have explored the incorporation of lanthanide atoms within borospherenes; $La_3B_{18}^-$ was the first observed metalloborospherene [18], and exhibits distinctive bonding interactions due to charge transfer and significant (d–p)$\pi$ and (d–p)$\delta$ bonding [19,20]. These interactions lead to novel structural motifs, including inverse-sandwich and trihedral configurations, which are not commonly observed in other types of boron clusters [18,21]. Similarly, transition-metal-based and actinide-based metalloborospherenes have been theoretically predicted, revealing complex electronic structures and unique bonding patterns that differ from those of their lanthanide counterparts [22,23]. These theoretical predictions indicate the possibility of forming stable complexes with actinides, which could have implications for both fundamental chemistry and potential applications. Although the study of metalloborospherenes is still in its early stages and lacks extensive experimental evidence, the initial findings suggest that both fullerene and borospherene cages have the potential to stabilize unconventional metal clusters, paving the way for new materials with advanced properties. These materials could find applications in various fields, including catalysis, electronic devices, and materials science, due to their unique structural and electronic characteristics. The exploration of these novel metalloborospherenes and metallofullerenes opens up exciting avenues for future research, highlighting the need for further experimental and theoretical investigations to fully understand their potential and to harness their properties for practical applications.

## 2. Structures of Fullerene and Borospherene

Fullerenes are enclosed cages of carbon molecules composed of many adjacent rings, with rings of size n = 5 or n = 6 being more prevalent than other ring sizes, suggesting that fullerenes consisting of pentagons and hexagons exhibit increased stability [1,2]. The smallest fullerene that satisfies these requirements is $C_{60}$, as shown in Figure 1 [3]. This cage structure is a network of $sp^2$-hybridized carbon atoms, where paired valence electrons from the 2 s orbitals collectively inhabit molecular orbitals arranged in triangular units centered around each carbon atom individually, forming a delocalized $\sigma$ network [4]. The triangular $sp^2$ hybrid orbitals of every adjacent carbon atom then overlap from end to end to form pentagonal rings, which constitute the faces of the "soccer ball" structure of the fullerene [4]. This concept is referred to as the Isolated Pentagon Rule (IPR), which asserts that the most stable state of a purely carbon or metal-doped fullerene is a network with the minimum number of separate pentagons required for the carbon atom count, with every bonding electron being fully delocalized over the $\sigma$ molecular orbital [5]. The remaining two valence 2p electrons in carbon inhabit a fully occupied 2p state, overlapping in a $\pi$ formation, where the lobes (amplitudes) of the 2p wavefunctions conjoin to cover a continuous $\pi$ (2p) orbital over the full carbon cage surface [5]. In these carbon fullerenes, the presence of two electrons in the 2p state enhances stability by resulting in occupancy of one electron in both the degenerate $2p_x$ and $2p_y$ orbitals of carbon in accordance with Hund's rule. This results in a sufficient number of electrons being distributed throughout both delocalized molecular orbitals [5].

Boron fullerenes, known as borospherenes [6], are analogous to carbon fullerenes. Figure 1 summarizes the reported borospherene structures validated by theoretical studies and photoelectron spectroscopy (PES) [24]. Unlike carbon, boron is electron-deficient, possessing only one 2p electron, which leads to the formation of delocalized $\sigma$ and $\pi$ bonds [6–9]. This electron deficiency necessitates a more complex bonding scheme to achieve stability. Boron clusters often adopt planar or quasi-planar structures (up to $B_{38}^-$) [7] to maximize electron delocalization and minimize electronic repulsion, a stark contrast to the spherical structures of carbon fullerenes that adhere to a $2(n+1)^2$ $\pi$-electron counting rule to determine spherical aromaticity [25]. In borospherenes, more electrons need to be delocalized over the structure to mitigate the electron deficiency effect, resulting in unique bonding characteristics and stability [26]. The presence of delocalized $\sigma$ and $\pi$ bonds in borospherenes allows for a double delocalization pattern, which is observed less frequently in carbon fullerenes. This double delocalization leads to increased stability and distinct chemical properties. Additionally, boron's propensity for forming multicenter bonds contributes to the diversity in structural motifs observed in borospherenes, ranging from planar to tubular and cage-like structures, depending on the size and composition of the cluster [8,9].

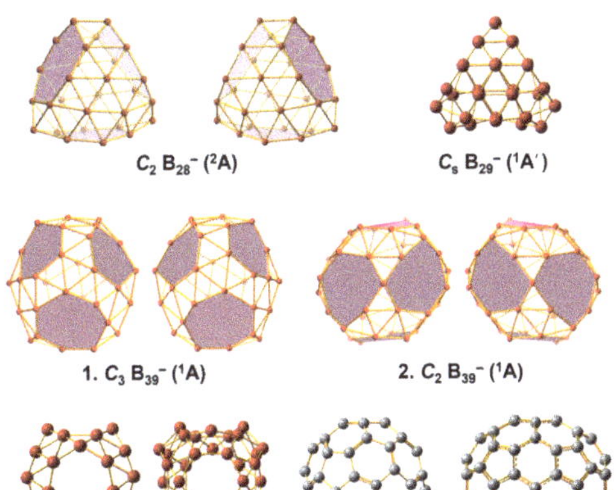


**Figure 1.** Known borospherene structures of $B_{28}^-$ [11,27], $B_{29}^-$[28], $B_{39}^-$ [29], and $B_{40}^-$ [6] validated by PES and quantum chemistry calculations compared with $I_h$-$C_{60}$ and $D_{5h}$-$C_{70}$. Here, $B_{28}^-$ and $B_{39}^-$ systems demonstrate chiral enantiomers.

The first borospherene structure experimentally and theoretically reported was neutral $B_{40}$ [6], as shown in Figure 1. This structure consists of eight $B_6$ quasi-planar triangle faces with two $B_6$ rings on the top and bottom and two $B_7$ rings on the sides in between the triangle faces. The anion counterpart, $B_{40}^-$, has two coexisting degenerate isomers with planar and cage-like shapes, respectively. In the $B_{39}^-$ cluster [29], two distinct chiral borospherene isomers with $C_2$ and $C_3$ symmetry were discovered through PES and theoretical studies, both derived from the $B_{40}$ borospherene. The $C_3$ isomer can be formed by replacing a $B_7$ heptagon in the $B_{40}$ cage with a $B_6$ hexagon, followed by structural reorganization. The $C_2$ $B_{39}^-$ isomer, on the other hand, can be created by removing a single boron atom from the waist of the $B_{40}$ borospherene, resulting in a narrowed boron double chain and the formation of a "defect" site. The seashell-like $B_{28}^-$ has been identified as the smallest borospherene, featuring chiral enantiomers with $C_2$ point-group symmetry [11,27], as shown in Figure 1. The PES of $B_{29}^-$ [28] exhibits a complex pattern, indicating the presence of low-lying isomers. Global minimum searches and extensive theoretical calculations have revealed a complicated potential energy surface for $B_{29}^-$, with multiple low-lying isomers contributing to the experimental spectrum. Among these, the global minimum is a three-dimensional seashell-like $C_s$ ($^1A'$) borospherene isomer, characterized by two heptagons on the waist and one octagon at the bottom. This is followed by a planar $C_1$ ($^1A$) isomer with a hexagonal hole and a stingray-shaped planar $C_s$ ($^1A'$) isomer featuring a pentagonal hole.

Ab initio calculations suggest that $B_{44}^-$ may be the next negatively charged borospherene, although extensive global minima searching has not been conducted [30]. Additionally, theoretical studies have predicted a variety of borospherene structures. The global minimum structures of several positively charged boron clusters have been identified as borospherene configurations, including $B_{29}^+$, $B_{31}^+$, $B_{35}^+$, $B_{38}^+$, $B_{38}^{2+}$, $B_{39}^+$, $B_{40}^+$, $B_{41}^+$, and $B_{42}^+$ [26,30–35]. To date, no borospherene clusters larger than $B_{42}$ have been identified through global minimum searches [36]. Nevertheless, numerous stable structures for large borospherenes have been predicted with high stability [36,37]. For larger clusters such as $B_{80}$, a buckminsterfullerene template was used as the starting structure, adding one boron atom to the center of each hexagon face [38]. Subsequent studies have revealed that core–shell structures, where additional atoms are incorporated inside the cage, are more stable than their hollow counterparts [39].

## 3. Metallofullerene and Metalloborospherene

Metallofullerenes are formed by trapping metal atoms inside the hollow cages of fullerenes [12,13], such as $C_{60}$ or larger fullerenes like $C_{70}$. The encapsulation process can occur through methods like arc discharge, laser ablation, and chemical vapor deposition. Their ability to encapsulate metal atoms opens up numerous possibilities for advancements in materials science, medicine, and catalysis, making them a significant area of study in nanotechnology and materials chemistry [14–16]. Metalloborospherenes, unlike metallofullerenes, involve significant interactions between the metal atoms and the boron cage structures [7,17]. These interactions often result in substantial changes to the boron framework, leading to the formation of unique metalloborospherene structures through metal-boron ring chemical bonding.

### *3.1. Lanthanide-Doped Fullerene and Borospherene Clusters*

Endohedral metallofullerenes (EMFs) offer a unique platform for studying otherwise unstable metallic clusters due to the protective environment provided by fullerene cages. Lanthanide atoms can be encapsulated within these cages, forming various structures such as monometallofullerenes, dimetallofullerenes, carbide clusterfullerenes with an $M_2C_2$ unit encapsulated inside the fullerene cages, and nitride clusterfullerenes [40,41]. The interaction between the lanthanide atoms and the coordinated carbon rings facilitates charge transfer, leading to unusual valence states and chemical bonding patterns. For example, single-electron lanthanide–lanthanide bonds can form within $C_{80}$ fullerenes, resulting in robust redox-active molecular magnets [42,43]. Crystallographic analysis of typical carbide clusterfullerenes has revealed a relationship between the ionic radius of encapsulated metal ions and the size of the fullerene cage. Using Shannon ionic radii [44], it has been observed that larger cages accommodate larger metal ions. For example, $Sc_2C_2$ is typically found within $C_{72–88}$ cages, $Er_2C_2$ within $C_{80–92}$ cages, and $La_2C_2$ within $C_{90–104}$ cages, respectively (Figure 2) [40].

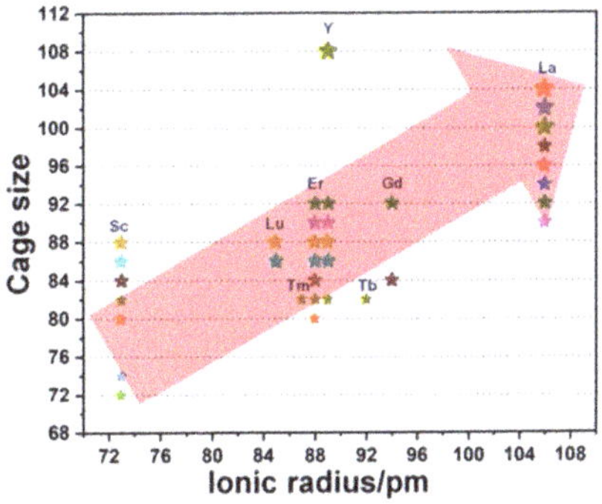


**Figure 2.** The relationship between the ionic radius of encapsulated metal ions and the fullerene cage size for typical $Ln_2C_2$ carbide clusterfullerenes. Each star in the diagram may represent more than one compound [40].

Due to the electron deficiency of boron, lanthanide elements usually form dative bonds with boron group ligands. Recent findings show that lanthanide-doped boron clusters exhibit unique structures due to charge transfer interactions and significant (d–p)$\pi$ and (d–p)$\delta$ bonding [19,20]. Unlike transition metal-doped boron clusters, which often form borometallic molecular wheels, lanthanide-doped clusters create distinct inverse-sandwich-type structures, particularly evident in $Ln_2B_n^-$ clusters (n = 7–9) [21]. Furthermore, the latest research suggests that these inverse-sandwich configurations can extend to form lanthanide–boron nanowires. The first experimentally observed tri-metal-doped inverse triple-decker, $La_3B_{14}^-$ with $C_{2v}$ point group, exhibits a unique structure where three lanthanum atoms are positioned between two conjoined $\eta^8$-$B_8$ rings that share a B–B unit on one edge, resulting in a tilted La–$B_8$–La–$B_8$–La configuration [45]. With increasing cluster size, recent studies have identified the first perfect spherical trihedral metalloborospherenes, $D_{3h}$ $Ln_3B_{18}^-$ (Ln = La, Tb) [18], through joint PES and first-principles theory investigations. These metalloborospherenes are notable for their distinct geometry, featuring three deca-coordinate lanthanide centers that are integral to the cage surface. These centers are coordinated within three equivalent $\eta^{10}$-$B_{10}$ decagons, which share two eclipsed triangular $B_6$ units connected by three $B_2$ units, as shown in Figure 3. Due to the 4f electrons of lanthanides, this configuration gives these metalloborospherenes unique magnetic and catalytic properties.

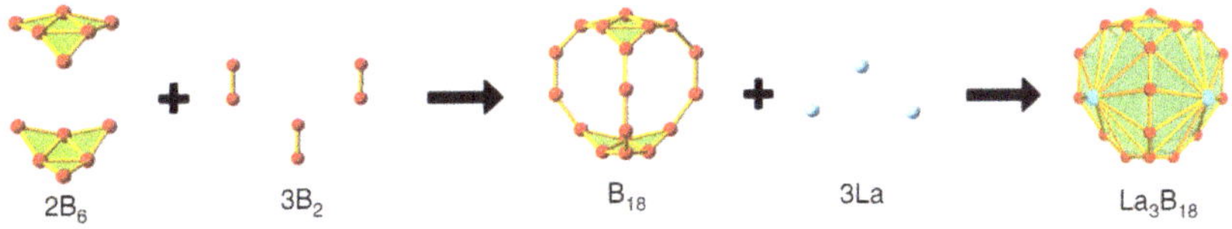


**Figure 3.** Preferable pathway for the $B_{18}$ construction and its bonding scheme with three La atoms to form $D_{3h}$ $Ln_3B_{18}^-$ metalloborospherene [18].

Following the discovery of the first lanthanide-doped borospherene, $La_3B_{18}^-$, a series of theoretical studies have emerged, predicting the existence of other lanthanide-doped borospherene structures with varying numbers of lanthanide atoms. Two core–shell-like metalloborospherenes, $C_{2v}$ $La_3\&[B_2@B_{17}]^-$ and $D_{3h}$ $La_3\&[B_2@B_{18}]^-$ have been identified, featuring a transition-metal-like $B_2$ core at the center of the $La_3B_{17}^-$ and $La_3B_{18}^-$ cages [46]. These structures exhibit unique donor–acceptor duality within the $La_3\&B_n^-$ spherical trihedral shells (n = 17, 18). The $B_2$ core acts as a donor–acceptor entity, contributing to the stability and electronic properties of the metalloborospherenes. Unlike the tri-lanthanide-doped metalloborospherene, which typically exhibits in-plane coordination with boron atoms, the tri-lanthanide-doped metallofullerene $Ln_3@C_{80}^+$ cation features an endohedral structure [47]. In this configuration, the $Ln_3$ cluster is situated inside the carbon cage and interacts through a three-center two-electron (3c-2e) $\sigma$ bond. Lu et al. theoretically proposed a $T_d$ $La_4B_{24}$ cluster featuring a tetrahedral cage structure [48]. This cluster consists of four interconnected $B_6$ triangles on the surface and four nona-coordinate lanthanum atoms embedded within four conjoined $\eta^9$-$B_9$ nonagonal ligands. This configuration represents a boron-based counterpart to the experimentally observed $T_d$ $C_{28}$ tetrahedral carbon fullerene [49].

Theoretically, as the number of boron atoms increases, the resulting boron cages become sufficiently large to accommodate endohedral doping. For instance, $M@B_n$ (M = Eu, Gd; n = 38, 40) clusters have been predicted to form spherically symmetric endohedral metalloborospherenes [50], which is a derivative of the $B_{40}$ borospherene structure [6].

### 3.2. *Transition Metal-Containing Fullerene and Borospherene*

In this section, we focus on transition metal-containing EMFs to compare their structures with metalloborospherenes. Over the past two decades, rare-Earth metals, such as Group-3 elements (Sc, Y) and many lanthanides, have been successfully encapsulated within fullerene cages. The exploration of EMFs with d-block transition metals began in 1992 with the detection of $Ti@C_{28}$ [51]. The macroscopic synthesis of Ti-containing EMFs, such as $Ti_2C_{80}$ and $Ti_2C_{84}$, was first reported by Shinohara et al. [52,53]. In 2013, Echegoyen et al. identified $Ti_2S@C_{78}$, which features a near-linear $Ti_2S$ cluster inside the fullerene cage [54]. Recently, Lu and colleagues prepared $Ti_3C_3@I_h(7)$-$C_{80}$ [55], which displays a unique cyclopropane-like geometry. The use of Group-3 metals in mixed-metal clusters has also led to the creation of nitride and carbide clusterfullerenes [56]. Synthesizing V-only EMFs has proven to be challenging. However, inspired by the successful synthesis of the M-Ti system, researchers used Sc to facilitate the encapsulation of vanadium. This approach led to the synthesis of mixed-metal nitride and carbide cluster fullerenes, such as $VSc_2N@I_h(7)$-$C_{80}$ and $VSc_2C@I_h(7)$-$C_{80}$ [57]. Unlike Ti-clusterfullerenes, V-containing clusters can host one or two V atoms, highlighting their unique chemical properties. This distinction underscores the complexity and diversity of transition metal-containing EMFs, as the incorporation of different metals can lead to varied structural and electronic characteristics.

Since the isolation of the first EMFs in the early 1990s, their electronic structures have been of significant interest due to their correlation with stability. The high electron affinity of fullerenes suggests an ionic structure for EMFs, with metal cations encapsulated within negatively charged carbon cages (e.g., $Y^{3+}@C_{82}^{3-}$) [58]. This was confirmed by ESR spectroscopy and molecular orbital analysis, which indicated that spin density and frontier MOs are localized on the fullerene cage. This ionic model applies to monometallofullerenes and complex species like dimetallofullerenes, nitride clusterfullerenes, and carbide clusterfullerenes, including Ti-containing EMFs such as $Ti_2C_{80}$ and $Ti_2C_{84}$. Various spectroscopic and computational studies [59–61] support this ionic nature, with XPS, UV-vis-NIR, and electrochemistry showing similar properties for isostructural EMFs. However, studies reveal a covalent contribution to metal–cage bonding, evident in smaller computed atomic charges and detailed electron density analyses [62].

Compared with metallofullerene, doping boron nanoclusters with transition metals results in significant structural transformations. As illustrated in Figure 4 and reported in our recent review paper [7], joint experimental and computational studies have observed structures including wheel-like clusters such as $D_{8h}$ $Co©B_8^-$, $D_{9h}$ $M©B_9^-$ (M = Rh, Ir, Re), and $D_{10h}$ $M©B_{10}^-$ (M = Ta, Nb) [63–65], half-sandwich clusters like $C_{3v}$ $CoB_{12}^-$ and $IrB_{12}^-$ [66], and tubular clusters such as $D_{8d}$ $CoB_{16}^-$ [67], $MnB_{16}^-$ [68], $RhB_{18}^-$ [69], $C_s$ $B_2$–$Ta@_{18}^-$, and $D_{10d}$ $Ta@B_{20}^-$, with $Ta@B_{20}^-$ exhibiting the highest coordination number (CN = 20) among tubular species [70]. The metal-centered tubular boron clusters exhibit multiple aromaticities due to the delocalized bonding, which results from the efficient overlap between the orbitals of the metal center and the surrounding boron group orbitals. It should be noted that no experimental evidence for the transition metal-containing metalloborospherene structures has been reported. Table 1 lists the predicted structures along with their corresponding doping patterns and symmetries, which await validation through experimental characterization. More recent theoretical predictions include spherical trihedral metalloborospherenes and endohedral complexes $B_{20}TM_n$ (TM = Sc, Y; n = 3, 4) [71], Ta-centered metalloborospherenes such as $Ta@B_{22}^-$ and $Ta@B_n^q$ (n = 23–28, q = −1–3), the smallest trihedral metalloborospherene $D_{3h}$ $Ta_3B_{12}^-$ with $\sigma + \pi + \delta$ aromaticity, and the spherical tetrahedral metalloborospherene $T_d$ $Ta_4B_{18}$ with four nonacoordinate Ta centers [72,73]. For larger boron sizes, theoretical studies have previously proposed the existence of metalloborospherenes, specifically for $MB_{40}$ (M = Be, Mg) and $Ni_nB_{40}$ (n = 1–4) [74–76], which feature seven-coordinate metal centers on the $B_{40}$ cage surface. These structures are related to the corresponding metalloborophenes, serving as precursors with similar coordination environments for the metal centers.

**Table 1.** Theoretically Predicted Transition Metal-Containing Borospherenes.

| Species | Symmetry | Doping Mechanism |
|---|---|---|
| $Ta_3B_{12}^-$ [75] | $D_{3h}$ | exohedral |
| $Zr_3B_{15}^+$ [77] | $D_{3h}$ | exohedral |
| $Zr_3B_{15}$ [77] | $D_{3h}$ | exohedral |
| $Zr_3B_{15}^-$ [77] | $D_{3h}$ | exohedral |
| $Hf_3B_{15}^+$ [77] | $D_{3h}$ | exohedral |
| $Hf_3B_{15}$ [77] | $D_{3h}$ | exohedral |
| $Hf_3B_{15}^-$ [77] | $D_{3h}$ | exohedral |
| $V_3B_{16}^-$ [78] | $C_{2v}$ | endohedral |
| $Cr_3B_{16}^-$ [78] | $C_{4v}$ | endohedral |
| $Mn_3B_{16}^-$ [78] | $C_{4v}$ | endohedral |
| $Fe_3B_{16}^-$ [78] | $D_{8d}$ | endohedral |
| $Co_3B_{16}^-$ [78] | $D_{8d}$ | endohedral |
| $Ni_3B_{16}^-$ [78] | $C_s$ | endohedral |
| $Ta_4B_{18}^-$ [79] | $T_d$ | exohedral |
| $Nb_4B_{18}^-$ [79] | $T_d$ | exohedral |
| $Sc_3B_{20}^-$ [71] | $D_{3h}$ | exohedral |
| $Y_3B_{20}^-$ [71] | $D_{3h}$ | exohedral |
| $Ti_3B_{20}^-$ [71] | $D_{3h}$ | exohedral |
| $Zr_3B_{20}^-$ [71] | $D_{3h}$ | exohedral |
| $Hf_3B_{20}^-$ [71] | $D_{3h}$ | exohedral |
| $Sc_3B_{20}^{3-}$ [71] | $D_{3h}$ | exohedral |
| $Y_3B_{20}^{3-}$ [71] | $D_{3h}$ | exohedral |
| $Sc_4B_{20}^-$ [71] | $D_{3h}$ | exohedral |
| $Y_4B_{20}^-$ [71] | $D_{3h}$ | exohedral |
| $ScB_{24}^-$ [80] | $D_2$ | endohedral |
| $TiB_{24}^-$ [80]<br>$VB_{24}^-$ [80] | $D_{3h}$<br>$C_{2v}$ | endohedral<br>endohedral |
| $CrB_{24}^-$ [80] | $C_s$ | endohedral |
| $MnB_{24}^-$ [80] | $C_s$ | endohedral |

**Table 1.** *Cont.*

| Species | Symmetry | Doping Mechanism |
| --- | --- | --- |
| $FeB_{24}^-$ [80] | $C_s$ | endohedral |
| $CoB_{24}^-$ [80] | $C_s$ | endohedral |
| $SrB_{40}^-$ [76,81] | $D_{2d}$ | endohedral |
| $CaB_{40}^-$ [76,81] | $C_{2v}$ | endohedral |
| $BeB_{40}^-$ [76,81] | $C_s$ | exohedral |
| $MgB_{40}^-$ [76,81] | $C_s$ | exohedral |
| $LiB_{40}$ [82] | $C_s$ | exohedral |
| $NaB_{40}$ [82] | $C_{2v}$ | endohedral |
| $KB_{40}$ [82] | $C_s$ | exohedral |
| $BaB_{40}$ [82] | $D_{2d}$ | endohedral |
| $TlB_{40}$ [82] | $C_s$ | exohedral |
| $TiB_{40}$ [83] | $C_1$ | endohedral |
| $Ti_2B_{40}$ [83] | $C_s$ | endohedral |

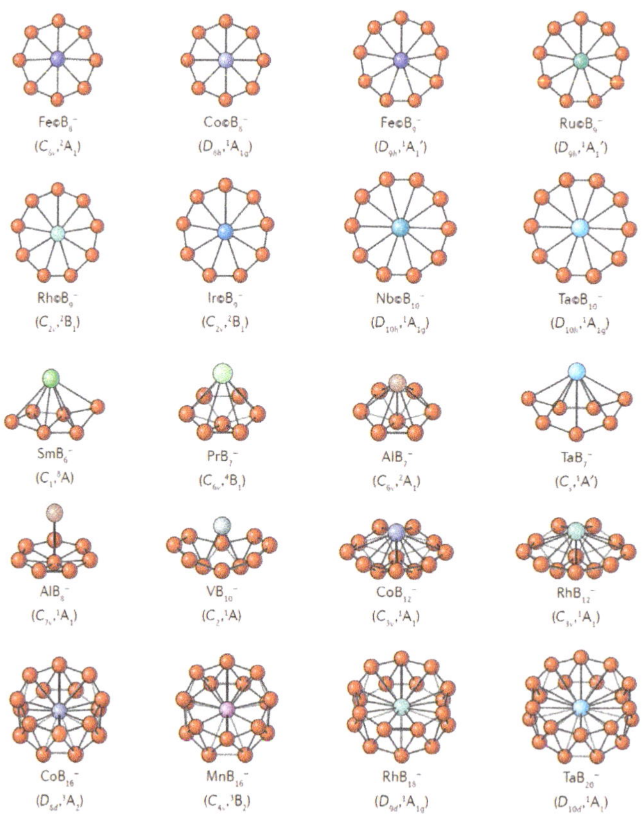


**Figure 4.** Three classes of metal-doped boron clusters have been observed experimentally: metal-centered monocyclic rings, half-sandwich type structures, and metal-centered tubular structures. The point group symmetry and electronic state are reported for each structure [7].

### 3.3. Actinide-Containing Fullerene and Borospherene

Actinide-based EMFs have recently garnered interest, compared with the extensively researched lanthanide-based EMFs [22]. Due to the diverse valence electron shells of actinide elements, actinide EMFs have revealed intricate electronic structures and novel bonding configurations not present in lanthanide counterparts [23], suggesting that the unique actinide clusters encapsulated within fullerene cages offer a valuable platform to examine the nature of actinide–ligand interactions within a controlled, inert environment, thereby unveiling new dimensions of actinide chemistry.

A defining characteristic of actinide EMFs is the substantial charge transfer from the metal to the fullerene cage, influencing both their stability and reactivity. Early actinides, such as thorium (Th), plutonium (Pu), and uranium (U), are notably redox-active and exhibit a wide range of oxidation states [84]. For instance, U can be present in various oxidation states, from U(III) to U(VI), in contrast to the limited oxidation states of lanthanides [85,86]. The high chemical activity of the 5f orbitals in actinides results in unique cage isomer preferences and novel electronic structures [85,87]. Although the radioactivity of actinides has traditionally restricted their study to theoretical calculations and limited spectroscopy, recent experimental advancements have significantly deepened our understanding, exemplified by the first crystallographic analysis of $Th@C_{3v}(8)\text{-}C_{82}$ and the identification of a formal $U^{5+}$ charge state in $U_2C@I_h(7)\text{-}C_{80}$ [49,88]. Additionally, $UCU@I_h(7)\text{-}C_{80}$ represents the first actinide metallic cluster EMF, showcasing unsupported U=C double bonds [89]. Recent advancements have expanded our knowledge of actinide clusterfullerene families with the synthesis and characterization of $U_2C_2@I_h(7)\text{-}C_{80}$ and $UCN@C_s(6)\text{-}C_{82}$ [90–92]. $U_2C_2@I_h(7)\text{-}C_{80}$ features two uranium atoms bridged by a carbon–carbon triple bond ($C \equiv C$), while $UCN@C_s(6)\text{-}C_{82}$ contains a triangular UCN cluster with $\eta^2$ (side-on) coordination of uranium by cyanide. These discoveries underscore the potential of fullerene cages to stabilize unconventional actinide clusters, offering profound insights into actinide bonding and paving the way for the development of advanced materials.

Although various metal-doped boron clusters have been reported, actinide (An)-doped boron clusters remain relatively unexplored due to significant challenges in both experimental and computational approaches. Similar to other metal atoms, doping boron clusters with actinides can enhance their stability and lead to diverse geometrical structures. Notable computational examples include the half-sandwich structures $AnB_{12}$ (An = Th to Cm) [93], the double-ring tubular structures $An@B_{20}$ [94], $An@B_{24}$ (An = Th, Pu, and Am) [95], and the actinide-centered borane $An(BH)_{24}$ (An = Th to Cm) [96]. Inspired by the discovery of monovalent lanthanides in the $LnB_8^-$ system through PES and quantum chemistry calculations [97], we investigated the $AnB_8^-$ system and identified the potential for monovalent actinide elements in later actinides such as $BkB_8^-$, $MdB_8^-$, and $NoB_8^-$ [98]. Despite significant interest in metalloborospherenes, there is limited evidence for the stabilization of boron cages by actinide metal atoms, with the notable exception of the recently theoretically reported boron cluster $U@B_{40}$. This discovery raises intriguing questions about the viability of other actinide metal-doped borospherenes. Theoretical calculations have demonstrated the existence of actinide endohedral borospherenes such as $An@B_n$ (An = U and Th; n = 36, 38, and 40) [99,100] and the chiral actinide endohedral borospherenes $Ac@B_{39}$, $[Ac@B_{39}]^{2+}$, $Th@B_{39}$, and $[Th@B_{39}]^{3+}$ [101]. Additionally, global searches have revealed that $U@B_{24}$ possesses a unique half-cage structure [96]. These findings highlight the potential for a broader range of actinide-doped boron clusters, suggesting new avenues for research into their stability, electronic properties, and geometric configurations. The continued exploration of these systems could provide valuable insights into the bonding characteristics and chemical behavior of actinides, further expanding the understanding of actinide chemistry in novel boron-based materials. Actinide metallofullerenes and actinide metalloborospherenes exhibit significant differences in their structural, electronic, and stability characteristics. Actinide metallofullerenes, composed of actinide atoms encapsulated within highly symmetric carbon-based fullerene cages, rely on substantial metal-to-cage charge transfer interactions for stability that are rarely observed in other systems where a single metal ion is positioned within a carbon cage, and exhibit unique redox properties influenced by the interaction of actinide 5f orbitals with the carbon $\pi$-system [22,87,88]. In contrast, actinide metalloborospherenes, consisting of actinide atoms within boron-based cages [102], present a broader range of geometrical configurations due to boron's ability to form multicenter bonds. The study of these actinide-doped clusters provides valuable insights into the complex chemistry of actinides, with metallofullerenes offering stable environments and metalloborospherenes presenting versatile and reactive platforms for further exploration.

## 4. Theoretical and Computational Methods

Theoretical and computational methods are crucial for studying metalloborospherene, as they enable the prediction of stable geometries and electronic structures that are challenging to investigate experimentally. Additionally, simulating spectroscopy and comparing it to experimental observations can provide evidence of the true global minimum structures. These methods offer detailed insights into the bonding patterns and aromaticity of metalloborospherene. We recently proposed the novel concept of "cluster-assembly materials" [98], wherein highly symmetric stable clusters serve as the foundational building blocks for derivative materials. This approach guides experimental efforts and enhances our understanding of these advanced materials.

### *4.1. Global Minimum Search*

Global minimum searching can be performed using several different algorithms. One of the earliest methods for this purpose is the Basin Hopping (BH) algorithm [103]. This method randomly perturbs the current structure, performs a geometry optimization, and then accepts the new structure if the energy is lowered, or based on a probability determined from the ratio of the population of the previous structure to that of the new structure as determined by the canonical ensemble partition function (Equation (1)). This strategy was implemented in the TGMin package, which allows acceptance of higher energy structures, enabling the BH algorithm to avoid becoming trapped in local minima that are not the global minimum [103–106]. The pure BH algorithm has been successfully used to identify the structures of $B_{40}^-$ [6], $MnB_{16}^-$ [68], etc.

$$p < e^{\frac{E(X_0)-E(X_{new})}{k_B T}} \tag{1}$$

A modified version of the BH method was developed in TGMin 2.0 [107], enhancing efficiency through several modifications: checking displaced structures for low-coordination numbers, freezing some atoms to reduce search dimensions, adjusting unreasonable structures before ab-initio calculations, and preventing the revisiting of prior structures. TGMin 2.0 has successfully identified the global minimum structures of $La_3B_{18}^-$, $Tb_3B_{18}^-$, $MnB_{16}^-$, and $La_3B_{14}^-$ [45,68,77].

In addition to the BH method, the Stochastic Surface Walking (SSW) method [108,109] enhances global minimum searching by adding random displacements to the current local minima structure and performing a biased dimer rotation to incorporate information about transition states between isomers, followed by geometry optimization [110]. This method has been successfully used to determine the global minimum structure of $B_{40}^-$ [6]. Similarly, the Coalescence Kick algorithm [111–113], which was employed to find the global minimum structure of $MnB_{16}^-$, involves displacing each atom within a spherical radius R of its prior position, akin to the displacement strategies used in BH, TGMin, and SSW.

Another advanced technique is the CALYPSO (Crystal Structure Analysis by Particle Swarm Optimization) package [114], which predicts structures by generating random structures with symmetry constraints followed by geometry optimization. Each structure is assigned a geometric parameter similar to a genetic code, and position adjustments are made through particle swarm optimization [115,116]. This method was used to predict the structures of $AnB_{36}$ (An = Pa, Np, Pu, Am, Cm, Bk, and Cf) clusters [102].

For the energy calculation, ab initio methods need to be employed using quantum chemistry software such as Gaussian (16) [117] and ADF (2024.1) [118,119]. These calculations are essential for accurately determining the electronic structure and properties of the systems under investigation. However, as the number of atoms in a system increases, the computational cost of ab initio calculations becomes a significant bottleneck. This necessitates the use of faster, less accurate basis sets and functionals for initial global minimum searches, followed by more precise calculations at higher levels of theory for the most promising structures.

To address these computational challenges, the development of machine learning (ML) potentials is becoming increasingly important. ML potentials can significantly accelerate energy and geometry optimizations by learning from a subset of high-quality ab initio calculations and then predicting the properties of new configurations with high accuracy. This approach has been successfully applied in various fields, demonstrating its potential to handle large systems efficiently [120–123]. For example, Behler–Parrinello neural networks [120] and Gaussian approximation potentials [124] are among the techniques that have been developed to create accurate and transferable potential energy surfaces based on machine learning.

Furthermore, integrating ML potentials with traditional quantum chemistry methods can create hybrid approaches that balance accuracy and computational efficiency. This integration allows for initial rapid screening of large configurational spaces using ML models, followed by detailed ab initio calculations on a select number of configurations. This hybrid strategy has shown promise in materials science and chemistry, making it a valuable tool for studying complex systems such as metalloborospherenes and other large clusters [125–127]. By leveraging the strengths of both ab initio methods and machine learning, researchers can more effectively explore the vast configurational landscapes of complex molecular systems, leading to a deeper understanding and more accurate predictions of their properties.

### *4.2. Photoelectron Spectroscopy Simulation*

The PES can be simulated using the ΔSCF-TDDFT method [128]. In this approach, the adiabatic detachment energy (ADE), which also refers to the electron affinity of the neutral species, is calculated as the energy difference between the neutral molecule at its optimized geometry and the anion at its optimized geometry, defining the onset of the first band in the PES. The peak of the first band, known as the first vertical detachment energy (VDE1), is determined as the energy difference between the neutral molecule and the anion, both at the anion geometry [129]. Higher VDEs are obtained by simulating the excitations of the neutral molecule at the anion geometry using time-dependent DFT (TDDFT) methods for systems without strong electron correlations, while for systems with strong electron correlation or spin–orbit (SO) coupling effects, single-electron VDEs from the anion ground state to the corresponding ground and excited states of the neutral cluster are calculated using the CASSCF/CCSD(T)/SO approach [130], where CCSD(T) energy values replace the diagonal elements of the CASSCF matrix with SO effects treated as a perturbation. Excitations with the electron detached from occupied molecular orbitals to the corresponding molecular orbital of the anion are selected, and the first VDE energy is added to these excitations to produce higher VDEs, corresponding to the peaks of subsequent bands in the PES.

The method of statistical average of different orbital model potentials (SAOP) [131] is typically preferred for the calculation of the excitation energies in metalloborophenes due to its accurate long-range behavior, which is essential for calculating charge transfer excitations in systems with significant charge separation in excited states, such as between metal fragments and boron clusters [18]. In pure boron fullerenes, such as $B_{40}^-$, the time-dependent DFT formalism [6,129] is used at the PBE0 level to achieve similar accuracy.

*4.3. Molecular Orbitals and Chemical Bonding Analysis*

The analysis of molecular orbitals (MOs) in metalloborophenes is vital for understanding their structural stability and the interactions between the metal atoms and the borospherene framework [132]. Moreover, determining the oxidation states of the encapsulated metals is crucial, which can be achieved through appropriate fragment analysis. As illustrated in Figure 5, the MO plot for $La_3B_{18}^-$ reveals that the 6e″, 12e′, and $2a_1''$ irreducible representations associated with the $La_3$ moiety exhibit La 5d character, whereas the 4e″, 6e′, and $1a_1''$ irreducible representations associated with the $B_{18}$ framework exhibit B 2p character. Consequently, the 5e″, 9e′, and $1a_1''$ MOs demonstrate significant bonding interactions between La 5d and $B_{18}$ 2p orbitals. This bonding pattern indicates that the La atoms are in their favored +III oxidation state in $La_3B_{18}^-$, which can be approximately represented as $(La^{3+})_3[B_{18}^{10-}]$.

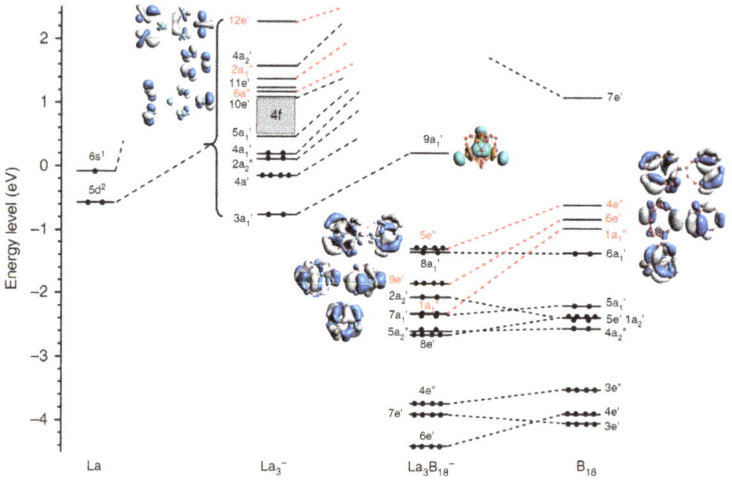


**Figure 5.** Kahn–Sham molecular orbital correlation diagram for the spherical $La_3B_{18}^-$ metalloborospherene with the fragment of $La_3^-$ and $B_{18}$ [18].

To further understand the charge and energy distribution between metal and borophene fragments, the EDA-NOCV (Energy Decomposition Analysis–Natural Orbital for Chemical Valence) method [133–135] can be applied by decomposing the internal energy into several distinct contributions: steric repulsion, orbital interactions, and preparation energy. Steric repulsion encompasses both Pauli repulsion, which arises from the overlap of electron densities, and electrostatic repulsion, which results from the Coulombic interactions between charged particles. Orbital energy, on the other hand, reflects the interactions between occupied and unoccupied molecular orbitals, capturing the essence of bonding and antibonding interactions within the system. Preparation energy accounts for the energy required to distort the fragments from their equilibrium geometries to the configurations they adopt within the complete molecule. This energy component is crucial for understanding the structural adjustments necessary for bonding. By applying the EDA-NOCV method, one can detect the flow of electrons between different fragments and quantify the contribution of each orbital to the total orbital interaction energy [135]. This detailed analysis allows for a deeper insight into the nature of the interactions, providing a comprehensive understanding of the electronic structure and bonding characteristics in metalloborophenes.

For instance, in the case of the $Be_2B_{24}^+$ metalloborospherene cluster, as shown in Figure 6 [136], EDA-NOCV reveals that the $Be_2$–$B_{24}$ interaction is significantly covalent in nature, with notable contributions from highly polarized covalent bonds formed between Be and the boron atoms. The analysis shows how these interactions stabilize the unique four-ring tubular structure of the boron framework, which consists of two $B_{12}$ tubular subunits connected by $B_6$ rings and capped by beryllium atoms. This method also illustrates the charge transfer and orbital mixing between the fragments, highlighting the role of Be to form strong covalent bonds, which is crucial for the stability and unusual geometry of the metalloborospherene.

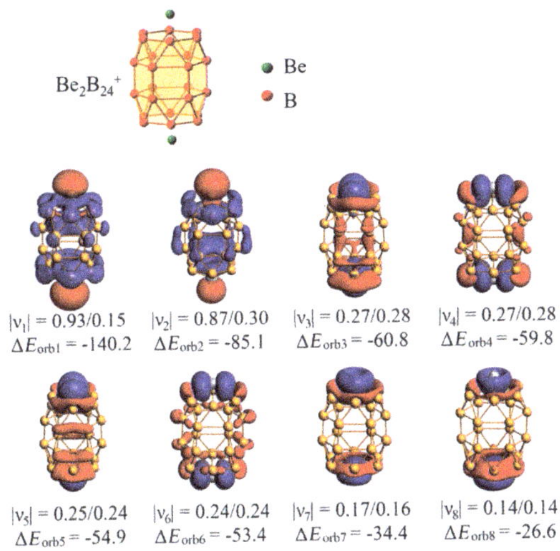


**Figure 6.** $Be_2B_{24}^+$ structure alongside plots of deformation densities $\Delta\rho_{1\text{–}8}$ for the pairwise orbital interactions in $Be_2B_{24}^+$ [136]. The associated interaction energies ($\Delta E_{orb}$ in kcal/mol) and eigenvalues ($\nu$) are also shown. Eigenvalues ($\nu$) represent the magnitude of charge transfer, with the direction of charge transfer illustrated from red to blue.

Bader's Quantum Theory of Atoms in Molecules (QTAIM) [137] provides a refined method for analyzing electron density and thereby understanding metal–cage interactions in EMFs [62]. Instead of focusing on individual molecular orbitals, QTAIM assesses the overall electron density, identifying natural boundaries between atoms via surfaces with zero flux in the gradient vector field. These boundaries, or atomic basins, allow for the integration of scalar properties, such as electron density, to specific atoms. Critical points (CPs) where the density gradient vanishes are classified by their rank (number of nonzero Hessian eigenvalues) and signature (sum of Hessian eigenvalue signs). Different types of CPs (nuclear, bond, ring, and cage) provide insights into bonding characteristics. Bond critical points (BCPs) are particularly crucial, as their presence between atoms signifies bonding. The electron density at BCPs ($\rho_{bcp}$) and its Laplacian ($\nabla^2\rho_{bcp}$) help distinguish between covalent (high $\rho_{bcp}$, negative $\nabla^2\rho_{bcp}$) and noncovalent (low $\rho_{bcp}$, positive $\nabla^2\rho_{bcp}$) interactions. Studies on EMFs using QTAIM, such as those by Kobayashi and Nagase [138,139], reveal the ionic character of metal–cage bonds through small $\rho_{bcp}$ values and positive $\nabla^2\rho_{bcp}$, alongside high bond ellipticity, indicating delocalized metal–cage interactions. Despite QTAIM's robustness, its application in EMF studies has been limited, primarily focusing on atomic charges rather than a comprehensive analysis of bonding indicators. This approach can provide a more detailed understanding of bonding in various EMF classes, including mono- and dimetallofullerenes, nitride cluster fullerenes, and carbide cluster fullerenes (Figure 7).

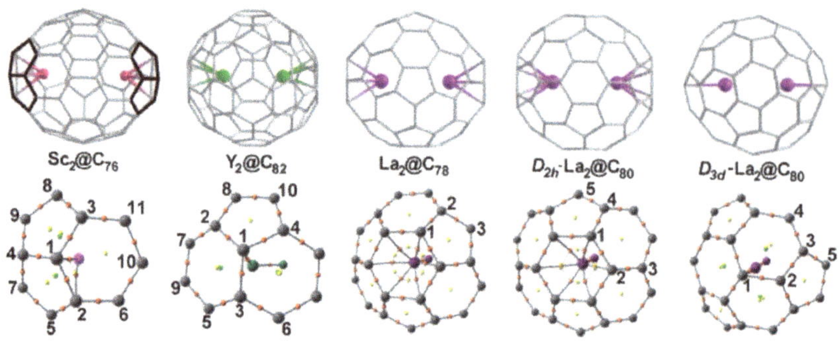


**Figure 7.** Molecular structures and fragments of molecular graphs in the vicinity of the metal atoms of $Sc_2@C_{76}$-Cs(17,490), $Y_2@C_{82}$-$C_{3v}$(8), $La_2@C_{78}$-$D_{3h}$(5), $D_{2h}$-$La_2@C_{80}$-$I_h$(7), and $D_{3d}$-$La_2@C_{80}$-$I_h$(7). Atoms are colored as follows: carbon (gray), bond critical points (BCP, red), ring critical points (RCP, yellow), and cage critical points (CCP, green) [62].

QTAIM has also been widely used to analyze the chemical bonding in metalloborospherenes. For example, in $Th@B_{36}$ metalloborospherene as shown in Figure 8 [140], QTAIM provides a detailed topological analysis of the electron density, highlighting critical points that reveal the nature of boron–boron (B-B) and metal–boron (Th-B) interactions. The analysis identifies BCPs for B-B bonds, indicating a significant covalent character due to high $\rho_{bcp}$ and negative $\nabla^2\rho_{bcp}$. Additionally, ring critical points (RCPs) and cage critical points (CCPs) indicate the presence of multicenter bonding, a characteristic feature of boron clusters. For Th-B interactions, QTAIM reveals smaller electron density and positive Laplacian values at the BCPs, suggesting a closed-shell interaction with partial covalent character, further supported by negative values of the energy density (H(r)) at these points.

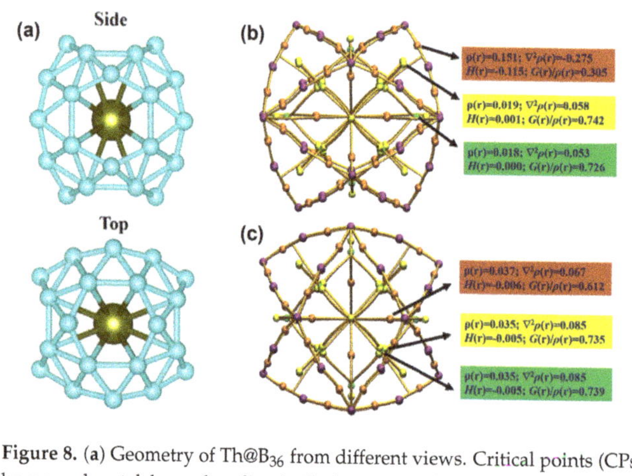


**Figure 8.** (**a**) Geometry of $Th@B_{36}$ from different views. Critical points (CPs) associated with boron–boron and metal–boron bonding in (**b**,**c**), respectively. The color scheme is purple for atoms, orange for bond critical points (BCPs), yellow for ring critical points (RCPs), and green for cage critical points (CCPs). Key parameters for different critical points are also provided [140].

The Adaptive Natural Density Partitioning (AdNDP) [141] analysis, which integrates Lewis theory with canonical molecular orbital theory, provides a powerful approach for describing the bonding in complex molecular systems. This method assigns each molecular orbital as an n-center (1,2)–electron bond, where n can range from 1 to the total number of atoms in the molecule, indicating a spectrum of bonding scenarios from localized to fully delocalized electrons.

AdNDP analysis is particularly effective in visualizing bonding patterns that are not easily represented by traditional two-center, two-electron bonds. For instance, in systems with delocalized bonding, such as the $La_3B_{18}^-$ system as shown in Figure 9, AdNDP can reveal multicenter bonds that involve three or more atoms sharing electron pairs. This capability is essential for understanding the electronic structure of such molecules, where conventional bonding models fall short.

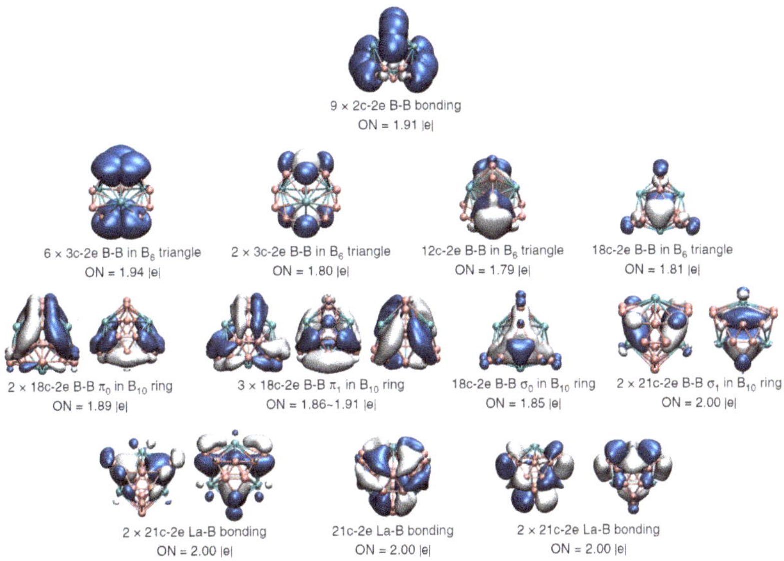


**Figure 9.** AdNDP analyses of $La_3B_{18}^-$ cluster [18].

Moreover, AdNDP analysis can also provide insights into the aromaticity of certain compounds by demonstrating the presence of delocalized electron clouds that contribute to the stability and reactivity of the molecule. All these advanced computational techniques collectively provide a comprehensive understanding of the electronic structure, bonding, and stability of metalloborophenes, facilitating the design and synthesis of new materials with tailored properties.

## 5. Conclusions and Perspectives

In summary, metalloborospherenes represent a fascinating and rapidly evolving class of materials that parallel the well-studied metallofullerenes. The unique electron-deficient nature of boron leads to diverse and complex bonding patterns, enabling the formation of stable boron cages when doped with various metals. These metalloborospherenes exhibit a range of intriguing structural motifs and electronic properties that differ significantly from their carbon-based counterparts.

Recent advancements have demonstrated the ability of metalloborospherenes to stabilize novel metal clusters, particularly those involving transition metals, lanthanides, and actinides. These studies have provided deep insights into the electronic structure, bonding characteristics, and potential applications of these materials. For instance, the stabilization of actinide clusters within borospherenes and fullerenes has revealed complex electronic configurations and bonding interactions not observed in lanthanide-doped systems.

The potential applications of metalloborospherenes are vast, spanning catalysis, materials science, and nanotechnology. The unique bonding and structural properties of these clusters offer opportunities for designing new materials with tailored electronic and catalytic properties. Moreover, the ability to encapsulate and stabilize actinide elements opens up new avenues for research in actinide chemistry, with implications for nuclear science and advanced materials.

Future research should continue to explore the synthesis, characterization, and theoretical modeling of metalloborospherenes. Advanced computational methods, such as global minimum searches and machine learning potentials, will play a crucial role in predicting and understanding the properties of these complex systems. Additionally, experimental techniques such as photoelectron spectroscopy and crystallography will be essential for validating theoretical predictions and uncovering new structural motifs. The practical application of metalloborospherenes relies heavily on the feasibility of their macroscale synthesis. Current research in this area is predominantly theoretical and experimental at the nanoscale. However, advancements in synthetic techniques for boron-based nanostructures and metallofullerenes provide a promising foundation for scaling up production. One potential method for macroscale synthesis involves chemical vapor deposition (CVD), which has been successfully employed for producing other boron-rich nanomaterials [142,143]. By optimizing the parameters of CVD, such as temperature, pressure, and precursor composition, it may be possible to achieve controlled synthesis of metalloborospherenes on a larger scale. Another promising approach is the use of solution-based synthesis methods, including solvothermal and hydrothermal processes [144,145]. These methods can facilitate the formation of boron cages and their subsequent doping with metal atoms in a controlled environment. The development of suitable solvents and reaction conditions will be crucial for this approach. Additionally, template-assisted synthesis [146], where preformed templates guide the assembly of boron and metal atoms into the desired structures, could provide a pathway for producing metalloborospherenes at scale. This method has been used successfully for other complex nanostructures and could be adapted for metalloborospherenes. Overall, while the macro-scale synthesis of metalloborospherenes presents challenges, ongoing advancements in nanomaterial synthesis techniques hold promise for overcoming these obstacles, making the large-scale production of these materials feasible in the near future.

Overall, the study of metalloborospherenes is a dynamic and promising field that bridges the gap between boron chemistry and nanomaterials science. The insights gained from this research will not only advance our understanding of boron-based clusters but also pave the way for the development of innovative materials with unique properties and wide-ranging applications.

**Author Contributions:** Writing—original draft preparation, J.B. and H.P.; writing—review and editing, W.-L.L.; supervision, W.-L.L.; project administration, W.-L.L. All authors have read and agreed to the published version of the manuscript.

**Funding:** This research received no external funding.

**Acknowledgments:** W.-L. Li acknowledges startup funding from the Jacob School of Engineering, UCSD.

**Conflicts of Interest:** The authors declare no conflicts of interest.

## References

1. Kroto , H. The stability of the fullerenes $C_n$, with $n$ = 24, 28, 32, 36, 50, 60 and 70. *Nature* **1987**, *329*, 529–531. [CrossRef]
2. Lu, X.; Chen, Z. Curved pi-conjugation, aromaticity, and the related chemistry of small fullerenes ($<C_{60}$) and single-walled carbon nanotubes. *Chem. Rev.* **2005**, *105*, 3643–3696. [PubMed]
3. Hebard, A. Buckminsterfullerene. *Annu. Rev. Mater. Sci.* **1993**, *23*, 159–191. [CrossRef]
4. Kroto, H.W.; Heath, J.R.; O'Brien, S.C.; Curl, R.F.; Smalley, R.E. $C_{60}$: Buckminsterfullerene. *Nature* **1985**, *318*, 162–163. [CrossRef]
5. Yadav, B.; Kumar, R. Structure, properties and applications of fullerenes. *Int. J. Nanotechnol.* **2008**, *2*, 15–24.
6. Zhai, H.J.; Zhao, Y.F.; Li, W.L.; Chen, Q.; Bai, H.; Hu, H.S.; Piazza, Z.A.; Tian, W.J.; Lu, H.G.; Wu, Y.B.; et al. Observation of an all-boron fullerene. *Nat. Chem.* **2014**, *6*, 727–731. [CrossRef]
7. Li, W.L.; Chen, X.; Jian, T.; Chen, T.T.; Li, J.; Wang, L.S. From planar boron clusters to borophenes and metalloborophenes. *Nat. Rev. Chem.* **2017**, *1*, 71. [CrossRef]

8. Sergeeva, A.P.; Popov, I.A.; Piazza, Z.A.; Li, W.L.; Romanescu, C.; Wang, L.S.; Boldyrev, A.I. Understanding boron through size-selected clusters: Structure, chemical bonding, and fluxionality. *Acc. Chem. Res.* **2014**, *47*, 1349–1358. [CrossRef] [PubMed]
9. Popov, I.A.; Piazza, Z.A.; Li, W.L.; Wang, L.S.; Boldyrev, A.I. A combined photoelectron spectroscopy and ab initio study of the quasi-planar $B_{24}^-$ cluster. *J. Chem. Phys.* **2013**, *139*, 144307. [CrossRef]
10. Zhai, H.J.; Kiran, B.; Li, J.; Wang, L.S. Hydrocarbon analogues of boron clusters—Planarity, aromaticity and antiaromaticity. *Nat. Mater.* **2003**, *2*, 827–833. [CrossRef]
11. Zhao, J.; Huang, X.; Shi, R.; Liu, H.; Su, Y.; King, R.B. $B_{28}$: The smallest all-boron cage from an ab initio global search. *Nanoscale* **2015**, *7*, 15086–15090. [CrossRef] [PubMed]
12. Popov, A.A.; Yang, S.; Dunsch, L. Endohedral fullerenes. *Chem. Rev.* **2013**, *113*, 5989–6113. [CrossRef] [PubMed]
13. Lu, X.; Akasaka, T.; Nagase, S. Chemistry of endohedral metallofullerenes: The role of metals. *Chem. Commun.* **2011**, *47*, 5942–5957. [CrossRef] [PubMed]
14. Shinohara, H. Endohedral metallofullerenes. *Rep. Prog. Phys.* **2000**, *63*, 843. [CrossRef]
15. Wilson, L.J.; Cagle, D.W.; Thrash, T.P.; Kennel, S.J.; Mirzadeh, S.; Alford, J.M.; Ehrhardt, G.J. Metallofullerene drug design. *Coord. Chem. Rev.* **1999**, *190-192*, 199–207. [CrossRef]
16. Xu, J.; Li, Y.K.; Janssens, E.; Hou, G.L. Multifacets of fullerene–metal clusters: From fundamental to application. *Acc. Chem. Res.* **2024**, *57*, 1670–1683. [CrossRef] [PubMed]
17. Barroso, J.; Pan, S.; Merino, G. Structural transformations in boron clusters induced by metal doping. *Chem. Soc. Rev.* **2022**, *51*, 1098–1123. [CrossRef] [PubMed]
18. Chen, T.T.; Li, W.L.; Chen, W.J.; Yu, X.H.; Dong, X.R.; Li, J.; Wang, L.S. Spherical trihedral metallo-borospherenes. *Nat. Commun.* **2020**, *11*, 2766. [CrossRef] [PubMed]
19. Jian, T.; Chen, X.; Li, S.D.; Boldyrev, A.I.; Li, J.; Wang, L.S. Probing the structures and bonding of size-selected boron and doped-boron clusters. *Chem. Soc. Rev.* **2019**, *48*, 3550–3591. [CrossRef]
20. Li, W.L.; Chen, T.T.; Xing, D.H.; Chen, X.; Li, J.; Wang, L.S. Observation of highly stable and symmetric lanthanide octa-boron inverse sandwich complexes. *Proc. Natl. Acad. Sci. USA* **2018**, *115*, E6972–E6977. [CrossRef]
21. Chen, T.T.; Li, W.L.; Li, J.; Wang, L.S. $[La(\eta^x\text{-}B_x)La]^-$ ($x$ = 7–9): A new class of inverse sandwich complexes. *Chem. Sci.* **2019**, *10*, 2534–2542. [CrossRef] [PubMed]
22. Cai, W.; Chen, C.H.; Chen, N.; Echegoyen, L. Fullerenes as nanocontainers that stabilize unique actinide species inside: Structures, formation, and reactivity. *Acc. Chem. Res.* **2019**, *52*, 1824–1833. [CrossRef] [PubMed]
23. Cai, W.; Zhang, M.; Echegoyen, L.; Lu, X. Recent advances in endohedral metallofullerenes. *Fundam. Res.* 2023, *in press*. [CrossRef]
24. Wang, L.S. Photoelectron spectroscopy of size-selected boron clusters: From planar structures to borophenes and borospherenes. *Int. Rev. Phys.* **2016**, *35*, 69–142. [CrossRef]
25. Hirsch, A.; Chen, Z.; Jiao, H. Spherical aromaticity in $I_h$ symmetrical fullerenes: The $2(N+1)^2$ rule. *Angew. Chem. Int. Ed.* **2000**, *39*, 3915–3917. [CrossRef]
26. Pei, L.; Li, H.R.; Yan, M.; Chen, Q.; Mu, Y.W.; Lu, H.G.; Wu, Y.B.; Li, S.D. Charge-induced structural transition between seashell-like $B_{29}^-$ and $B_{29}^+$ in 18 $\pi$-electron configurations. *Phys. Chem. Chem. Phys.* **2018**, *20*, 15330–15334. [CrossRef]
27. Wang, Y.J.; Zhao, Y.F.; Li, W.L.; Jian, T.; Chen, Q.; You, X.R.; Ou, T.; Zhao, X.Y.; Zhai, H.J.; Li, S.D.; et al. Observation and characterization of the smallest borospherene, $B_{28}^-$ and $B_{28}$. *J. Chem. Phys.* **2016**, *144*, 064307. [CrossRef] [PubMed]
28. Li, H.R.; Jian, T.; Li, W.L.; Miao, C.Q.; Wang, Y.J.; Chen, Q.; Luo, X.M.; Wang, K.; Zhai, H.J.; Li, S.D.; et al. Competition between quasi-planar and cage-like structures in the $B_{29}^-$ cluster: Photoelectron spectroscopy and ab initio calculations. *Phys. Chem. Chem. Phys.* **2016**, *18*, 29147–29155. [CrossRef] [PubMed]
29. Chen, Q.; Li, W.L.; Zhao, Y.F.; Zhang, S.Y.; Hu, H.S.; Bai, H.; Li, H.R.; Tian, W.J.; Lu, H.G.; Zhai, H.J.; et al. Experimental and theoretical evidence of an axially chiral borospherene. *ACS Nano* **2015**, *9*, 754–760. [CrossRef]
30. Chen, Q.; Zhang, S.Y.; Bai, H.; Tian, W.J.; Gao, T.; Li, H.R.; Miao, C.Q.; Mu, Y.W.; Lu, H.G.; Zhai, H.J.; et al. Cage-like $B_{41}^+$ and $B_{42}^{2+}$ : New chiral members of the borospherene family. *Angew. Chem. Int. Ed.* **2015**, *54*, 8160–8164. [CrossRef]
31. Pei, L.; Yan, M.; Zhao, X.Y.; Mu, Y.W.; Lu, H.G.; Wu, Y.B.; Li, S.D. Sea-shell-like $B_{31}^+$ and $B_{32}$: Two new axially chiral members of the borospherene family. *RSC Adv.* **2020**, *10*, 10129–10133. [CrossRef]
32. Liu, H.; Chen, Q.; Li, H.R.; Zhao, X.Y.; Tian, X.X.; Mu, Y.W.; Lu, H.G.; Li, S.D. Aromatic cage-like $B_{34}$ and $B_{35}^+$ : New axially chiral members of the borospherene family. *Phys. Chem. Chem. Phys.* **2018**, *20*, 15344–15349. [CrossRef]
33. Liu, H.; Mu, Y.W.; Li, S.D. Axially Chiral Cage-like $B_{38}^+$ and $B_{38}^{2+}$: New aromatic members of the borospherene family. *J. Clust. Sci.* **2022** *33*, 81–87. [CrossRef]
34. Zhao, X.Y.; Chen, Q.; Li, H.R.; Mu, Y.W.; Lu, H.G.; Li, S.D. Cage-like $B_{39}^+$ clusters with the bonding pattern of $\sigma + \pi$ double delocalization: New members of the borospherene family. *Phys. Chem. Chem. Phys.* **2017**, *19*, 10998–11003. [CrossRef] [PubMed]
35. Li, H.R.; Chen, Q.; Tian, X.X.; Lu, H.G.; Zhai, H.J.; Li, S.D. Cage-like $B_{40}^+$ : A perfect borospherene monocation. *J. Mol. Model.* **2016**, *22*, 1–6. [CrossRef]
36. Zhang, M.; Lu, H.G.; Li, S.D. $B_{111}$, $B_{112}$, $B_{113}$, and $B_{114}$: The most stable core-shell borospherenes with an icosahedral $B_{12}$ core at the center exhibiting superatomic behaviors. *Nano Res.* **2021**, *14*, 4719–4724. [CrossRef]
37. Pei, L.; Yan, Q.Q.; Li, S.D. Predicting the structural transition in medium-sized boron nanoclusters: From bilayer $B_{64}$, $B_{66}$, $B_{68}$, $B_{70}$, and $B_{72}$ to core-shell $B_{74}$. *Eur. J. Inorg. Chem.* **2021**, *2021*, 2618–2624. [CrossRef]

38. Gonzalez Szwacki, N.; Sadrzadeh, A.; Yakobson, B.I. $B_{80}$ fullerene: An ab initio prediction of geometry, stability, and electronic structure. *Phys. Rev. Lett.* **2007**, *98*, 166804. [CrossRef] [PubMed]
39. Mukhopadhyay, S.; He, H.; Pandey, R.; Yap, Y.K.; Boustani, I. Novel spherical boron clusters and structural transition from 2D quasi-planar structures to 3D double-rings. *J. Phys. Conf. Ser.* **2009**, *176*, 012028. [CrossRef]
40. Shen, W.; Hu, S.; Lu, X. Endohedral metallofullerenes: New structures and unseen phenomena. *Chem. Eur. J.* **2020**, *26*, 5748–5757. [CrossRef]
41. Lu, X.; Feng, L.; Akasaka, T.; Nagase, S. Current status and future developments of endohedral metallofullerenes. *Chem. Soc. Rev.* **2012**, *41*, 7723–7760. [CrossRef] [PubMed]
42. Liu, F.; Spree, L.; Krylov, D.S.; Velkos, G.; Avdoshenko, S.M.; Popov, A.A. Single-electron lanthanide-lanthanide bonds inside fullerenes toward robust redox-active molecular magnets. *Acc. Chem. Res.* **2019**, *52*, 2981–2993. [CrossRef] [PubMed]
43. Hu, Z.; Yang, S. Endohedral metallofullerene molecular nanomagnets. *Chem. Soc. Rev.* **2024**, *53*, 2863–2897. [CrossRef] [PubMed]
44. Shannon, R.D. Revised effective ionic radii and systematic studies of interatomic distances in halides and chalcogenides. *Acta Crystallogr. A* **1976**, *32*, 751–767. [CrossRef]
45. Chen, T.T.; Li, W.L.; Chen, W.J.; Li, J.; Wang, L.S. $La_3B_{14}^-$ : An inverse triple-decker lanthanide boron cluster. *Chem. Commun.* **2019**, *55*, 7864–7867. [CrossRef] [PubMed]
46. Zhao, X.Y.; Yan, M.; Wei, Z.; Li, S.D. Donor–acceptor duality of the transition-metal-like B2 core in core–shell-like metallo-borospherenes $La_3\&[B_2@B_{17}]^-$ and $La_3\&[B_2@B_{18}]^-$. *RSC Adv.* **2020**, *10*, 34225–34230. [PubMed]
47. Jiang, Y.; Li, Z.; Wu, Y.; Wang, Z. Ln3@C80+ (Ln = lanthanide): A new class of stable metallofullerene cations with multicenter metal–metal bonding in the sub-nanometer confined space. *Inorg. Chem. Front.* **2022**, *9*, 2173–2181. [CrossRef]
48. Lu, X.Q.; Gao, C.Y.; Wei, Z.; Li, S.D. Cage-like $La_4B_{24}$ and Core-Shell $La_4B_{29}^{0/+/-}$: Perfect spherically aromatic tetrahedral metallo-borospherenes. *J. Mol. Model.* **2021**, *27*, 1–7. [CrossRef]
49. Guo, T.; Diener, M.D.; Chai, Y.; Alford, M.J.; Haufler, R.E.; McClure, S.M.; Ohno, T.; Weaver, J.H.; Scuseria, G.E.; Smalley, R.E. Uranium stabilization of $C_{28}$: A tetravalent fullerene. *Science* **1992**, *257*, 1661–1664. [CrossRef]
50. Xi, C.; Yang, L.; Liu, C.; You, P.; Li, L.; Jin, P. Lanthanide metals in the boron cages: Computational prediction of $M@B_n$ (M = Eu, Gd; n = 38, 40). *Int. J. Quantum Chem.* **2018**, *118*, e25576. [CrossRef]
51. Dunlap, B.I.; Haeberlen, O.D.; Roesch, N. Asymmetric localization of titanium in carbon molecule ($C_{28}$). *J. Phys. Chem.* **1992**, *96*, 9095–9097. [CrossRef]
52. Cao, B.; Hasegawa, M.; Okada, K.; Tomiyama, T.; Okazaki, T.; Suenaga, K.; Shinohara, H. EELS and $^{13}C$ NMR characterization of pure $Ti_2@C_{80}$ metallofullerene. *J. Am. Chem. Soc.* **2001**, *123*, 9679–9680. [CrossRef] [PubMed]
53. Cao, B.; Suenaga, K.; Okazaki, T.; Shinohara, H. Production, Isolation, and EELS Characterization of $Ti_2@C_{84}$ Dititanium Metallofullerenes. *J. Phys. Chem. B* **2002**, *106*, 9295–9298. [CrossRef]
54. Li, F.F.; Chen, N.; Mulet-Gas, M.; Triana, V.; Murillo, J.; Rodríguez-Fortea, A.; Poblet, J.M.; Echegoyen, L. $Ti_2S@D_{3h}(24109)$-$C_{78}$: A sulfide cluster metallofullerene containing only transition metals inside the cage. *Chem. Sci.* **2013**, *4*, 3404–3410. [CrossRef]
55. Yu, P.; Shen, W.; Bao, L.; Pan, C.; Slanina, Z.; Lu, X. Trapping an unprecedented $Ti_3C_3$ unit inside the icosahedral $C_{80}$ fullerene: A crystallographic survey. *Chem. Sci.* **2019**, *10*, 10925–10930. [CrossRef] [PubMed]
56. Yu, P.; Bao, L.; Yang, L.; Hao, D.; Jin, P.; Shen, W.; Fang, H.; Akasaka, T.; Lu, X. Crystallographic characterization of $Ti_2C_2@D_{3h}(5)$-$C_{78}$, $Ti_2C_2@C_{3v}(8)$-$C_{82}$, and $Ti_2C_2@C_s(6)$-$C_{82}$: Identification of unsupported $Ti_2C_2$ cluster with cage-dependent configurations. *Inorg. Chem.* **2020**, *59*, 9416–9423. [CrossRef] [PubMed]
57. Guan, R.; Chen, Z.C.; Huang, J.; Tian, H.R.; Xin, J.; Ying, S.W.; Chen, M.; Zhang, Q.; Li, Q.; Xie, S.Y.; et al. Self-driven carbon atom implantation into fullerene embedding metal–carbon cluster. *Proc. Natl. Acad. Sci. USA* **2022**, *119*, e2202563119. [CrossRef]
58. Johnson, R.D.; De Vries, M.S.; Salem, J.; Bethune, D.S.; Yannoni, C.S. Electron paramagnetic resonance studies of lanthanum-containing $C_{82}$. *Nature* **1992**, *355*, 239–240. [CrossRef]
59. Alvarez, L.; Pichler, T.; Georgi, P.; Schwieger, T.; Peisert, H.; Dunsch, L.; Hu, Z.; Knupfer, M.; Fink, J.; Bressler, P.; et al. Electronic structure of pristine and intercalated $Sc_3N@C_{80}$ metallofullerene. *Phys. Rev. B* **2002**, *66*, 035107. [CrossRef]
60. Campanera, J.M.; Bo, C.; Olmstead, M.M.; Balch, A.L.; Poblet, J.M. Bonding within the endohedral fullerenes $Sc_3N@C_{78}$ and $Sc_3N@C_{80}$ as determined by density functional calculations and reexamination of the crystal structure of $Sc_3N@C_{78}Co(OEP)$ 1.5($C_6H_6$)0.3($CHCl_3$). *J. Phys. Chem. A* **2002**, *106*, 12356–12364. [CrossRef]
61. Macchi, P.; Sironi, A. Chemical bonding in transition metal carbonyl clusters: Complementary analysis of theoretical and experimental electron densities. *Coord. Chem. Rev.* **2003**, *238–239*, 383–412. [CrossRef]
62. Popov, A.A.; Dunsch, L. Bonding in endohedral metallofullerenes as studied by quantum theory of atoms in molecules. *Chem. Eur. J.* **2009**, *15*, 9707–9729. [CrossRef] [PubMed]
63. Romanescu, C.; Galeev, T.R.; Li, W.L.; Boldyrev, A.I.; Wang, L.S. Aromatic metal-centered monocyclic boron rings: $Co©B_8^-$ and $Ru©B_9^-$. *Angew. Chem. Int. Ed.* **2011**, *50*, 9334–9337. [CrossRef] [PubMed]
64. Chen, T.T.; Li, W.L.; Bai, H.; Chen, W.J.; Dong, X.R.; Li, J.; Wang, L.S. $Re©B_8^-$ and $Re©B_9^-$: New members of the transition-metal-centered borometallic molecular wheel family. *J. Phys. Chem. A* **2019**, *123*, 5317–5324. [CrossRef] [PubMed]
65. Romanescu, C.; Galeev, T.R.; Li, W.L.; Boldyrev, A.I.; Wang, L.S. Transition-metal-centered monocyclic boron wheel clusters ($M©B_n$): A new class of aromatic borometallic compounds. *Acc. Chem. Res.* **2013**, *46*, 350–358. [CrossRef] [PubMed]

66. Popov, I.A.; Li, W.L.; Piazza, Z.A.; Boldyrev, A.I.; Wang, L.S. Complexes between planar boron clusters and transition metals: A photoelectron spectroscopy and ab initio study of $CoB_{12}^-$ and $RhB_{12}^-$. *J. Phys. Chem. A* **2014**, *118*, 8098–8105. [CrossRef] [PubMed]
67. Popov, I.A.; Jian, T.; Lopez, G.V.; Boldyrev, A.I.; Wang, L.S. Cobalt-centred boron molecular drums with the highest coordination number in the $CoB_{16}^-$ cluster. *Nat. Commun.* **2015**, *6*, 8654. [CrossRef] [PubMed]
68. Jian, T.; Li, W.L.; Popov, I.A.; Lopez, G.V.; Chen, X.; Boldyrev, A.I.; Li, J.; Wang, L.S. Manganese-centered tubular boron cluster–$MnB_{16}^-$: A new class of transition-metal molecules. *J. Chem. Phys.* **2016**, *144*, 154310. [CrossRef] [PubMed]
69. Jian, T.; Li, W.L.; Chen, X.; Chen, T.T.; Lopez, G.V.; Li, J.; Wang, L.S. Competition between drum and quasi-planar structures in $RhB_{18}^-$: Motifs for metallo-boronanotubes and metallo-borophenes. *Chem. Sci.* **2016**, *7*, 7020–7027. [CrossRef]
70. Li, W.L.; Jian, T.; Chen, X.; Li, H.R.; Chen, T.T.; Luo, X.M.; Li, S.D.; Li, J.; Wang, L.S. Observation of a metal-centered $B_2$-$Ta@B_{18}^-$ tubular molecular rotor and a perfect $Ta@B_{20}^-$ boron drum with the record coordination number of twenty. *Chem. Commun.* **2017**, *53*, 1587–1590. [CrossRef]
71. Yan, L. Large $B_7$ triangles in hollow spherical trihedral metallo-borospherenes and their endohedral complexes of $B_{20}TM_n$ (TM = Sc, Y; n = 3, 4): A theoretical characterization. *Inorg. Chem.* **2022**, *61*, 10652–10660. [CrossRef] [PubMed]
72. Li, H.R.; Liu, H.; Lu, X.Q.; Zan, W.Y.; Tian, X.X.; Lu, H.G.; Wu, Y.B.; Mu, Y.W.; Li, S.D. Cage-like $Ta@B_n^q$ complexes (n = 23–28, q = −1 − +3) in 18-electron configurations with the highest coordination number of twenty-eight. *Nanoscale* **2018**, *10*, 7451–7456. [CrossRef] [PubMed]
73. Li, H.R.; Liu, H.; Tian, X.X.; Zan, W.Y.; Mu, Y.W.; Lu, H.G.; Li, J.; Wang, Y.K.; Li, S.D. Structural transition in metal-centered boron clusters: From tubular molecular rotors $Ta@B_{21}$ and $Ta@B_{22}^+$ to cage-like endohedral metalloborospherene $Ta@B_{22}^-$. *Phys. Chem. Chem. Phys.* **2017**, *19*, 27025–27030. [CrossRef] [PubMed]
74. Li, H.R.; Tian, X.X.; Luo, X.M.; Yan, M.; Mu, Y.W.; Lu, H.G.; Li, S.D. Heteroborospherene clusters $Ni_n{\in}B_{40}$ (n = 1–4) and heteroborophene monolayers $Ni_2{\in}B_{14}$ with planar heptacoordinate transition-metal centers in $\eta^7$-$B_7$ heptagons. *Sci. Rep.* **2017**, *7*, 5701. [CrossRef] [PubMed]
75. Zhang, Y.; Zhao, X.Y.; Yan, M.; Li, S.D. From inverse sandwich $Ta_2B_7^+$ and $Ta_2B_8$ to spherical trihedral $Ta_3B_{12}^-$: Prediction of the smallest metallo-borospherene. *RSC Adv.* **2020**, *10*, 29320–29325. [CrossRef] [PubMed]
76. Bai, H.; Chen, Q.; Zhai, H.J.; Li, S.D. Endohedral and exohedral metalloborospherenes: $M@B_{40}$ (M = Ca, Sr) and $M\&B_{40}$ (M = Be, Mg). *Angew. Chem. Int. Ed.* **2015**, *54*, 941–945.
77. Yan, L. Expanded spherical trihedral metallo-borospherenes of transition-metal doped boron clusters: $TM_3B_{15}^q$ (TM = Zr, Hf; q = −1, 0, +1). *Results Phys.* **2022**, *33*, 105214. [CrossRef]
78. Li, C.G.; Shen, Z.G.; Zhang, J.; Cui, Y.Q.; Li, J.J.; Xue, H.Y.; Li, H.F.; Ren, B.Z.; Hu, Y.F. Analysis of the structures, stabilities and electronic properties of $MTa@B_{16}^-$ (M = V, Cr, Mn, Fe, Co, Ni) clusters and assemblies. *New J. Chem.* **2020**, *44*, 5109–5119. [CrossRef]
79. Zhang, Y.; Lu, X.Q.; Yan, M.; Li, S.D. Perfect spherical tetrahedral metallo-borospherene $Ta_4B_{18}$ as a superatom following the 18-electron rule. *ACS Omega* **2021**, *6*, 10991–10996. [CrossRef]
80. Li, S.X.; Yang, Y.J.; Chen, D.L.; Long, Z.W. Structures, and electronic and spectral properties of single-atom transition metal-doped boron clusters $MB_{24}^-$ (M = Sc, Ti, V, Cr, Mn, Fe, Co, and Ni). *RSC Adv.* **2022**, *12*, 16706–16716. [CrossRef]
81. An, Y.; Zhang, M.; Wu, D.; Fu, Z.; Wang, T.; Xia, C. Electronic transport properties of the first all-boron fullerene $B_{40}$ and its metallofullerene Sr@ $B_{40}$. *Phys. Chem. Chem. Phys.* **2016**, *18*, 12024–12028. [CrossRef] [PubMed]
82. Fa, W.; Chen, S.; Pande, S.; Zeng, X.C. Stability of metal-encapsulating boron fullerene $B_{40}$. *J. Phys. Chem. A* **2015**, *119*, 11208–11214. [CrossRef] [PubMed]
83. Jin, P.; Yang, L.; Liu, C.; Hou, Q.; Li, L. Computational prediction of the endohedral metalloborofullerenes $Ti_n@B_{40}$ (n = 1, 2). *Theor. Chem. Acc.* **2017**, *136*, 1–12. [CrossRef]
84. Mikheev, N. Lower oxidation states of lanthanides and actinides. *Inorg. Chim. Acta.* **1984**, *94*, 241–248. [CrossRef]
85. MacDonald, M.R.; Fieser, M.E.; Bates, J.E.; Ziller, J.W.; Furche, F.; Evans, W.J. Identification of the +2 oxidation state for uranium in a crystalline molecular complex, [K(2.2.2-Cryptand)][$(C_5H_4SiMe_3)_3U$]. *J. Am. Chem. Soc.* **2013**, *135*, 13310–13313. [CrossRef] [PubMed]
86. La Pierre, H.S.; Scheurer, A.; Heinemann, F.W.; Hieringer, W.; Meyer, K. Synthesis and characterization of a uranium(II) monoarene complex supported by $\delta$ backbonding. *Angew. Chem. Int. Ed.* **2014**, *53*, 7158–7162. [CrossRef]
87. Hayton, T.W. Recent developments in actinide–ligand multiple bonding. *Chem. Commun.* **2013**, *49*, 2956–2973. [CrossRef] [PubMed]
88. Akiyama, K.; Zhao, Y.; Sueki, K.; Tsukada, K.; Haba, H.; Nagame, Y.; Kodama, T.; Suzuki, S.; Ohtsuki, T.; Sakaguchi, M.; et al. Isolation and characterization of light actinide metallofullerenes. *J. Am. Chem. Soc.* **2001**, *123*, 181–182. [CrossRef] [PubMed]
89. Zhang, X.; Li, W.; Feng, L.; Chen, X.; Hansen, A.; Grimme, S.; Fortier, S.; Sergentu, D.C.; Duignan, T.J.; Autschbach, J.; et al. A diuranium carbide cluster stabilized inside a $C_{80}$ fullerene cage. *Nat. Commun.* **2018**, *9*, 2753. [CrossRef]
90. Zhuang, J.; Abella, L.; Sergentu, D.C.; Yao, Y.R.; Jin, M.; Yang, W.; Zhang, X.; Li, X.; Zhang, D.; Zhao, Y.; et al. Diuranium(IV) carbide cluster $U_2C_2$ stabilized inside fullerene cages. *J. Am. Chem. Soc.* **2019**, *141*, 20249–20260. [CrossRef]
91. Yao, Y.R.; Qiu, J.; Zheng, L.; Jiang, H.; Xia, Y.; Chen, N. Endohedral Fullerenes. In *Atomically Precise Nanochemistry*; John Wiley & Sons, Ltd.: Hoboken, NJ, USA, 2023; Chapter 11, pp. 331–371.

92. Meng, Q.; Abella, L.; Yang, W.; Yao, Y.R.; Liu, X.; Zhuang, J.; Li, X.; Echegoyen, L.; Autschbach, J.; Chen, N. UCN@$C_s(6)$-$C_{82}$: An encapsulated triangular UCN Cluster with ambiguous U oxidation state [U(III) versus U(I)]. *J. Am. Chem. Soc.* **2021**, *143*, 16226–16234. [CrossRef] [PubMed]
93. Hu, S.X.; Chen, M.; Ao, B. Theoretical studies on the oxidation states and electronic structures of actinide-borides: $AnB_{12}$ (An = Th–Cm) clusters. *Phys. Chem. Chem. Phys.* **2018**, *20*, 23856–23863. [CrossRef] [PubMed]
94. Wang, J.; Zhang, N.X.; Wang, C.Z.; Wu, Q.Y.; Lan, J.H.; Chai, Z.F.; Nie, C.M.; Shi, W.Q. Theoretical probing of twenty-coordinate actinide-centered boron molecular drums. *Phys. Chem. Chem. Phys.* **2021**, *23*, 26967–26973. [CrossRef] [PubMed]
95. Wang, J.; Wang, C.Z.; Wu, Q.Y.; Lan, J.H.; Chai, Z.F.; Nie, C.M.; Shi, W.Q. Construction of the largest metal-centered double-ring tubular boron clusters based on actinide metal doping. *J. Phys. Chem. A* **2022**, *126*, 3445–3451. [CrossRef] [PubMed]
96. Hu, S.X.; Zhang, P.; Zou, W.; Zhang, P. New theoretical insights into high-coordination-number complexes in actinides-centered borane. *Nanoscale* **2020**, *12*, 15054–15065. [CrossRef] [PubMed]
97. Li, W.L.; Chen, T.T.; Chen, W.J.; Li, J.; Wang, L.S. Monovalent lanthanide (I) in borozene complexes. *Nat. Commun.* **2021**, *12*, 6467. [CrossRef] [PubMed]
98. Wang, B.; Li, W.L. Revisiting the quasi-aromaticity in polynuclear metal chalcogenide clusters and their derivative "cluster-assembly" crystalline structures. *Phys. Chem. Chem. Phys.* **2024**, *26*, 17370–17382. [CrossRef]
99. Wang, C.Z.; Bo, T.; Lan, J.H.; Wu, Q.Y.; Chai, Z.F.; Gibson, J.K.; Shi, W.Q. Ultrastable actinide endohedral borospherenes. *Chem. Commun.* **2018**, *54*, 2248–2251. [CrossRef]
100. Yu, T.; Gao, Y.; Xu, D.; Wang, Z. Actinide endohedral boron clusters: A closed-shell electronic structure of U@$B_{40}$. *Nano Res.* **2018**, *11*, 354–359. [CrossRef]
101. Zhang, N.; Li, A.; Wang, C.; Wu, Q.; Lan, J.; Chai, Z.; Zhao, Y.; Shi, W. Theoretical prediction of chiral actinide endohedral borospherenes. *New J. Chem.* **2021**, *45*, 6803–6810. [CrossRef]
102. Zhang, N.X.; Wang, C.Z.; Lan, J.H.; Wu, Q.Y.; Chai, Z.F.; Shi, W.Q. Actinide-doped boron clusters: From borophenes to borospherenes. *Phys. Chem. Chem. Phys.* **2022**, *24*, 29705–29711. [CrossRef] [PubMed]
103. Wales, D.J.; Doye, J.P.K. Global optimization by basin-hopping and the lowest energy structures of Lennard-Jones clusters containing up to 110 atoms. *J. Phys. Chem. A* **1997**, *101*, 5111–5116. [CrossRef]
104. Zhao, Y.; Chen, X.; Li, J. TGMin: A global-minimum structure search program based on a constrained basin-hopping algorithm. *Nano Res.* **2017**, *10*, 3407–3420. [CrossRef]
105. White, R.P.; Mayne, H.R. An investigation of two approaches to basin hopping minimization for atomic and molecular clusters. *Chem. Phys. Lett.* **1998**, *289*, 463–468. [CrossRef]
106. Liberti, L.; Maculan, N. *Global Optimization: From Theory to Implementation;* Springer Science & Business Media: Berlin/Heidelberg, Germany, 2006; Volume 84.
107. Chen, X.; Zhao, Y.F.; Zhang, Y.Y.; Li, J. TGMin: An efficient global minimum searching program for free and surface-supported clusters. *J. Comput. Chem.* **2019**, *40*, 1105–1112. [CrossRef]
108. Shang, C.; Liu, Z.P. Constrained Broyden minimization combined with the dimer method for locating transition state of complex reactions. *J. Chem. Theory Comput.* **2010**, *6*, 1136–1144. [CrossRef]
109. Shang, C.; Liu, Z.P. Constrained Broyden dimer method with bias potential for exploring potential energy surface of multistep reaction process. *J. Chem. Theory Comput.* **2012**, *8*, 2215–2222. [CrossRef] [PubMed]
110. Shang, C.; Liu, Z.P. Stochastic surface walking method for structure prediction and pathway searching. *J. Chem. Theory Comput.* **2013**, *9*, 1838–1845. [CrossRef]
111. Averkiev, B. *Geometry and Electronic Structure of Doped Clusters via the Coalescence Kick Method;* Utah State University: Logan, UT, USA, 2009.
112. Saunders, M. Stochastic search for isomers on a quantum mechanical surface. *J. Comput. Chem.* **2004**, *25*, 621–626. [CrossRef]
113. Bera, P.P.; Sattelmeyer, K.W.; Saunders, M.; Schaefer, H.F.; Schleyer, P.v.R. Mindless chemistry. *J. Phys. Chem. A* **2006**, *110*, 4287–4290. [CrossRef]
114. Wang, Y.; Lv, J.; Zhu, L.; Ma, Y. Crystal structure prediction via particle-swarm optimization. *Phys. Rev. B* **2010**, *82*, 094116. [CrossRef]
115. Kennedy, J.; Eberhart, R. Particle swarm optimization. In Proceedings of the ICNN'95-International Conference on Neural Networks, Perth, WA, Australia, 27 November–1 December 1995; Volume 4, pp. 1942–1948.
116. Yen, G.; Lu, H. Dynamic population strategy assisted particle swarm optimization. In Proceedings of the 2003 IEEE International Symposium on Intelligent Control, Houston, TX, USA, 8 October 2003; pp. 697–702.
117. Frisch, M.J.; Trucks, G.W.; Schlegel, H.B.; Scuseria, G.E.; Robb, M.A.; Cheeseman, J.R.; Scalmani, G.; Barone, V.; Petersson, G.A.; Nakatsuji, H.; et al. *Gaussian 16 Revision C.01;* Gaussian Inc.: Wallingford, CT, USA, 2016.
118. ADF 2024.1, SCM. Vrije Universiteit, Amsterdam, The Netherlands. Available online: http://www.scm.com (accessed on 25 June 2024).
119. te Velde, G.; Bickelhaupt, F.M.; Baerends, E.J.; Fonseca Guerra, C.; van Gisbergen, S.J.A.; Snijders, J.G.; Ziegler, T. Chemistry with ADF. *J. Comput. Chem.* **2001**, *22*, 931–967. [CrossRef]
120. Behler, J.; Parrinello, M. Generalized neural-network representation of high-dimensional potential-energy surfaces. *Phys. Rev. Lett.* **2007**, *98*, 146401. [CrossRef] [PubMed]
121. Rupp, M.; Tkatchenko, A.; Müller, K.R.; von Lilienfeld, O.A. Fast and accurate modeling of molecular atomization energies with machine learning. *Phys. Rev. Lett.* **2012**, *108*, 058301. [CrossRef] [PubMed]

122. Schütt, K.T.; Arbabzadah, F.; Chmiela, S.; Müller, K.R.; Tkatchenko, A. Quantum-chemical insights from deep tensor neural networks. *Nat. Commun.* **2017**, *8*, 13890. [CrossRef] [PubMed]
123. Smith, J.S.; Isayev, O.; Roitberg, A.E. ANI-1: An extensible neural network potential with DFT accuracy at force field computational cost. *Chem. Sci.* **2017**, *8*, 3192–3203. [CrossRef] [PubMed]
124. Bartók, A.P.; Payne, M.C.; Kondor, R.; Csányi, G. Gaussian approximation potentials: The accuracy of quantum mechanics, without the electrons. *Phys. Rev. Lett.* **2010**, *104*, 136403. [CrossRef]
125. Chmiela, S.; Tkatchenko, A.; Sauceda, H.E.; Poltavsky, I.; Schütt, K.T.; Müller, K.R. Machine learning of accurate energy-conserving molecular force fields. *Sci. Adv.* **2017**, *3*, e1603015. [CrossRef] [PubMed]
126. Zhang, L.; Han, J.; Wang, H.; Car, R.; E, W. Deep potential molecular dynamics: A scalable Model with the accuracy of quantum mechanics. *Phys. Rev. Lett.* **2018**, *120*, 143001. [CrossRef]
127. Deringer, V.L.; Caro, M.A.; Csányi, G. Machine learning interatomic potentials as emerging tools for materials science. *Adv. Mat.* **2019**, *31*, 1902765.
128. Li, J.; Li, X.; Zhai, H.J.; Wang, L.S. $Au_{20}$: A tetrahedral cluster. *Science* **2003**, *299*, 864–867. [CrossRef]
129. Bauernschmitt, R.; Ahlrichs, R. Treatment of electronic excitations within the adiabatic approximation of time dependent density functional theory. *Chem. Phys. Lett.* **1996**, *256*, 454–464. [CrossRef]
130. Li, W.L.; Liu, H.T.; Jian, T.; Lopez, G.V.; Piazza, Z.A.; Huang, D.L.; Chen, T.T.; Su, J.; Yang, P.; Chen, X.; et al. Bond-bending isomerism of $Au_2I_3^-$: Competition between covalent bonding and aurophilicity. *Chem. Sci.* **2016**, *7*, 475–481. [PubMed]
131. Gritsenko, O.; Schipper, P.; Baerends, E. Approximation of the exchange-correlation Kohn–Sham potential with a statistical average of different orbital model potentials. *Chem. Phys. Lett.* **1999**, *302*, 199–207. [CrossRef]
132. Stowasser, R.; Hoffmann, R. What do the Kohn- Sham orbitals and eigenvalues mean? *J. Am. Chem. Soc.* **1999**, *121*, 3414–3420. [CrossRef]
133. Mitoraj, M.P.; Michalak, A.; Ziegler, T. A combined charge and energy decomposition scheme for bond analysis. *J. Chem. Theory Comput.* **2009**, *5*, 962–975. [CrossRef]
134. Frenking, G.; Matthias Bickelhaupt, F. The EDA Perspective of Chemical Bonding. In *The Chemical Bond*; John Wiley & Sons, Ltd.: Hoboken, NJ, USA, 2014; Chapter 4, pp. 121–157.
135. Zhao, L.; Hermann, M.; Schwarz, W.E.; Frenking, G. The Lewis electron-pair bonding model: Modern energy decomposition analysis. *Nat. Rev. Chem.* **2019**, *3*, 48–63. [CrossRef]
136. Dong, X.; Liu, Y.Q.; Tiznado, W.; Cabellos-Quiroz, J.L.; Zhao, J.; Pan, S.; Cui, Z.H. Designing a Four-Ring Tubular Boron Motif through Metal Doping. *Inorg. Chem.* **2022**, *61*, 14553–14559. [CrossRef] [PubMed]
137. Poater, J.; Duran, M.; Sola, M.; Silvi, B. Theoretical evaluation of electron delocalization in aromatic molecules by means of atoms in molecules (AIM) and electron localization function (ELF) topological approaches. *Chem. Rev.* **2005**, *105*, 3911–3947. [CrossRef]
138. Kobayashi, K.; Nagase, S. Bonding features in endohedral metallofullerenes. Topological analysis of the electron density distribution. *Chem. Phys. Lett.* **1999**, *302*, 312–316. [CrossRef]
139. Kobayashi, K.; Nagase, S. A stable unconventional structure of $Sc_2@C_{66}$ found by density functional calculations. *Chem. Phys. Lett.* **2002**, *362*, 373–379. [CrossRef]
140. Du, J.; Jiang, G. Theoretical characterization of the endohedral metalloborospherenes $M@B_{36}$ (M = Ti, Zr, Hf, Ce, Th, $Pa^+$, $U^{2+}$, $Np^{3+}$, and $Pu^{4+}$). *J. Mol. Liq.* **2020**, *319*, 114088. [CrossRef]
141. Zubarev, D.Y.; Boldyrev, A.I. Developing paradigms of chemical bonding: Adaptive natural density partitioning. *Phys. Chem. Chem. Phys.* **2008**, *10*, 5207–5217. [CrossRef] [PubMed]
142. Tian, J.; Xu, Z.; Shen, C.; Liu, F.; Xu, N.; Gao, H.J. One-dimensional boron nanostructures: Prediction, synthesis, characterizations, and applications. *Nanoscale* **2010**, *2*, 1375–1389. [CrossRef] [PubMed]
143. Yin, J.; Li, J.; Hang, Y.; Yu, J.; Tai, G.; Li, X.; Zhang, Z.; Guo, W. Boron Nitride Nanostructures: Fabrication, Functionalization and Applications. *Small* **2016**, *12*, 2942–2968. [CrossRef] [PubMed]
144. Yang, M.; Jin, H.; Sun, Z.; Gui, R. Experimental synthesis, functionalized modifications and potential applications of monoelemental zero-dimensional boron nanomaterials. *J. Mater. Chem. A* **2022**, *10*, 5111–5146. [CrossRef]
145. Tan, C.; Cao, X.; Wu, X.J.; He, Q.; Yang, J.; Zhang, X.; Chen, J.; Zhao, W.; Han, S.; Nam, G.H.; et al. Recent Advances in Ultrathin Two-Dimensional Nanomaterials. *Chem. Rev.* **2017**, *117*, 6225–6331. [CrossRef]
146. Rycenga, M.; Camargo, P.H.C.; Xia, Y. Template-assisted self-assembly: A versatile approach to complex micro- and nanostructures. *Soft Matter* **2009**, *5*, 1129–1136. [CrossRef]

*inorganics* MDPI

*Review*

# Endometallofullerenes in the Gas Phase: Progress and Prospect

**Yameng Hou and Xianglei Kong ***

State Key Laboratory of Elemento-Organic Chemistry, Frontiers Science Center for New Organic Matter, Tianjin Key Laboratory of Biosensing and Molecular Recognition, College of Chemistry, Nankai University, Tianjin 300071, China; 2120210890@mail.nankai.edu.cn
* Correspondence: kongxianglei@nankai.edu.cn

**Abstract:** This review describes the progress of the gas-phase study of endometallofullerenes (EMFs) by mass spectrometry and theoretical calculation over the past 15 years. The attention herein focuses on the gas-phase syntheses, reactions, and generation mechanisms of some novel EMF ions, along with their structures and properties. The highlighted new species include EMFs with small-size carbon cages of $C_{2n}$ ($n < 60$), multiple metal atoms ($M_x@C_{2n}$, $x \geq 3$), late transition metals, and encaged ionic bonds. Furthermore, the gas-phase experimental and calculational supports for top-down or bottom-up models are summarized and discussed. These gas-phase results not only provide experimental evidence for the existence of related novel EMF species and possible synthesis methods for them, but they also provide new insights about chemical bonds in restricted space. In addition, the opportunities and further development directions faced by gas-phase EMF study are anticipated.

**Keywords:** endometallofullerenes; gas phase; electronic properties; mass spectrometry

## 1. Introduction

Interestingly, the history of endometallofullerenes (EMFs) is almost as long as that of fullerene [1,2]. Just one week after the wonderful experiment on the magic cluster ion of $C_{60}^+$, ions of $LaC_{2n}^+$ (2n = 46–70) were observed in the laser ablation mass spectrum of the La-loaded graphite by Smalley and his co-workers. The endohedral structure proposal was further supported by the "shrink wrapping" experiment and reaction test with oxygen or ammonia gas [3,4]. About six years later, which is one year after the success of the synthetization of solid $C_{60}$ [5], solid EMFs of $La@C_{82}$ and $La_2@C_{80}$ were also synthesized by the laser ablation method and the Krätschmer–Huffman method [4,6]. The number of known EMFs has been rapidly and constantly growing in the past twenty years, which is mainly attributed to the improvement of synthesis, separation, and analysis methods [7–10]. Carbon cages are now able to encapsulate metal atoms in different forms, such as metallic nitrides, oxides, sulfides, and carbides. Because the cage can serve as an ideal nanoscale container for research on exceptional structures and unique clusters or chemical bonds, EMFs have great potential for showcasing new physical and chemical properties, as well as molecular devices [11–14].

However, due to the great success and rapid development of solid-state synthesis technology for EMFs, there was a time when interest in gas-phase EMF studies decreased. Notably, progress in new materials such as graphene and mass spectrometry technology bring new opportunities for gas-phase research on EMFs [15]. It can be seen that in the past 15 years, gas-phase EMF study has mainly focused on two aspects. The first one is the discovery of novel EMFs, which are difficult to prepare using existing methods and are often considered very difficult or impossible to generate, such as EMFs with small cages, with multiple metal atoms, and with transition metals or ionic bonds. This type of research can inform researchers about the stability of these novel EMFs and provide methodological recommendations for their synthesis. The second one is the generation mechanism of EMFs, which helps us to understand the generation process of EMF in principle and provides guidance for new methods aimed at EMF synthesis. Both aspects will be summarized and discussed in this review. In summary, we focus on the discovery of new endohedral species in gas-phase experiments, which is typically related to the method of laser ablation mass spectrometry. The structures and properties of important species, mostly based on theoretical calculations, are also included. The mechanisms of generation of these novel EMFs are summarized and discussed in the last section, and an outlook on the prospects of gas-phase EMF studies is presented, too.

## 2. Experimental Method

From both historical and practical perspectives, gas-phase EMF study is mainly performed with mass spectrometry technology, combined with different cluster ion sources. Considering the diversity of ion species generated and the isotopic distribution of ions, high-resolution mass spectrometry is necessary in these experiments. Although high-resolution time-of-flight mass spectrometers are also applied, more experiments are conducted on Fourier-transform ion cyclotron resonance (FT-ICR) mass spectrometers, which have higher resolution and stronger capabilities for ion trapping and manipulation [16]. For the generation of the EMF ions to be studied, the Smalley-type laser-vaporization supersonic expansion cluster beam source is widely used, and there are also many kinds of variations of this type of source [1,2,17]. Generally, a rotatable solid target disk is vaporized by a focused beam from a YAG laser at 532 nm. In the source, pulsed helium gas should be adjusted in timing and pressure to coincide with the laser shot. However, there are also other cluster sources based on laser ablation in vacuum without using the accompanying pulsed helium gas. Two typical experimental setups applied by Dunk et al. in Florida State University [18–20] and Kong et al. in Nankai University [21–23] are shown in Figure 1a,b, respectively. In Dunk et al.'s setup, the cluster source block (shown in the upper corner of Figure 1a) includes vertical channels for helium gas and the laser beam. The other channel, located downstream from the target, is aligned with the gas channel to fulfill the free jet expansion and introduction of clusters into octopoles after the skimmer [18]. Dissimilarly, Kong et al. directly applied a commercial-matrix-assisted laser desorption/ionization (MALDI) [24] source for the laser ablation experiments without modifying any hardware [23]. In their experiments, the precursor was deposited on a metal plate (the MALDI plate) in turn. A focused 355 nm YAG laser beam was applied. Although a pulsed valve is also used in the source chamber (Figure 1b), its function is mainly to help in the accumulation of the generated ions in the hexapole. It is worth noting that although both experimental devices were implemented on FT-ICR mass spectrometers, the EMF species produced and the dependence of their yields on experimental conditions vary in the two setups due to their different cluster source designs and precursor materials used. This will be reflected in the examples mentioned in the following sections. For the generated EMF ions, they can be mass selected in the cell with the method of stored waveform inverse Fourier transform (SWIFT) [25] and can be further studied by other tandem mass spectrometric methods, including collision-induced dissociation (CID), ultraviolet photodissociation (UVPD), or reaction tests with reactive gas molecules (for example, $O_2$) [3,4,26].

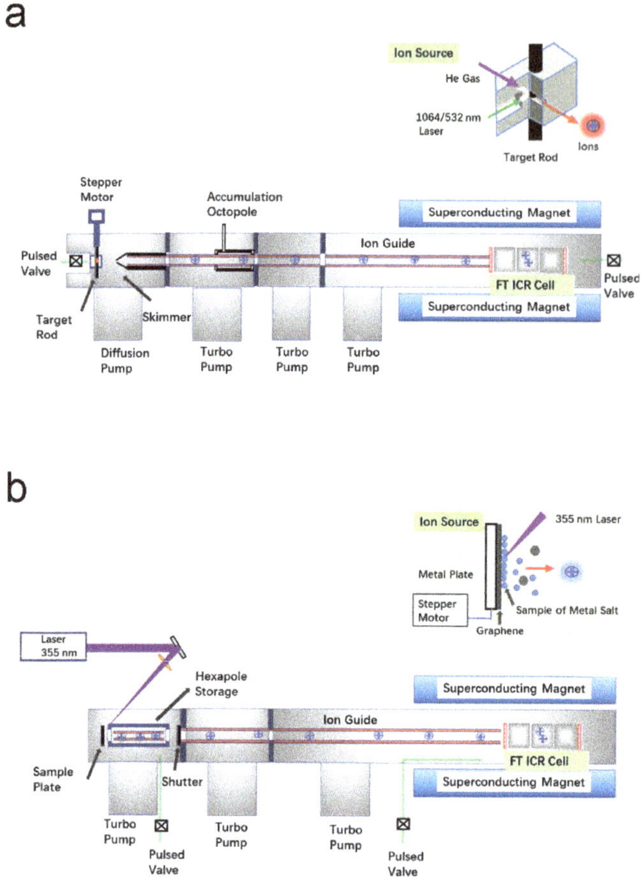


**Figure 1.** (**a**) Experimental setup for gas-phase EMF studies applied at Florida State University [18–20]. A 9.4 T FT-ICR mass spectrometer combined with a self-designed cluster source was applied. The cut-away view of the source block is shown in the upper right of the graph. The vaporization of target rods is achieved by laser shots at 532 nm, in conjunction with the opening of the pulsed valve. (**b**) Experimental setup for gas-phase EMF study applied in Nankai University [21–23]. A 7.0 T FT-ICR mass spectrometer with a commercial ion source (Varian IonSpec ProMALDI) was applied in these experiments without any modification. The precursor is deposited on the metal plate and irradiated by the 355 nm laser. The metal plate is controlled by a stepper motor and moves in the x–y plane during experiments.

## 3. Novel EMFs in the Gas Phase

### *3.1. EMFs with Small-Size Carbon Cages of $C_{2n}$ ($n < 60$)*

As closed-cage carbon molecules, the theoretically predicted smallest fullerene is $C_{20}$. However, for a very long time, questions regarding the existence of this structure have been raised due to its high curvature and reactivity. Prinzbach et al. reported the short-lived species $C_{20}Br_{20}$ in 1993 [27] and showed in 2000 that the fullerene $C_{20}$ with a cage structure can be produced from $C_{20}HBr_{13}$ by gas-phase debromination [28]. Calculation examples show that doping the metal atoms in $C_{20}$ can change its electronic structure [29–35]. On the other hand, Tomanek and Schluter predicted that closed carbon cages show a larger stability than other structures for clusters with sizes larger than 20, based on their theoretical calculations [36], and later Kent et al. predicted that the smallest stable fullerenes should be the $C_{26}$ and $C_{28}$ clusters [37]. Thus, the possible stabilization of $C_{28}$ by an endohedral species has been a very intriguing topic and has attracted the interest of many researchers [33,38–43]. The first experimental evidence in support of the idea was reported by Smalley et al. in 1992, in which they investigated the mass spectra of soot prepared by laser furnace and arc discharge methods [43]. Shrink-wrapping experiments indicated their endohedral structures, in which the fragmentation of larger endohedral uranofullerenes appeared to dissociate by only $C_2$ loss. The authors claimed that a tetravalent internal metal atom could donate four electrons to the fullerene to afford a closed-shell $M^{4+}@C_{28}{}^{4-}$ species with a large HOMO−LUMO gap and permit sufficient stabilization for such species [43].

By using a pulsed laser vaporization cluster source, Kroto and Marshal et al. have also reported that $C_{28}$ fullerene could be stabilized by encapsulation with appropriate metal forms [44]. Impressively, for the first time, they reported the high-resolution mass spectrum of Ti@$C_{28}$ (Figure 2). For a long time, titanium has been one of the most attractive tetravalent candidates for $C_{28}$ stabilization by encapsulation. However, its major isotope, $^{48}$Ti, which is just 52 mDa different in mass from $C_4$, makes the ion of Ti@$C_{28}$ hard to resolve from the signal of empty-cage $C_{32}$ with conventional mass spectrometers. Using a 9.0-T Fourier-transform ion cyclotron resonance mass spectrometry (FT-ICR) mass spectrometer, Kroto and Marshal et al. clearly identified the species, along with ions of Ti@$C_{2n}$ (n = 19–23), in the mass spectrum [44]. Collision-induced dissociation (CID) experiments on Ti@$C_{28}$ show that the ion can remain completely intact with neither the loss of Ti nor the loss of $C_3$ or C after many collisions, indicating the fact that the metal is not directly attached to the skeleton, which is neither linear nor cyclic. Further CID experiments on Ti@$C_{30}$ show that Ti@$C_{30}$ can fragment to Ti@$C_{28}$ by $C_2$ loss, demonstrating that both species are endohedral. They also investigated other metal elements in the periodic table and found that both Zr@$C_{28}$ and U@$C_{28}$ formed in abundance, but Hf@$C_{28}$ was much weaker.

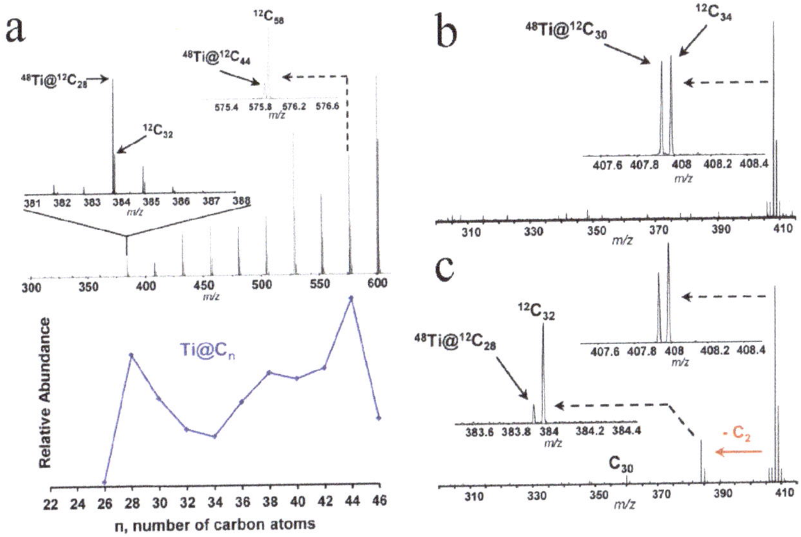


**Figure 2.** (**a**) Laser vaporization mass spectrum of a Ti-doped graphite target. Among the observed cations in the Ti@$C_{2n}$ family, Ti@$C_{28}$ is observed in great abundance without the accompany of the empty-cage $C_{28}$. (**b**) Isolated ions of Ti@$C_{30}$ and $C_{34}$, and (**c**) fragmentation pathway of Ti@$C_{30}$. Taken with permission from [44].

Dunk et al. have synthesized a series of uranium-endohedral fullerenes using the method of Krätschmer–Huffman arc discharge [45]. Several uranofullerenes were able to be observed in experiments utilizing solvents with different polarities. Via extractions using pyridine and 1,2-Dichlorobenzene (o-DCB), species of U@$C_{2n}$ (2n = 28, 60, 66, 68, and 70) were observed in the FT-ICR mass spectra. The distribution of the series of uranium-endohedral fullerenes was found to be different from that obtained in the early experiment reported by Smalley and his colleagues [43]. Some missing species of uranium-endohedral fullerenes have been observed for the first time in Dunk et al.'s experiment [45].

Due to its primary oxidation state at +4, it was predicted that thorium could form Th@$C_{28}$ via the method described above. However, Dunk et al. found that the species Th@$C_{28}$ was absent and replaced by the species Th@$C_{36}$ in the resulting mass spectrum, which could be attributed to the larger ionic radius of $Th^{4+}$ relative to $U^{4+}$ [20]. For the purpose of creating mono-metallofullerene from metal-incorporated graphite, more than 80 components were examined separately in their experiment. The detectable species with the smallest carbon cages are M@$C_{50}$ for Group 1 elements (K, Rb, Cs), M@$C_{42}$ or M@$C_{44}$ for Group 2 elements (Ca, Sr, Ba), M@$C_{36}$ for trivalent rare earth metals (Sc, Y, La, Ce, Pr, Nd, Gd, Tb, Dy Ho, Er, Lu), M@$C_{44}$ for divalent lanthanides (Sm, Eu, Yb), and M@$C_{50}$ for Ga and In. The results indicate that the oxidation states of the encapsulated metals (corresponding to the charge transfer to the carbon cage outside) and the ionic radius are determining factors that govern the size limits of the observed EMFs and the corresponding EMF formation.

Alternatively, Rodríguez-Fortea et al. recently reported the cation species of U@$C_{2n-1}$B (2n − 1 = 27–59) using their laser vaporization source [46]. As shown in Figure 3, mass spectrometric results identified the boron-doped U@$C_{27}$B as the most prevalent borafullerene within the U@$C_{2n-1}$B family. According to density functional theory (DFT) calculations, the U atom is positioned in the middle of the cage and interacts with every other atom there as though it were a 28-hapto ligand. This exotic radical shows us an intriguing example of unique host–guest chemistry in metallofullerenes.

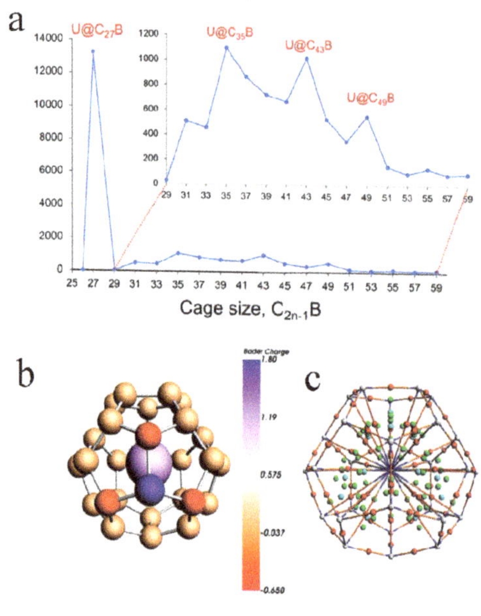


**Figure 3.** (**a**) Distribution of cations of U@$C_{2n-1}B^{+}$ formed by laser vaporization of U-containing graphite. The region 2n − 1 = 29 to 59 is expanded in the inset for clarity. (**b**) Bader charges and (**c**) AIM critical points (CPs) and bond paths for U@$C_{27}$B. Atom CP in white, bond CP in red, ring CP in green, and cage CP in light blue. Taken with permission from [46].

The observed new EMF species in small carbon cages in the gas phase were also simulated using theoretical methods on their structure and properties [31–35,47–49]. For example, Rodrñguez-Fortea, Kroto, and Pobletand have reported their theoretically calculated results of the Ti@$C_{2n}$ (2n = 26–50) family [47]. The stability and unique electronic characteristics of the two main species, Ti@$C_{28}$ and Ti@$C_{44}$, account for their abundance. It is demonstrated that all of the optimum isomers in the family are connected by a straightforward $C_2$ insertion, with the exception of a few cages that need an extra $C_2$ rearrangement. Da et al. studied the ground-state electronic structures of U@$C_{28}$ in its neutral, monovalent, and bivalent cationic forms [41]. The findings indicate that, rather than the $(5f)^1$ $(cage)^1$ ground state with a $D_2$ symmetry, the neutral U@$C_{28}$ is characterized by a $(cage)^2$ ground state with a *Td* symmetry [41]. Meng et al. studied the geometric and bonding properties of EMFs, including an f-block metal atom or ion enclosed in a $C_{20}$ cage [48]. Munoz-Castro investigated the availability of $U_2$@$C_{36}$ as an example for the small endohedral dimetallofullerene. The result indicates its neutral and dicationic forms can behave as plausible candidates with a U-U distance of 2.27Å [49]. Zhou et al. reported the structures and properties of TM@$C_n$ (n = 20, 24, 28) based on their DFT calculations, showing that after the embedding of TM atoms, the stability of TM@$C_{20}$ is greatly enhanced [33]. Their recent results show that the general species of $Mg^{2+}$@$[C_{20}]^{n-2}$ can be classified as superatoms [35].

### 3.2. EMFs including Multiple Metal Atoms ($M_x$@$C_{2n}$, $x \geq 3$)

The Coulomb repulsion among the metal ions in EMFs with multiple metal atoms creates a barrier to their stabilities. As the number and size of metal atoms increase, this situation becomes more serious. There are two ways to solve the problem [50]. One is to introduce electronegative elements (such as N, S, and O) to compensate for the Coulomb repulsion; the other is to adopt fullerene cages with large sizes. For the generation and structure study of EMFs that include multiple metal atoms, gas-phase experiments provide a convenient and effective method, especially when combined with DFT calculations. Graphene has been found to be a powerful precursor for the generation of EMF ions that encage multiple metal atoms in the gas phase [23,51–58]. For example, the laser ablation of $LaCl_3$/graphene can generate EMF ions of $La_n$@$C_{2m}^+$ with a long accumulation time in the cell of the FT-ICR mass spectrometer [53]. As shown in Figure 4a, compared to the wide distributions of La@$C_{58–194}^+$ and $La_2$@$C_{74–168}^+$, the distribution of trimetallofullerene ions is narrower (2n = 96–160), with a center located at $La_3C_{110}$. The primary products are the monometallofullerene ions; nonetheless, the species ranked second in abundance is $La_3C_{2m}$, instead of $La_2C_{2m}$. Other EMF ions of $La_4C_{112–160}$, $La_5C_{2m}$ ($132 \leq 2$ m $\leq 170$), and $La_6C_{2m}$ ($152 \leq 2$ m $\leq 180$) ions were also identified.

Although the diverse species generated in these experiments make the clear isolation of a special EMF ion difficult, the CID experiments have also been applied to some EMF ions, such as $CaC_{60}^+$, $SrC_{60}^+$, $BaC_{60}^+$, $BaC_{92}^+$, and $BaC_{130}^+$ [23]. For other EMF ions with more metal atoms or larger carbon cages, the gas-phase oxygen reaction test was applied to verify their endohedral structures. Nevertheless, since endohedral carbide cluster fullerene structures cannot be ruled out, it is not possible to properly identify their structures based solely on these mass spectrometric experiments.

The laser ablation of the mixture of graphene and $LuCl_3$ generated EMF ions with more metal atoms encaged (Figures 4b and 5) [54]. As shown in Figure 5, EMF ions of $Lu_nC_{2m}^+$ (n = 1–9, $50 \leq 2m \leq 198$) were observed in the mass spectrum. The gas-phase reaction test experiments confirmed their endohedral architectures. A few mono-EMF ions, such as $LuC_{50}^+$ and $LuC_{66}^+$, exhibit appreciable relative intensities, suggesting that the encaged metal atom significantly stabilizes relative IPR-violating cages. Similar to the cases of La and Y, the yield of trimetallofullerene ions, $Lu_3C_{2m}^+$, is the highest among all of the multimetallic metallofullerene ions. The intensities of $Lu_4C_{2m}^+$ and $Lu_5C_{2m}^+$ are still strong, suggesting the potential of their practical synthesis. The more Lu atoms contained in the EMFs, the larger the required carbon cage size. It has been found that the cage's average size rises by 12 carbon atoms for every additional Lu atom within the fullerene.

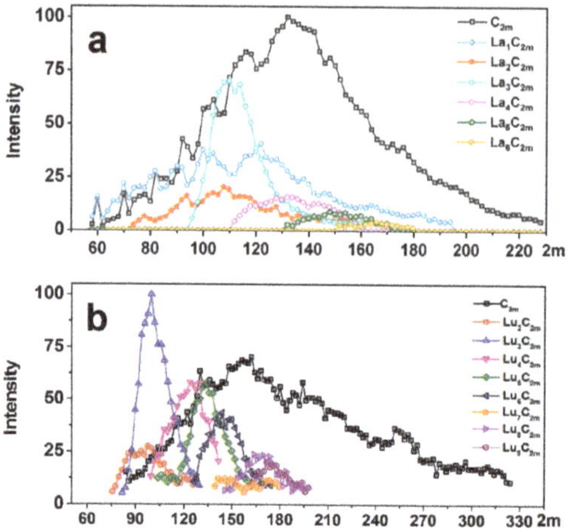


**Figure 4.** Intensity distributions of (**a**) $La_{1\text{–}6}C_{2m}^+$ and (**b**) $Lu_{2-9}C_{2m}^+$. (Figure 4b is taken with permission from [54]).

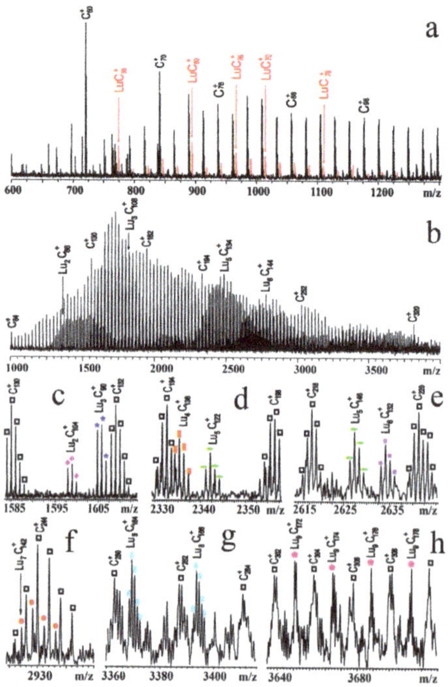


**Figure 5.** Laser ablation mass spectra of $LuCl_3$/graphene (**a**) obtained in the *m/z* region of 600–1300, and (**b**) obtained in the *m/z* region of 1000–4000. Some metallofullerene ions of $Lu_{2-9}C_{2n}^+$ are further shown in (**c**–**h**) in detail. Taken with permission from [54].

Other rare earth metals were also considered in order to obtain a general understanding of the multi-metallofullerene ions generated by laser ablation. Together with the previous data for alkali metal and alkali earth metal species, Table 1 summarizes all the multi-metallofullerene species created under identical experimental conditions in such studies [23,52–55]. As shown here, EMF ions with numerous metal atoms (n > 4) are only available in the cases of Y, La, Tb, Ho, and Lu. The increased number of encaged metal atoms also creates requirements for the size of carbon cages, but these requirements vary for different metals. For instance, the distribution of $La_5C_{2m}$ falls between 132 and 170 in terms of 2m, whereas the ranges for Tb, Ho, and Lu are 124 and 150, 114 and 160, and 104 and 160, respectively. The outcomes show the impacts of the metals' ionic radii in addition to the interaction of metal clusters with carbon cages. For the majority of these rare earth metals, it is possible to generate distinct multi-EMF ions ($n \geq 2$); however, Sm and Eu exhibit divergent outcomes, with no metallofullerene ions reported in the studies. For alkali metals and alkaline earth metals, only dimetallofullerene ions were observed in the examples of calcium and barium.

**Table 1.** Observed multi-metallofullerene ions of different metals in gas-phase laser ablation experiments.

| M | Oxide States | Observed Multi-Metallofullerene Ions | Ref. |
|---|---|---|---|
| La | +3 | $La_2C^+_{74-168}$, $La_3C^+_{96-160}$, $La_4C^+_{112-160}$, $La_5C^+_{132-170}$, $La_6C^+_{152-180}$ | [53] |
| Ce | +3 +4 | $Ce_2C^+_{88-106}$, $Ce_3C^+_{94-136}$, $Ce_4C^+_{100-150}$ | [55] |
| Pr | +3 +4 | $Pr_2C^+_{88-102}$, $Pr_3C^+_{94-134}$, $Pr_4C^+_{108-138}$ | [55] |
| Nd | +3 | $Nd_3C^+_{96-130}$ | [55] |
| Sm | +2 +3 | None | [55] |
| Eu | +2 +3 | None | [55] |
| Gd | +3 | $Gd_2C^+_{86-108}$, $Gd_3C^+_{90-128}$, $Gd_4C^+_{106-134}$ | [55] |
| Tb | +3 +4 | $Tb_2C^+_{80-112}$, $Tb_3C^+_{86-130}$, $Tb_4C^+_{100-144}$, $Tb_5C^+_{124-150}$, $Tb_6C^+_{134-164}$, $Tb_7C^+_{148-176}$ | [55] |
| Dy | +3 | $Dy_2C^+_{82-108}$, $Dy_3C^+_{90-128}$ | [55] |
| Ho | +3 | $Ho_2C^+_{74-104}$, $Ho_3C^+_{86-124}$, $Ho_4C^+_{98-112}$, $Ho_5C^+_{114-154}$, $Ho_6C^+_{132-162}$ | [55] |
| Er | +3 | $Er_2C^+_{82-96}$, $Er_3C^+_{90-114}$ | [54] |
| Tm | +2 +3 | $Tm_2C^+_{80-100}$, $Tm_3C^+_{88-122}$, $Tm_4C^+_{110-132}$ | [54] |
| Yb | +2 +3 | $Yb_3C^+_{90-110}$ | [54] |
| Lu | +3 | $Lu_2C^+_{76-128}$, $Lu_3C^+_{82-130}$, $Lu_4C^+_{100-142}$, $Lu_5C^+_{104-160}$, $Lu_6C^+_{126-174}$, $Lu_7C^+_{140-180}$, $Lu_8C^+_{146-198}$, $Lu_9C^+_{172-198}$ | [54] |
| Y | +3 | $Y_2C^+_{74-130}$, $Y_3C^+_{88-128}$, $Y_4C^+_{102-142}$, $Y_5C^+_{128-154}$, $Y_6C^+_{130-154}$ | [53] |
| Li | +1 | None | [52] |
| Na | +1 | None | [52] |
| Rb | +1 | None | [52] |
| Cs | +1 | None | [52] |
| Mg | +2 | None | [23] |
| Ca | +2 | $Ca_2C^+_{96-160}$ | [23] |
| Sr | +2 | None | [23] |
| Ba | +2 | $Ba_2C^+_{80-182}$ | [23] |

### *3.3. EMFs including Late Transition Metals*

Although EMFs with rare earth metals and their adjacent elements (groups I–IV) have been investigated in the gas phase, those containing post-transition metals are very rarely experimentally studied. Entrapping these transition metals in fullerene cages is still a challenge [56]. There may be many reasons for this, but one of them is that these metal species usually exhibit high reactivity with carbon atoms during the generative process. Thus, it is suggested that the most performable way to make EMFs containing post-transition metals is to use heterometal clusters to reduce the reactivity of these metal species.

Recently, Kong's group successfully synthesized metal fullerene ions containing platinum atoms in the gas phase for the first time [56]. By laser ablation of a mixed sample of graphene/$LaCl_3$/$PtCl_4$ in vacuum, the EMF ions of $La_mPtC_{2n}{}^+$ were observed (Figure 6a) and verified with oxygen reaction tests [50,56]. An example is shown in Figure 6. Apart from the EMF ions of $La_2@C_{2n}{}^+$ and $La_3@C_{2n}{}^+$, $La_mPtC_{2n}{}^+$ (m = 2–3; n = 43–65) was also observed. The distributions of the observed EMF ions are compared and shown in Figure 6b. It can be found that EMF ions of $La_2PtC_{2n}{}^+$ have a distribution from 2n = 86 to 114 centered on the $La_2PtC_{98}{}^+$, and those of $La_3PtC_{2n}{}^+$ have a distribution from 2n = 98 to 130 centered on the $La_3PtC_{112}{}^+$.

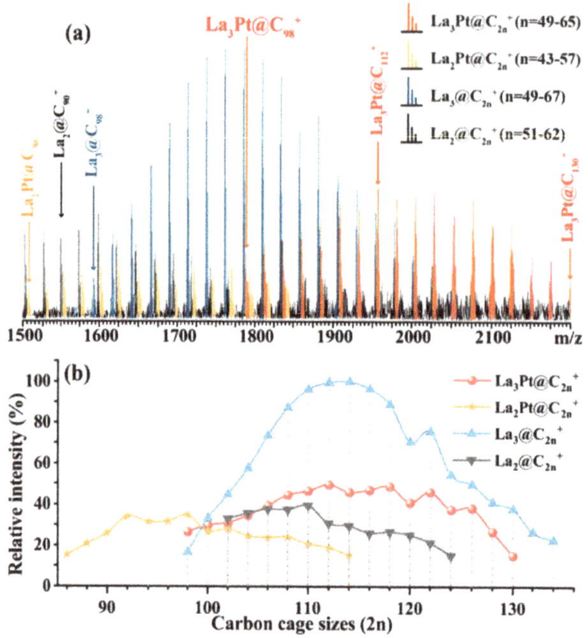


**Figure 6.** (**a**) Laser ablation MS spectrum of the mixture of graphene/$LaCl_3$/$PtCl_4$ in the region of 1500–2200 *m/z*. For convenience, the abundant fullerene ions of $C_{2n}^+$ (n = 63–91) have been manually removed. (**b**) Intensity distributions of $La_mPt@C_{2n}^+$ (m = 2, 3) and $La_m@C_{2n}^+$(m = 2, 3) observed in the upper figure. Taken with permission from [50].

Extensive theoretical studies have been performed for the neutral forms of these two selected species, $La_2PtC_{90}^+$ and $La_3PtC_{98}^+$ [50,56]. For $La_2PtC_{90}$ (the neutral species of $La_2PtC_{90}^+$), the two most stable structures are $La_2Pt@C_2(99915)$-$C_{90}$ and $La_2Pt@C_2(99916)$-$C_{90}$, with the energy of the latter being 8.1 kcal/mol greater than that of the former. Moreover, large HOMO-LUMO gaps of 1.28 and 1.13 eV have been found for the two isomers. Impressively, the most stable isomer of the corresponding metal carbide EMFs, which has been identified to be $La_2PtC_2@D2(81738)$-$C_{88}$, has a much higher energy (>89 kcal/mol) than $La_2Pt@C_2(99915)$-$C_{90}$. As shown in Figure 7a,b, the internal $La_2Pt$ clusters in both structures exhibit a "V" shape, with La coordinating with the carbon cage and Pt located at the center of the $La_2Pt$ cluster, serving as a connecting point to stabilize the EMFs. Within the triatomic cluster, the distances between La–La and La−Pt are ~4.01 and 2.73 Å, respectively, indicating the existence of the La–Pt bond.

The two most stable isomers of $La_3Pt@C_{98}$ were found to be $La_3Pt@C_2(231010)$-$C_{98}$ and $La_3Pt@C_1(231005)$-$C_{98}$ [50]. Their structures are shown in Figure 7c,d. In both isomers, the $La_3Pt$ clusters form pyramidal tetrahedrons in carbon cages. The distances between two nearby La atoms are 4.18~4.40 Å, a little longer than those in the top two isomers of $La_2Pt@C_{90}$. The encaged clusters of $La_3Pt$ in both isomers have typical La–Pt bonds with a length of about 2.80 Å, and La–Pt–La bond angles of ~100°, with a shape similar to the metal nitride clusters in $M_3N@C_{80}$ (M = Y, Tb, Gd), but different from planar triangles in the cases of $M_3N@C_{80}$ (M = Sc, Lu, Dy).

In order to explore the role of the encaged Pt atom, Mulliken charge distribution calculations were carried out for these structures [50,56]. Results show that the carbon cages and the endohedral La atoms are negatively and positively charged, respectively. Very interestingly, it is noted that Pt atoms within $La_2Pt$ or $La_3Pt$ clusters appear to have a negative valence and may mediate the interactions between two La or among three La atoms. For $La_2Pt@C_2(99915)\text{-}C_{90}$, the charges of the two La atoms are +1.82 and +1.81, and the Pt atom is negatively charged (with a charge of −0.64). This role is similar to that of Pt in the embedded tetrametallic fullerene $La_3Pt@C_{98}$. For both $La_3Pt@C_1(231005)\text{-}C_{98}$ and $La_3Pt@C_2(231010)\text{-}C_{98}$, the La atoms present positive states (from 0.49 to 0.52), while the Pt atoms are negatively charged (around −0.24). In the case of $La_3Pt@C_{98}$, all three of the La atoms create 2c–2e bonds with the negatively charged Pt atom, which serves as a mediator. Interestingly, the entire molecule is stabilized by the 4c–2e tetrametallic bond, with the highest occupancy number in the unit of $La_3Pt$. It is noteworthy that seven electrons have transferred from the inner metal cluster to the outside fullerene cage in this instance, which is quite rare and has not been documented before. This is further evidenced by the adaptive natural density partitioning (adNDP) analysis, where Pt takes two electrons to fill its 5d-shell from La atoms and shares its $6s^2$ pair as a back-donation with La atoms, resulting in its unique electron distribution in a form of $[(La^{3+})_3Pt^{2-}]^{7+}@C_{98}{}^{7-}$.

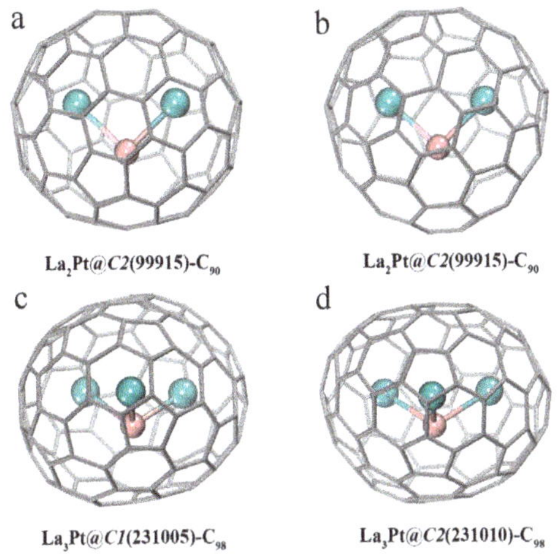


**Figure 7.** The calculated structures of two most stable isomers of $La_2Pt@C_{90}$: (**a**) $La_2Pt@C_2(99915)\text{-}C_{90}$, (**b**) $La_2Pt@C_2(99916)\text{-}C_{90}$. The calculated structures of the two most stable isomers of $La_3Pt@C_{98}$: (**c**) $La_3Pt@C_1(231005)\text{-}C_{98}$ and (**d**) $La_3Pt@C_2(231010)\text{-}C_{98}$.

### *3.4. EMFs including Ionic Bonds*

There are still big challenges to encapsulating various ionic bonds in fullerenes (except for metal oxides) [57–61]. The confinement of an individual ionic bond is very difficult, since it has neither sharing electrons nor a fixed pairing relationship between the two ions. Due to the difficulty in generation, they have only very rarely been the subject of experimental research. Using a Smalley-type ion source, Martin et al. observed such fullerene ions in the gas phase in their pioneering works. They identified ions of $CaCl@C_{66}{}^+$ and $HoCl@C_{60}{}^+$ in the gas phase and gave a justification for the species' threshold sizes using the spherical approximation [61].

Recently, Kong's group reported the identification of endohedral ions of $LuClC_{2n}{}^+$ (2n = 90−190) with high-resolution laser ablation mass spectrometry [57]. In the experiments, a mixture of graphene/$LuCl_3$/LiCl was applied as the precursor, and oxygen gas was used as the reaction gas to verify their endohedral structures. Similarly, mixtures of graphene and sodium halides (NaX) (X = Cl, Br, I), EMF ions of $NaCl@C_{2n}{}^+$ (2n = 120–244), $NaBr@C_{2n}{}^+$ (2n = 110–240), and $NaI@C_{2n}{}^+$ (2n = 116–198) were also observed by the same group [58]. Interestingly, Li or K could not substitute for the encapsulated metal ion in these studies, indicating that the metal cation's effects on the EMFs outweigh those of halide anions.

To figure out the structures and properties of the endohedral ionic bond fullerenes, the species of LuCl@$C_{90}$ was selected and studied systematically using DFT methods [57]. LuCl@$C_2$(99917)-$C_{90}$ and LuCl@$C_2$(99914)-$C_{90}$ were found to be the top two isomers (Figure 8), which have significant HOMO−LUMO gaps of 1.61 and 1.82 eV, respectively. Statistical thermodynamic analysis shows that the former prevails inside a small temperature range of 0 to 500 K. With the temperature increasing, it is exceeded by LuCl@$C_2$(99914)-$C_{90}$ at 650 K and by LuCl@$C_2$(99913)-$C_{90}$ at 3000 K. For all optimized isomers, it was found that the Lu atoms are situated between the Cl atoms and the cages, and the Cl atoms are found in the hubs of the cages. The potential curve of an encaged ionic bond is distinct from that of isolated ionic bonds, which are identified by their usual Morse or Morse long-range potentials. As shown in Figure 7, infinite potential wells exist due to the restriction of the carbon cages, and the wells are deeper than those of unprotected ionic species, clearly showing that they have different characteristics. In addition, the calculation shows that the equilibrium distances of the Lu–Cl bonds in both isomers is 2.49 Å, which is larger than that of $LuCl_2^+$ (2.30 Å) and a little smaller than that of LuCl (2.55 Å).

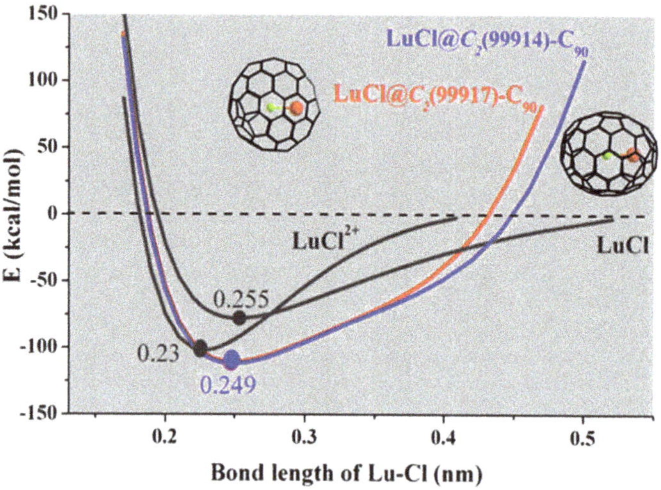


**Figure 8.** Potential curves of $LuCl^{2+}$, LuCl, LuCl@$C_2$(99917)-$C_{90}$, and LuCl@$C_2$(99914)-$C_{90}$. Taken with permission from Ref. [57].

An interesting effect of endohedral ionic bond fullerenes is that it is possible to rotate the bond inside the carbon cages. The energy barrier mainly depends on the interaction between the metal ion and the cage [57]. For instance, seven energy minima were discovered during the rotation of the Cl−Lu bond in the lowest energy isomer of LuCl@$C_2$(99917)-$C_{90}$, and the energy barriers were found in the range of 20−220 kcal/mol. However, the results depend very much on the selected plane for rotation. If the perpendicular rotation plane is selected, there are only four energy minima and the rotation energy barrier is considerably larger, reaching up to 600 kcal/mol.

In fact, the properties of encaged ionic bonds have previously been noticed by theoretical chemists and have always been a focus of their attention [62–65]. Foroutan-Nejad et al. proposed a single-molecule switch based on the neutral dipolar molecule that uses alkali metal halides contained in a $C_{70}$ cage for data storage [65]. The encaged ionic bonds have two local energy minima separated by an energy barrier relative to the fullerene. Attractively, an external electric field can be used to manipulate the dipolar system inside the fullerene. This is possible with contemporary instruments like scanning tunneling microscopes. The ionic bond aligns along the main axis of $C_{70}$ to minimize the steric clash. The conformations with the dipole moment along the electric field are more stable than the others when the electric field is on. With an increasing electric field, the stabilization should lead to the reorientation of the encaged ionic bond. The same group has also proposed using a two- or four-electrode system to speed up the switching procedure, showing their potential in the application of random-access memories and memristors [64]. As illustrated in Figure 9, the dipolar MX causes the system to operate like a diode by polarizing the fullerene shell electrons. A series of species of MX@$C_{70}$ (M = a metal or hydrogen, X = a halogen or a chalcogen) have been studied to be identified as possible molecular memristors by the same research group.

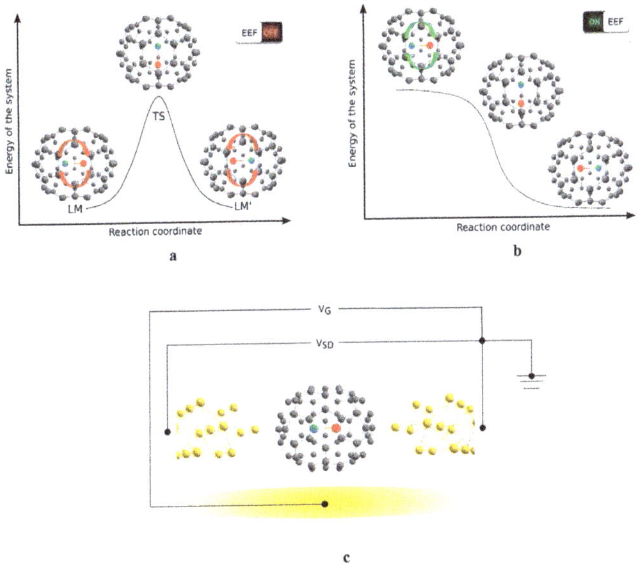


**Figure 9.** The principle of an MX@$C_{70}$ molecular memristor (EEF stands for the external electric field). (**a**) MX@$C_{70}$ with two local minima (LM and LM′) separated by a transition state (TS) with an energy barrier. (**b**) Upon application of external voltage, one of the local minima becomes unstable. This leads to the rotation of MX between LM and LM′ (or LM′ and LM), and thus to the switching of the system's polarity and conductivity. (**c**) The four-electrode setup in which the source–drain voltage (VSD) is applied horizontally and the gate voltage (VG) is perpendicular to the VSD. Taken with permission from Ref. [64].

Impressively, Gan et al. recently reported the preparation of an open-cage fullerene as a selective molecular trap for ionic bonds of LiF/$[BeF]^+$ [60]. They made use of a hydroxy group-containing open-cage $C_{60}$ derivative. LiF and $[BeF]^+$ were encapsulated as a result of the hydroxy group's ability to coordinate lithium or beryllium cations, functioning as a bait or hook. The whole structures were clearly identified by single-crystal X-ray diffraction. A bright future lies ahead for the development of more cage-opening reactions for the prospective uses of fullerene-based molecular containers, especially in light of the recent advancements in the logic-in-memory operations of a two-terminal single-metallofullerene device at room temperature [62].

## 4. Mechanism Study

Similar to those of fullerenes, the suggested mechanisms for the formation of EMFs can be generally classified into two types: bottom-up or top-down [20,44,47,66–82]. For example, Dunk et al. demonstrated that U@$C_{28}$ acts as a precursor for larger uranofullerenes while itself being formed through a bottom-up process [44]. They have identified the structures of Ti@$C_{2n}$ (2n = 26–50) and uncovered essential features in their growth processes by DFT calculations and molecular dynamics (MD) simulations [47]. Further studies by the same group show more proofs of the suggested bottom-up formation process of giant mono-EMFs from Pr@$C_{2v}$-$C_{82}$ [20]. They proposed that, under EMF synthesis circumstances, mono-metallofullerenes with small to giant sizes might form in the gas phase via a bottom-up mechanism. The extent of charge transferred to carbon cages appears to be determined by the elements' ionization energies during graphite/metal evaporation. It has been suggested that the unconventional small-sized EMFS could stabilize by charge transfer, although the existing strain energy resulted in the impeding of $C_2$ insertion and the inhibition of overall bottom-up formation of large-sized metallofullerenes. MD simulation also plays an important role in understanding the formation mechanisms of EMFs [83–87]. Using constructed multi-body potential functions, Yamaguchi et al. found that the clustering process of a system containing 500 isolated carbon atoms and five La atoms under a controlled temperature of 3000 K can result in lanthanum-containing caged clusters. For the Ni–C system, the Ni atom ultimately stayed on the face of caged structures [83]. Deng et al. investigated the effects of the inert cooling gas helium in the Sc-EMF forming process with the method of quantum chemical molecular dynamics. Their results show that Sc encapsulation involves nucleation of carbon clusters with Sc atoms at early stages, and the existence of He gas reduces the size of the self-assembled cages [84]. An extended simulation with an Fe atom shows no formation of such Fe-EMF, which is consistent with experimental results [71]. Fan et al. found that the growth patterns of Sc-EMFs resemble those of hollow cages, and the He gas could promote both the formation of carbon cages and the encapsulation of scandium atoms [86].

The bottom-up mechanism is suitable for the formation of mono-metallofullerenes from metal-incorporated graphite [20]. However, the generation of multiple metallofullerenes in some experiments, such as the examples shown in Figures 3–5, may have different mechanisms. The top-down mechanism offers another kind of possibility [71–78]. There is evidence to support the theory that giant fullerenes formed from large-size polycyclic aromatic hydrocarbons (PAHs), or graphene sheets shrunk into smaller ones, through interstellar and laboratory observations as well as theoretical calculations [71,77,78]. For example, Irle et al. use a 'shrinking hot giant fullerene' mechanism in which giant carbon cages occur first, followed by multiple $C_2$ losses with self-assembly to appropriate symmetric fullerene cages, such as $I_h$-$C_{60}$ and $D_{5h}$-$C_{70}$ [71]. Moreover, the transformation of graphene to a fullerene cage was demonstrated by Chuvilin et al., based on their transmission electron microscopy and theoretical studies [74].

For EMFs, the top-down mechanism is also applied in some cases [23,48–56,79–82]. Dorn et al. isolated and characterized metal carbide metallofullerenes $M_2C_2$@$C_1$(51383)-$C_{84}$ (M = Y, Gd) and proposed that the cage with fused pentagons is a preserved 'missing link' in the top-down mechanism [80]. The unique asymmetric structure can form many symmetric fullerenes that account for some well-known solvent-extractable metallofullerenes using limited, well-established rearrangement steps. Lu et al. presented the identified $La_2C_2$@$C_2$(816)-$C_{104}$ as an origin formation of other known ideal tubular cages by removing the pyracylene motif, providing support for the top-down formation mechanism of such fullerenes [79]. Using the method of transmission electron spectroscopy, Sinitsa et al. proposed a method for synthesis of nickel EMFs, in which the formation was initiated by electron irradiation of a metal cluster surrounded by amorphous carbon inside a carbon nanotube [87]. Their MD simulations also rationalize the experimental observations. On the other hand, Kong et al. suggested the top-down mechanism to explain the multiple EMF ions produced in the process of laser ablation of graphene flakes mixed with metal salts [23,48–56]. It is proposed that the instantaneous process of wrapping metal atoms or clusters inside curved graphene flakes produced by laser irradiation can lead to the production of EMFs with large-sized cages and multi-metal atoms. Moreover, the mechanism can be used to interpret the generation of fullerenes with encaged ionic bonds in a similar way. It is important to notice the difference between the use of graphene and graphite as precursors in these experiments, since their differently reported mass spectra indicate their different laser ablation processes and mechanisms.

## 5. Prospect

The gas-phase EMF study provides valuable information in many ways. The importance of gas-phase research on EMFs in their discovery can never be overstated, and they are going to remain essential in interdisciplinary investigations of new species in this big family. In this review, we summarized the progress of gas-phase EMF studies in the past 15 years, focusing on novel EMF species identified in the gas phase and the relative mechanism study, which provided valuable information for the structural research, synthesis, and design of such new molecules. Although the research undertaken in gas-phase EMF research is much less widespread compared to those engaged in the synthesis, separation, and application of EMFs, the gas-phase results still provide valuable clues for new members in the family of EMF. On the one hand, these novel EMF species observed in the gas phase provide direct experimental evidence for their existence and valuable clues for their solid-phase synthesis methods, which can promote the exploration and discovery of new synthesis methods for EMFs. On the other hand, the structural study of these new EMF species can provide insights into their novel structures and new chemical bonds, different properties, and potential applications.

Experimentally, high-resolution mass spectrometry is widely applied to qualitatively identify charged EMF species and their molecular compositions. The distributions of EMF ions related to different experimental conditions, precursors, and metal elements can provide valuable information on the mechanism of the generation of important and novel EMF species, especially when combined with theoretical calculations. Among them, the selection of precursors is found to be important, and it is worth noting that, in future research, the synthesis of specific novel EMFs can be more effectively achieved by selecting appropriate precursors. However, it should be pointed out that a great deal of essential work in gas-phase EMF study still needs to be done. First, new methods for the generation of new or novel EMF ions are necessary. Considering that the bottom-up mechanism dominates most laser ablation experiments, based on the Smalley-type source, different generation and ionization methods based on the top-down mechanism are needed for large-size or special EMF ions. Different precursors and new sample preparation methods are important, too. Second, the formation of EMFs encaging transition metals is one of the most fascinating topics in gas-phase EMF chemistry. It is also an important means of discovering novel EMFs. Considering the examples of $La_2Pt@C_{90}$, and $La_3Pt@C_{98}$, it is natural to assume that more types of embedded metal clusters or ionic bonds can be discovered or synthesized. On the other hand, the ionic bonds composed of different elements can provide new members for the EMF family and may be applied as crucial components of molecular devices. Third, it is expected that more methods of ion chemistry should be applied to the investigation of the structure and characteristics of gas-phase fullerene ions. In addition to the widely used method of CID, other methods, such as reaction mass spectrometry, collisional section measurement, UV photodissociation spectroscopy, etc., should be combined in tandem mass spectrometric study. Fourth, the gas-phase study of EMF anions, which may open up a different path for discovering new EMFs, is still very limited.

**Author Contributions:** Conceptualization, X.K.; writing—original draft preparation, Y.H.; writing—review and editing, X.K. All authors have read and agreed to the published version of the manuscript.

**Funding:** This research was funded by National Natural Science Foundation of China (No. 22174076).

**Conflicts of Interest:** The authors declare no conflict of interest.

## References


1. Kroto, H.W.; Heath, J.R.; Obrien, S.C.; Curl, R.F.; Smalley, R.E. $C_{60}$: Buckminsterfullerene. *Nature* **1985**, *318*, 162–163. [CrossRef]
2. Heath, J.R.; O'Brien, S.C.; Zhang, Q.; Liu, Y.; Curl, R.F.; Kroto, H.W.; Tittel, F.K.; Smalley, R.E. Lanthanum complexes of spheroidal carbon shells. *J. Am. Chem. Soc.* **1985**, *107*, 7779. [CrossRef]
3. Weiss, F.D.; O'Brien, S.C.; Elkind, J.L.; Curl, R.F.; Smalley, R.E. Photophysics of metal complexes of spheroidal carbon shells. *J. Am. Chem. Soc.* **1988**, *110*, 4464. [CrossRef]
4. Chai, Y.; Guo, T.; Jin, C.; Haufler, R.E.; Chibante, L.P.F.; Fure, J.; Wang, L.; Alford, J.M.; Smalley, R.E. Fullerenes with metal inside. *J. Phys. Chem.* **1991**, *95*, 7564–7568. [CrossRef]
5. Krätschmer, W.; Lamb, L.D.; Fostiropoulos, K.; Huffman, D.R. Solid $C_{60}$: A new form of carbon. *Nature* **1990**, *347*, 354. [CrossRef]
6. Alvarez, M.M.; Gillan, E.G.; Holczer, K.; Kaner, R.B.; Min, K.S.; Whetten, R.L. Lanthanum carbide ($La_2C_{80}$): A soluble dimetallofullerene. *J. Phys. Chem.* **1991**, *95*, 10561–10563. [CrossRef]
7. Lu, X.; Feng, L.; Akasaka, T.; Nagase, S. Current status and future developments of endohedral metallofullerenes. *Chem. Soc. Rev.* **2012**, *41*, 7723–7760. [CrossRef] [PubMed]
8. Popov, A.A.; Yang, S.F.; Dunsch, L. Endohedral fullerenes. *Chem. Rev.* **2013**, *113*, 5989–6113. [CrossRef] [PubMed]
9. Rodríguez-Fortea, A.; Balch, A.L.; Poblet, J.M. Endohedral metallofullerenes: A unique host–guest association. *Chem. Soc. Rev.* **2011**, *40*, 3551–3563. [CrossRef] [PubMed]
10. Shen, W.; Hu, S.; Lu, X. Endohedral metallofullerenes: New structures and unseen phenomena. *Chem. Eur. J.* **2020**, *26*, 5748–5757. [CrossRef] [PubMed]
11. Shen, W.; Bao, L.; Lu, X. Endohedral metallofullerenes: An ideal platform of sub-nano chemistry. *Chin. J. Chem.* **2022**, *40*, 275–284. [CrossRef]
12. Gan, L.B. Molecular containers derived from [60] fullerene through peroxide chemistry. *Acc. Chem. Res.* **2019**, *52*, 1793–1801. [CrossRef]
13. Yang, S.F.; Ioffe, I.N.; Troyanov, S.I. Chlorination-promoted skeletal transformations of fullerenes. *Acc. Chem. Res.* **2019**, *52*, 1783–1792. [CrossRef]
14. Li, M.; Zhao, R.; Dang, J.; Zhao, X. Theoretical study on the stabilities, electronic structures, and reaction and formation mechanisms of fullerenes and endohedral metallofullerenes. *Coord. Chem. Rev.* **2022**, *471*, 214762. [CrossRef]
15. Boltalina, O.V. Ions of Endometallofullerenes in the Gas Phase. In *Endohedral Fullerenes: Electron Transfer and Spin*; Springer: Berlin/Heidelberg, Germany, 2017; pp. 81–102.
16. Marshall, A.G.; Hendrickson, C.L.; Jackson, G.S. Fourier transform ion cyclotron resonance mass spectrometry: A primer. *Mass Spectrom. Rev.* **1998**, *17*, 1–35. [CrossRef]
17. Duncan, M.A. Invited review article: Laser vaporization cluster sources. *Rev. Sci. Instrum.* **2012**, *83*, 041101. [CrossRef] [PubMed]
18. Dunk, P.W.; Kaiser, N.K.; Hendrickson, C.L.; Quinn, J.P.; Ewels, C.P.; Nakanishi, Y.; Sasaki, Y.; Shinohara, H.; Marshall, A.G.; Kroto, H.W. Closed network growth of fullerenes. *Nat. Commun.* **2012**, *3*, 855. [CrossRef]
19. Dunk, P.W.; Adjizian, J.J.; Kaiser, N.K.; Quinn, J.P.; Blakney, G.T.; Ewels, C.P.; Marshall, A.G.; Kroto, H.W. Metallofullerene and fullerene formation from condensing carbon gas under conditions of stellar outflows and implication to stardust. *Proc. Natl. Acad. Sci. USA* **2013**, *110*, 18081–18086. [CrossRef] [PubMed]
20. Dunk, P.W.; Mulet-Gas, M.; Nakanishi, Y.; Kaiser, N.K.; Rodríguez-Fortea, A.; Shinohara, H.; Poblet, J.M.; Marshall, A.G.; Kroto, H.W. Bottom-up formation of endohedral mono-metallofullerenes is directed by charge transfer. *Nat. Commun.* **2014**, *5*, 5844. [CrossRef]
21. Kong, X.L.; Li, S.Q.; Zhang, S.; Huang, Y.; Cheng, Y.S. Large carbon cluster anions generated by laser ablation of graphene. *J. Am. Soc. Mass Spectrom.* **2011**, *22*, 2033–2041. [CrossRef]
22. Kong, X.L.; Huang, Y.; Chen, Y.S. Difference in Formation of Carbon Cluster Cations by Laser Ablation of Graphene and Graphene Oxide. *J. Mass. Spectrom.* **2012**, *47*, 523–528. [CrossRef]
23. Kong, X.L.; Bao, X.D. Generation of $M@C_{2n}^+$ (M = Ca, Sr, Ba, 2n = 50–230) by Laser Ablation of Graphene. *Int. J. Mass. Spectrom.* **2014**, *372*, 1–7. [CrossRef]
24. Karas, M.; Hillenkamp, F. Laser desorption ionization of proteins with molecular masses exceeding 10,000 daltons. *Anal. Chem.* **1988**, *60*, 2299–2301. [CrossRef] [PubMed]
25. Guan, S.; Marshall, A.G. Stored waveform inverse Fourier transform (SWIFT) ion excitation in trapped-ion mass spectrometry: Theory and applications. *Int. J. Mass Spectrom. Ion Proc.* **1996**, *158*, 5–37. [CrossRef]
26. Laskin, J.; Futrell, J.H. Collisional activation of peptide ions in FT-ICR mass spectrometry. *Mass Spectrom. Rev.* **2003**, *22*, 158–181. [CrossRef] [PubMed]
27. Wahl, F.; Wörth, J.; Prinzbach, H. The pagodane route to dodecahedranes: An improved approach to the $C_{20}H_{20}$ parent framework; partial and total functionalizations—Does $C_{20}$-fullerene exist? *Angew. Chem. Int. Ed.* **1993**, *32*, 1722–1726. [CrossRef]
28. Prinzbach, H.; Weiler, A.; Landenberger, P.; Wahl, F.; Wörth, J.; Scott, L.T.; Gelmont, M.; Olevano, D.; Issendorff, B.V. Gas-phase production and photoelectron spectroscopy of the smallest fullerene, $C_{20}$. *Nature* **2000**, *407*, 60–63. [CrossRef] [PubMed]
29. An, Y.P.; Yang, C.L.; Wang, M.S.; Ma, X.G.; Wang, D.H. Geometrical and electronic properties of the clusters of $C_{20}$ Cage doped with alkali metal atoms. *J. Clust. Sci.* **2011**, *22*, 31–39. [CrossRef]

30. Baei, M.T.; Soltani, A.; Torabi, P.; Hosseini, F. Formation and electronic structure of $C_{20}$ fullerene transition metal clusters. *Monatshefte für Chemie-Chem. Mon.* **2014**, *145*, 1401–1405. [CrossRef]
31. Rad, A.S.; Ayub, K. Nonlinear optical and electronic properties of Cr-, Ni-, and Ti- substituted $C_{20}$ fullerenes: A quantum-chemical study. *Mater. Res. Bull.* **2018**, *97*, 399–404. [CrossRef]
32. Rad, A.S.; Ayub, K. Substitutional doping of zirconium-, molybdenum-, ruthenium-, and palladium: An effective method to improve nonlinear optical and electronic property of $C_{20}$ fullerene. *Comput. Theor. Chem.* **2017**, *1121*, 68–75.
33. Zhou, Z.H.; Liu, Z.F.; Li, Z.; Wang, Z.Y. Structures, stability, electronic, magnetic properties of TM@$C_n$ (n = 20, 24, 28) and polymerization of cyclocarbon-24: A DFT study. *Mater. Chem. Phys.* **2021**, *260*, 124154. [CrossRef]
34. Yang, Y.F.; Gromov, E.V.; Cederbaum, L.S. Charge separated states of endohedral fullerene Li@$C_{20}$. *J. Chem. Phys.* **2019**, *151*, 114306. [CrossRef] [PubMed]
35. Li, J.R.; Wang, R.; Huang, W.R.; Zhu, Y.; Teo, B.K.; Wang, Z.G. Smallest endohedral metallofullerenes $[Mg@C_{20}]_n$ (n = 4, 2, 0, −2, and −4): Endo-ionic interaction in superatoms. *J. Phys. Chem. Lett.* **2023**, *14*, 2862–2868. [CrossRef] [PubMed]
36. Tománek, D.; Schluter, M.A. Growth regimes of carbon clusters. *Phys. Rev. Lett.* **1991**, *67*, 2331. [CrossRef] [PubMed]
37. Kent, P.R.C.; Towler, M.D.; Needs, R.J.; Rajagopal, G. Carbon clusters near the crossover to fullerene stability. *Phys. Rev. B* **2000**, *62*, 15394. [CrossRef]
38. Guo, T.; Smalley, R.E.; Scuseria, G.E. Ab initio theoretical predictions of $C_{28}$, $C_{28}H_4$, $C_{28}F_4$, $(Ti@C_{28})H_4$, and M@$C_{28}$ (M = Mg, Al, Si, S, Ca, Sc, Ti, Ge, Zr, and Sn. *J. Chem. Phys.* **1993**, *99*, 352–359. [CrossRef]
39. Garg, I.; Sharma, H.; Kapila, N.; Dharamvir, K.; Jindal, V.K. Transition metal induced magnetism in smaller fullerenes ($C_n$ for n $\leq$ 36). *Nanoscale* **2011**, *3*, 217–224. [CrossRef] [PubMed]
40. Koley, S.; Sen, S.; Chakrabarti, S. Role of molecule-electrode coupling strength in inducing inelastic transmission spectra of Hf@$C_{28}$. *Chem. Phys.* **2020**, *539*, 110930. [CrossRef]
41. Dai, X.; Gao, Y.; Jiang, W.R.; Lei, Y.Y.; Wang, Z.G. U@$C_{28}$: The electronic structure induced by the 32-electron principle. *Phys. Chem. Chem. Phys.* **2015**, *17*, 23308. [CrossRef]
42. Zhao, Y.X.; Yuan, K.; Li, M.Y.; Zhao, X. Oxidation states of gallium (infrequent i and common iii) tunable via medium-sized $C_{60}$ and small-sized $C_{28}$ fullerenes. *Inorg. Chem. Front.* **2020**, *7*, 4113. [CrossRef]
43. Guo, T.; Diener, M.D.; Chai, Y.; Alford, M.J.; Haufler, R.E.; McClure, S.M.; Ohno, T.; Weaver, J.H.; Scuseria, G.E.; Smalley, R.E. Uranium stabilization of $C_{28}$: A tetravalent fullerene. *Science* **1992**, *257*, 1661–1664. [CrossRef]
44. Dunk, P.W.; Kaiser, N.K.; Mulet-Gas, M.; Rodríguez-Fortea, A.; Poblet, J.M.; Shinohara, H.; Hendrickson, C.L.; Marshall, A.G.; Kroto, H.W. The smallest stable fullerene, M@$C_{28}$ (m = Ti, Zr, U): Stabilization and growth from carbon vapor. *J. Am. Chem. Soc.* **2012**, *134*, 9380–9389. [CrossRef]
45. Gómez-Torres, A.; Esper, R.; Dunk, P.W.; Morales-Martínez, R.; Rodríguez-Fortea, A.; Echegoyen, L.; Poblet, J.M. Small cage uranofullerenes: 27 years after their first observation. *Helv. Chim. Acta* **2019**, *102*, e1900046. [CrossRef]
46. Moreno-Vicente, A.; Alías-Rodríguez, M.; Dunk, P.W.; de Graaf, C.; Poblet, J.M.; Rodríguez-Fortea, A. Highly oxidized U(vi) within the smallest fullerene: Gas-phase synthesis and computational study of boron-doped U@$C_{27}$B. *Inorg. Chem. Front.* **2023**, *10*, 908. [CrossRef]
47. Mulet-Gas, M.; Abella, L.; Dunk, P.W.; Rodríguez-Fortea, A.; Kroto, H.W.; Poblet, J.M. Small endohedral metallofullerenes: Exploration of the structure and growth mechanism in the Ti@$C_{2n}$ (2n = 26–50) family. *Chem. Sci.* **2015**, *6*, 675–686. [CrossRef] [PubMed]
48. Meng, F.C.; Zhou, Z.W.; Zhang, P.L.; Jiang, M.; Xu, X.L.; Wang, Y.; Gou, J.H.; Huie, D.; Die, D. Encapsulation of an f-block metal atom/ion to enhance the stability of $C_{20}$ with the $I_h$ symmetry. *Phys. Chem. Chem. Phys.* **2015**, *17*, 4328. [CrossRef] [PubMed]
49. Muñoz-Castro, A. U@$C_{36}$. Is there enough room for a second uranium? *RSC Adv.* **2016**, *6*, 78176. [CrossRef]
50. Hou, Y.M.; Mu, L.; Zhou, S.Z.; Xu, Y.; Kong, X. Structure and bonding properties of the platinum-mediated tetrametallic endohedral fullerene $La_3Pt@C_{98}$. *Dalton Trans.* **2023**, *52*, 7021–7030. [CrossRef] [PubMed]
51. Kong, X.L.; Huang, Y. Applications of graphene in mass spectrometry. *J. Nanosci. Nanotechnol.* **2014**, *14*, 4719–4732. [CrossRef] [PubMed]
52. Bao, X.D.; Kong, X.L. Generation of M@$C_{2n}^+$ (M = Cs, Rb, K, 2n = 70–220) by Laser Ablation of Graphene. *Rapid Comm. Mass. Spectrom.* **2015**, *29*, 1774–1778. [CrossRef] [PubMed]
53. Kong, X.; Bao, X. Formation of Endohedral Metallofullerene (EMF) ions of $M_nC_{2m}^+$ (M = La, Y, n $\leq$ 6, 50 $\leq$ 2m $\leq$ 194) in the Laser Ablation Process with Graphene as Precursor. *Rapid Comm. Mass. Spectrom.* **2017**, *31*, 865–872. [CrossRef] [PubMed]
54. Wang, Y.Y.; Ren, J.; Mu, L.; Kong, X.L. Metallofullerene ions of $Lu_nC_{2m}^+$ (1 $\leq$ n $\leq$ 9, 50 $\leq$ 2m $\leq$ 198) generated by laser ablation of graphene/ $LuCl_3$. *Int. J. Mass. Spectrom.* **2017**, *422*, 105–110. [CrossRef]
55. Wang, Y.; Ma, L.; Ren, J.; Mu, L.; Kong, X.L. A systematic study on the generation of multimetallic lanthanide fullerene ions by laser ablation mass spectrometry. *Rapid Commun. Mass Spectrom.* **2018**, *32*, 1396–1402. [CrossRef] [PubMed]
56. Mu, L.; Yang, S.; Feng, R.X.; Kong, X.L. Encapsulation of Platinum in Fullerenes: Is that Possible? *Inorg. Chem.* **2017**, *56*, 6035–6038. [CrossRef] [PubMed]
57. Wang, Y.Y.; Shi, Y.Y.; Fan, X.T.; Ren, J.; Kong, X.L. Encapsulation of an Ionic Bond in Fullerenes: What is the Difference? *Inorg. Chem.* **2019**, *58*, 3601–3605. [CrossRef] [PubMed]
58. Fan, X.T.; Wang, Y.Y.; Kong, X.L. Generation of sodium halide endohedral metallofullerenes in the gas phase. *Rapid Comm. Mass. Spectrom.* **2020**, *34*, e8826. [CrossRef] [PubMed]

59. Krachmalnicoff, A.; Bounds, R.; Mamone, S.; Alom, S.; Concistrè, M.; Meier, B.; Kouril, K.; Light, M.E.; Johnson, M.R.; Rols, S.; et al. The dipolar endofullerene HF@$C_{60}$. *Nat. Chem.* **2016**, *8*, 953–957. [CrossRef] [PubMed]
60. Gao, R.; Liu, Z.; Liu, Z.; Liang, T.; Su, J.; Gan, L. Open-cage fullerene as a selective molecular trap for LiF/$[BeF]^+$. *Angew. Chem. Int. Ed.* **2023**, *62*, e202300151, *Erratum in Angew. Chem.***2023**, *135*, e202300151. [CrossRef] [PubMed]
61. Martin, T.P.; Heinebrodt, M.; Näher, U.; Göhlich, H.; Lange, T.; Schaber, H. Fullerenes doped with metal halides. *Int. J. Mod. Phys. B* **1992**, *6*, 3871–3877. [CrossRef]
62. Li, J.; Hou, S.J.; Yao, Y.R.; Zhang, C.Y.; Wu, Q.Q.; Wang, H.C.; Zhang, H.W.; Liu, X.Y.; Tang, C.; Wei, M.X.; et al. Room-temperature logic-in-memory operations in single-metallofullerene devices. *Nat. Mater.* **2022**, *21*, 917–923. [CrossRef] [PubMed]
63. Kang, D.; Yao, Y.; Su, Z.; Xu, H.L. Probing the structure–property relationships of Na+···Cl–@$C_{50}N_5H_5$ under the external electric field. *Inorg. Chem.* **2022**, *61*, 17646–17652. [CrossRef] [PubMed]
64. Jaros, A.; Bonab, E.F.; Straka, M.; Foroutan-Nejad, C. Fullerene-based switching molecular diodes controlled by oriented external electric fields. *J. Am. Chem. Soc.* **2019**, *141*, 19644–19654. [CrossRef] [PubMed]
65. Foroutan-Nejad, C.; Andrushchenko, V.; Straka, M. Dipolar molecules inside $C_{70}$: An electric field-driven room-temperature single-molecule switch. *Phys. Chem. Chem. Phys.* **2016**, *18*, 32673–32677. [CrossRef]
66. Sabirov, D.S.; Terentyev, A.O.; Shepelevich, I.S.; Bulgakov, R.G. Inverted thermochemistry of "norbornadiene–quadricyclane" molecular system inside fullerene nanocages. *Comput. Theor. Chem.* **2014**, *1045*, 86–92. [CrossRef]
67. Sabirov, D.S.; Terentyev, A.O.; Sokolov, V.I. Activation energies and information entropies of helium penetration through fullerene walls. Insights into the formation of endofullerenes $nX@C_{60/70}$ ($n$ = 1 and 2) from the information entropy approach. *RSC Adv.* **2016**, *6*, 72230–72237. [CrossRef]
68. Yu, P.; Li, M.; Shen, W.; Hu, S.; Yu, P.Y.; Tian, X.; Zhao, X.; Bao, L.; Lu, X. An unprecedented $C_{80}$ cage that violates the isolated pentagon rule. *Inorg. Chem. Front.* **2022**, *9*, 2264–2270. [CrossRef]
69. Brotsman, V.A.; Ioffe, I.N.; Troyanov, S.I. Crippling the $C_{70}$ fullerene: Non-classical $C_{68}Cl_{26}(OH)_2$ and $C_{68}Cl_{25}(OH)_3$ with three heptagons and only fused pentagons via chlorination-promoted skeletal transformations. *Chem. Commun.* **2022**, *58*, 6918–6921. [CrossRef]
70. Kirkpatrick, J.; McMorrow, B.; Turban, D.H.P.; Gaunt, A.L.; Spencer, J.S.; Matthews, A.G.D.G.; Obika, A.; Thiry, L.; Fortunato, M.; Pfau, D.; et al. Pushing the frontiers of density functionals by solving the fractional electron problem. *Science* **2021**, *374*, 1385–1389. [CrossRef]
71. Irle, S.; Zheng, G.; Wang, Z.; Morokuma, K. The $C_{60}$ formation puzzle "solved": QM/MD simulations reveal the shrinking hot giant road of the dynamic fullerene self-assembly mechanism. *J. Phys. Chem. B* **2006**, *110*, 14531–14545. [CrossRef]
72. Curl, R.F.; Lee, M.K.; Scuseria, G.E. $C_{60}$ buckminsterfullerene high yields unraveled. *J. Phys. Chem. A* **2008**, *112*, 11951–11955. [CrossRef] [PubMed]
73. Cross, R.J.; Saunders, M. Transmutation of fullerenes. *J. Am. Chem. Soc.* **2005**, *127*, 3044–3047. [CrossRef] [PubMed]
74. Chuvilin, A.; Kaiser, U.; Bichoutskaia, E.; Besley, N.A.; Khlobystov, A.N. Direct transformation of graphene to fullerene. *Nat. Chem.* **2010**, *2*, 450–453. [CrossRef] [PubMed]
75. Li, Z.; Cheng, Z.; Wang, R.; Li, Q.; Fang, Y. Spontaneous formation of nanostructures in graphene. *Nano Lett.* **2009**, *9*, 3599–3602. [CrossRef]
76. Shenoy, V.B.; Reddy, C.D.; Zhang, Y.W. Spontaneous curling of graphene sheets with reconstructed edges. *ACS Nano* **2010**, *4*, 4840–4844. [CrossRef] [PubMed]
77. Berné, O.; Tielens, A.G.G.M. Formation of buckminsterfullerene ($C_{60}$) in interstellar space. *Proc. Natl. Acad. Sci. USA* **2012**, *109*, 401–406. [CrossRef] [PubMed]
78. Berné, O.; Montillaud, J.; Joblin, C. Top-down formation of fullerenes in the interstellar medium. *Astron. Astrophys.* **2015**, *588*, C1. [CrossRef]
79. Cai, W.T.; Li, F.F.; Bao, L.P.; Xie, Y.P.; Lu, X. Isolation and crystallographic characterization of $La_2C_2$@Cs(574)-$C_{102}$ and $La_2C_2$@C2(816)-$C_{104}$: Evidence for the top-down formation mechanism of fullerenes. *J. Am. Chem. Soc.* **2016**, *138*, 6670–6675. [CrossRef]
80. Zhang, J.Y.; Bowles, F.L.; Bearden, D.W.; Ray, W.K.; Fuhrer, T.; Ye, Y.Q.; Dixon, C.; Harich, K.; Helm, R.F.; Olmstead, M.M.; et al. A missing link in the transformation from asymmetric to symmetric metallofullerene cages implies a top-down fullerene formation mechanism. *Nat. Chem.* **2013**, *5*, 880–885. [CrossRef]
81. Li, M.Y.; Zhao, Y.X.; Han, Y.B.; Yuan, K.; Nagase, S.; Ehara, M.; Zhao, X. Theoretical investigation of the key roles in fullerene-formation mechanisms: Enantiomer and enthalpy. *ACS Appl. Nano Mater.* **2020**, *3*, 547–554. [CrossRef]
82. Mulet-Gas, M.; Abella, L.; Cerón, M.R.; Castro, E.; Marshall, A.G.; Rodríguez-Fortea, A.; Echegoyen, L.; Poblet, J.M.; Dunk, P.W. Transformation of doped graphite into cluster-encapsulated fullerene cages. *Nat. Commun.* **2017**, *8*, 1222. [CrossRef] [PubMed]
83. Yamaguchi, Y.; Maruyama, S. A molecular dynamics study on the formation of metallofullerene. *Eur. Phys. J. D* **1999**, *9*, 385–388. [CrossRef]
84. Deng, Q.M.; Heine, T.; Irle, S.; Popov, A.A. Self-assembly of endohedral metallofullerenes: A decisive role of cooling gas and metal–carbon bonding. *Nanoscale* **2016**, *8*, 3796–3808. [CrossRef]
85. Gan, L.H.; Lei, D.; Fowler, P.W. Structural interconnections and the role of heptagonal rings in endohedral trimetallic nitride template fullerenes. *J. Comput. Chem.* **2016**, *37*, 1907–1913. [CrossRef]

86. Fan, H.; Liu, Z.; Gan, L.H.; Wang, C.R. The formation mechanism of Sc-based metallofullerenes: A molecular dynamics simulation study. *Phys. Chem. Chem. Phys.* **2024**, *26*, 5499–5507. [CrossRef] [PubMed]
87. Sinitsa, A.S.; Chamberlain, T.W.; Zoberbier, T.; Lebedeva, I.V.; Popov, A.M.; Knizhnik, A.A.; McSweeney, R.L.; Biskupek, J.; Kaiser, U.; Khlobystov, A.N. Formation of Nickel Clusters Wrapped in Carbon Cages: Toward New Endohedral Metallofullerene Synthesis. *Nano Lett.* **2017**, *17*, 1082–1089. [CrossRef] [PubMed]

*inorganics*

*Review*

# Capturing Unstable Metallofullerenes

**Fupin Liu * and Alexey A. Popov ***

Leibniz Institute for Solid State and Materials Research (IFW Dresden), 01069 Dresden, Germany
* Correspondence: liu_fupin@outlook.com (F.L.); a.popov@ifw-dresden.de (A.A.P.)

**Abstract:** Metallofullerenes are interesting molecules with unique structures and physicochemical properties. After they are formed in the arc-discharge process, they are first buried in the carbon soot, which requires solvent extraction to fish them out, normally followed by HPLC separation. In this minireview, we summarize the main procedures developed to obtain pure metallofullerenes, including well-established extraction with conventional fullerene solvents followed by HPLC (procedure (**I**) as well as several methods developed for isolation and purification of unstable fullerenes insoluble in conventional fullerene solvents, including chemical modification followed by dissolution (**II.1**), chemical functionalization *during* extraction followed by HPLC (**II.2**), and chemical functionalization of ionic EMFs *after* redox-extraction followed by HPLC (procedure **II.3**). The main focus here is on procedure **II.3**, for which the current status and future perspective are discussed.

**Keywords:** endohedral metallofullerenes; extraction; open-shell electronic structure; electrophilic trifluoromethylation; benzylation

## 1. Introduction

Endohedral metallofullerenes (EMFs) are molecules with metal ions encapsulated in a fullerene cage, which are interesting because of their unique coordination between metal ions and the fullerene cage, resulting in fascinating physicochemical properties [1–52]. The pursuit of novel EMF structures has been a strong impetus in the field, not only for the discovery of a bizarre interplay between metal ions and fullerene cages, but also for their appealing application potentials endowed by this unique interaction. Metallofullerenes are usually prepared by direct current arc discharge of graphite rods embedded with a metal source. However, this process simultaneously produces EMFs, empty fullerenes, as well as other carbonaceous species, and the EMFs are then fished out of the carbon soot. The general procedure uses extraction with some organic solvents to dissolve metallofullerenes from the carbon soot, followed by separation procedures to obtain pure compounds (Procedure **I**, Figure 1). $CS_2$ is the most commonly used solvent during the extraction, although toluene, chlorobenzene and other aromatic solvents are also applied occasionally; we will refer to this group of solvents as **fullerene solvents** in the following discussion. Separation of the dissolved fullerenes is usually performed by high-performance liquid chromatography (HPLC), but may also include some non-HPLC steps.

Procedure **I** prevails in the EMF research because many fullerenes/metallofullerenes are well soluble in these solvents, and the extraction procedure is relatively straightforward. However, there is a certain fraction of fullerenes that show moderate to negligible solubility in fullerene solvents. These fullerenes therefore remain in the soot after extraction with fullerene solvents, but their formation in the arc-discharge synthesis could be demonstrated by extraction with other solvents or chemical functionalization, as will be discussed further in this review. In the following, we define **stable fullerenes** as those fullerenes that can be routinely obtained by Procedure **I**, while **unstable fullerenes** are those that cannot be obtained by this procedure due to poor solubility in fullerene solvents at the extraction and separation stages. The boundary between stable and unstable fullerenes is not clearly defined. Fullerenes are usually considered stable if their solution in fullerene solvents remains intact for months, i.e., long enough to complete the separation and characterization procedures.

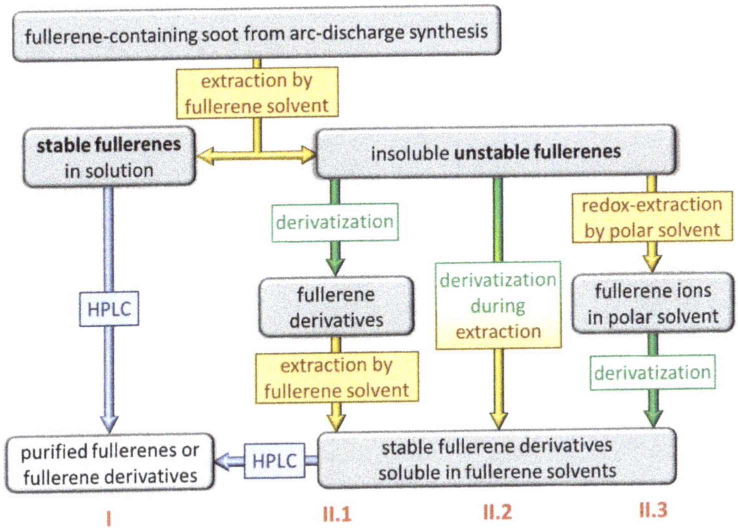


**Figure 1.** Schematic description of fullerene separation procedures discussed in this minireview.

## 2. Unstable Fullerenes

It should be emphasized that the stability discussed here is not stability in the thermodynamic sense. The **unstable fullerenes**, which are the focus of this review, are formed during synthesis, and if we assume that the yield of fullerenes in the arc-discharge process is controlled by thermodynamic factors, we should also assume **thermodynamic stability** of these unstable fullerenes. In other words, the relative energies of these structures with respect to different cage isomers should be low, and calculations of equilibrium concentration at the temperature of the synthesis will likely predict considerable concentration for these fullerenes. Therefore, it is their low **kinetic stability** or, equivalently, high reactivity, that poses the problem for their isolation and characterization. The high reactivity is usually related to the electronic structure and can be by deduced from the analysis of the frontier orbital levels. Examples of frontier orbital situations in three types of fullerenes, which often have unstable structures among them, are shown in Figure 2:

(1) Empty fullerenes. Soluble empty fullerenes have a large HOMO-LUMO gap and a closed-shell electronic structure with the singlet ground state. However, if the HOMO-LUMO gap is small, the triplet (biradical) state may be accessible or even become lower in energy than the singlet (Figure 2a). The molecule thus has an open-shell electronic structure, which is usually synonymous with enhanced reactivity. A typical example of an insoluble empty fullerene with a triplet ground state is $D_{3h}$-$C_{74}$ [53].

(2) Monometallofullerenes with trivalent metals, $M^{III}@C_{2n}$ (Figure 2b). Since endohedral metal atoms transfer their electrons to the fullerene cage, the $M^{III}$ state implies that the fullerene cage accepts three electrons and thus has a radical nature with one unpaired electron. Interestingly, some monometallofullerenes, such as $M^{III}@C_{2v}(9)$-$C_{82}$, behave as stable fullerenes and can be obtained in pure form by procedure **I**, while the others, such as $M^{III}@C_{60}$ or $M^{III}@C_{2v}(5)$-$C_{80}$, are insoluble in fullerene solvents and require special procedures for extraction and isolation. $M^{III}@C_s(6)$-$C_{82}$ is a good example of a dual behavior as it shows reasonable solubility and can be obtained by conventional procedure **I**, but tends to form dimers upon crystallization [54,55]. The boundary

between the two types is blurred, indicating that the peculiarities of the π-system distribution for a given carbon cage also play an important role.

(3) Dimetallofullerenes of trivalent metals (Figure 2c). Their situation in some way resembles that of empty fullerenes since the transfer of six electrons from the metal dimer to the host fullerene should result in a closed-shell electronic structure. Indeed, dimetallofullerenes of early lanthanides, such as $La_2@C_{2n}$ and $Ce_2@C_{2n}$, are soluble in fullerene solvents and can be purified using procedure **I**. The difference from empty fullerenes is that the LUMO of dimetallofullerenes is usually a metal–metal bonding orbital, and its energy depends on the metal. For heavier lanthanides, starting with Nd, the energy of the metal-based LUMO becomes so low that the triplet state, in which one unpaired electron occupies the metal-based MO and one occupies the fullerene MO, becomes more stable than the closed-shell singlet [56,57].

(4) Clusterfullerenes, i.e., metallofullerenes with endohedral clusters such as $M_3N$, $M_2C_2$, $M_2O$, $M_2S$, etc., usually have a closed-shell electronic structure of the carbon cage, and the vast majority of them can be obtained from the soot and separated by procedure **I**. It is certainly possible that some of them are unstable due to the small HOMO-LUMO gap of the carbon cage, similarly to insoluble empty fullerenes, but we are not aware of any experimentally characterized examples of clusterfullerenes that are not soluble in fullerene solvents but could be extracted and separated by other approaches. Clusterfullerenes are therefore not discussed further in this review.

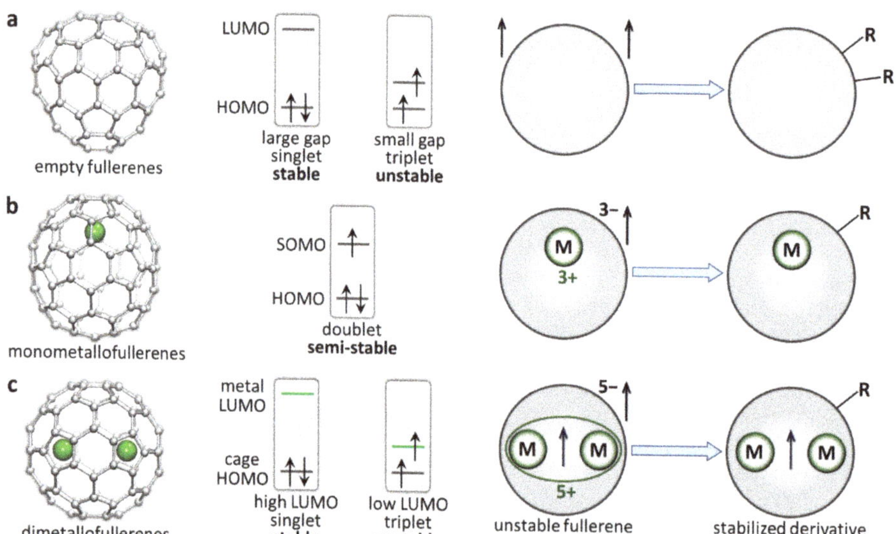


**Figure 2.** Major types of unstable fullerenes: (**a**) empty fullerenes, (**b**) monometallofullerenes with trivalent metals, (c) dimetallofullerenes with trivalent metals. For each type, the figure shows an example of a molecular structure (**left**), population of the frontier MOs (middle; fullerene MO levels are gray, metal-based MO levels are green), formal charge of the metal species and the cage (**b**,**c**) and stabilization of the open-shell electronic structure of the carbon cage by addition of radical groups (**right**). Vertical arrows denote unpaired electrons.

The low solubility of unstable fullerenes is likely caused by the tendency of fullerene cages to stabilize their open-shell structure by forming covalent bonds between fullerene molecules (dimerization/polymerization) or by forming bonds with other unidentified components of the soot. Either way, polymerization or binding to soot particles renders unstable fullerenes poorly soluble and causes them to remain in the soot during extraction with fullerene solvents. Importantly, it is the open-shell structure of the carbon cage that is critical for kinetic stability and separation of EMFs. Endohedral species can also have an open-shell electronic structure, but it does not affect the kinetic stability if the fullerene cage has closed-shell electronic structure. For instance, dimetallo-azafullerenes $M_2@C_{79}N$ (M = Y, Gd, Tb) [58,59], trimetal-carbide clusterfullerenes $M_3C_2@C_{80}$ (M = Sc, Dy, Er) [60–62] or dimetallofullerenes derivatives $M_2@C_{80}(R)$ (M = rare earth elements) [63–67] have unpaired electrons localized on their endohedral species, but this does not hinder their solubility in fullerene solvents and their overall kinetic stability.

## 3. Redox Extraction

Extensive studies of metallofullerene synthesis in the late 1990s revealed that while the solubility of some metallofullerenes in fullerene solvents is low, it can be greatly enhanced if extraction is performed using polar nitrogen-containing solvents such as pyridine, N,N-dimethylformamide (DMF), or aniline [68–79]. Subsequent studies found that EMFs in these solvents adopt the negative charge sate [80,81], suggesting that stabilization of insoluble EMFs, which allows their extraction, occurs via the electron transfer from the solvent. Indeed, since unstable metallofullerenes have an open-shell electronic structure in their pristine neutral form, the addition or removal of an electron can change the fullerene structure to the closed-shell, while endohedral metal species can still retain their spin structure. The process by which fullerenes are reduced or oxidized during extraction with a polar solvent is hereafter referred to as **redox extraction**. Redox extraction requires the presence of either reducing agents (amines [74–79], which can simultaneously be a solvent), or oxidizing agent (magic blue $(BrC_6H_4)_3NSbCl_6$, $AgSbF_6$) [82–85], or the use of electrolysis [70,86]. Ionic fullerenes are not soluble in most fullerene solvents and instead require the use of highly polar solvents, whereas uncharged fullerenes are conversely poorly soluble in many polar solvents. Therefore, redox extraction can be used not only for extraction of unstable fullerenes, but also for a partial separation of fullerenes based on their redox properties.

One of the most popular forms of redox extraction in EMF synthesis applies DMF [72,73,81,87–90], which undergoes partial thermal decomposition during extraction with the formation of amines. Hereby, amines act as mild reducing agents, facilitating the formation of EMF anions that are well soluble in DMF. Meanwhile, fullerenes with large HOMO-LUMO gap, including the majority of empty fullerenes, remain in the uncharged state. Since the solubility of neutral fullerenes in DMF is rather modest, extraction with hot DMF is quite efficient for dissolving of mono- and dimetallofullerenes and separating them from empty fullerenes [91,92].

Redox extraction helps to obtain unstable fullerenes from a soot as charged species, but their further separation is less straightforward. To date, $Li@C_{60}$ is the only unstable fullerene that has been isolated in a pure cationic form by redox methods [84,85]. HPLC procedures and column materials used for fullerene separation have been developed for uncharged fullerenes, and their efficiency for ionic species is not guaranteed. In addition, polar solvents normally used for unstable fullerenes, such as DMF or aniline, have high boiling points and are toxic, which adds obstacles for further separation by multi-step HPLC. Therefore, the separation of fullerenes extracted with DMF usually involves non-HPLC methods, such as electrochemical or chemical redox reactions, which change the charge state of the dissolved fullerenes and thus strongly affect their solubility [70,86,90,92,93]. The disadvantage of this strategy is that the redox reaction is not only controlled by the intrinsic redox potentials of the fullerenes, but also strongly affected by their concentrations. This means that very high purity separations can hardly be achieved by this method, and HPLC is still required in the final step. If dissolved anionic EMFs are also stable in the pristine form, they can be oxidized to the neutral form and then separated by HPLC [90]. However, if neutral EMFs are unstable, the oxidation of their anions will obviously produce open-shell species with ensuing low kinetic stability and solubility problems. For instance, attempts in our laboratory to obtain dimetallofullerenes of heavy lanthanides by oxidation of their DMF-soluble anions immediately formed insoluble precipitates that could not be further processed.

## 4. Chemical Functionalization of Unstable Fullerenes

### *4.1. General Remarks*

An alternative approach to stabilize fullerenes with open-shell electronic structure is their chemical derivatization by attachment of exohedral groups, which should quench unpaired electrons of the carbon cage. Obviously, each unpaired cage electron can be compensated by the addition of one radical group, and therefore, we will only consider derivatization with groups that form single bonds to the fullerene surface. Cycloaddition reactions, which usually introduce two bonds to a fullerene at once, are not suitable for stabilization of EMFs with the open-shell electronic structure of the fullerene cage. Figure 2 sketches the minimum number of groups that are required to stabilize different types of unstable fullerenes:

(1) Empty fullerenes with triplet ground state require the addition of at least two radical groups to quench the cage biradical. However, it is not guaranteed that the bisadduct $C_{2n}(R)_2$ will have a large HOMO-LUMO gap. Depending on the MO structure of the molecule, the bisadduct may also have a small gap, thus requiring the addition of another pair of radical groups. However, regardless of how many pairs of radical groups are ultimately required to obtain a stable derivative, it is certain that the number of groups will be even.
(2) Monometallofullerenes $M^{III}@C_{2n}$ have one unpaired electron, and therefore require the addition of at least one radical group. In general, the number of groups in the stable closed-shell $M^{III}@C_{2n}(R)_x$ derivative should be odd.
(3) Open-shell dimetallofullerenes of trivalent metals are peculiar in that of the two unpaired valence electrons, one is localized on the fullerene cage and one on the metal dimer. The latter does not seem to be relevant for the kinetic stability, which means that only the unpaired electron of the cage should be quenched. Therefore, dimetallofullerenes require the addition of at least one and generally odd number of groups, similarly to monometallofullerenes.

Once unstable fullerenes are stabilized by derivatization, they can be processed akin to stable fullerenes, since derivatives are soluble in fullerene solvents and can be separated by conventional HPLC to achieve high-purity separation. This strategy will be referred to as Procedure **II**.

### *4.2. Derivatization Followed by Dissolution*

Depending on the sequence of derivatization and extraction, three modifications of Procedure **II** can be distinguished, as shown in Figure 1. In method **II.1**, chemical derivatization is first applied to insoluble unstable fullerenes. A classic example is the thermal radical trifluoromethylation of a mixture of insoluble empty fullerenes (left after pre-extraction of stable fullerenes with *o*-dichlorobenzene) by $CF_3I$ gas [94]. This procedure resulted in a series of $C_{2n}(CF_3)_{12}$ derivatives that exhibited high HOMO-LUMO gaps and good solubility in fullerene solvents. Separation of the mixture by HPLC and subsequent $^{19}F$ NMR and/or single-crystal X-ray diffraction studies of purified derivatives revealed the formation of some previously unreported empty fullerenes, such as $D_{3h}$-$C_{74}$, $T_d$-$C_{76}$, $D_{3h}(5)$-$C_{78}$, $C_{2v}(5)$-$C_{80}$, or $C_2(5)$-$C_{82}$ [94,95]. More recently, Shinohara et al. modified the method by adding polytetrafluoroethylene into the arc-discharge generator, thus performing the reaction in situ, before any extraction step [96]. Applying this procedure in the synthesis of Y, La, and Gd EMFs, they were able to isolate $CF_3$ derivatives of several elusive monometallofullerenes, including $M@C_{60}$, $M@C_{70}$, $M@C_{72}$, $M@C_{74}$, etc., and determine their molecular structures by single-crystal X-ray diffraction [96–99]. In situ chlorination [100–102] and hydrogenation [103–105] during fullerene formation, which allowed the capture of many unstable and nonclassical empty fullerenes in the form of chlorides and hydrides, can also be considered as a version of Procedure **II.1**, but its application to metallofullerenes has not been very successful so far: halogenated EMF species have been detected in mass-spectra, but not yet isolated in bulk form [106].

*4.3. In Situ Derivatization during Exraction*

One notable breakthrough in the stabilization and isolation of unstable metallofullerenes by their chemical derivatization was made by Akasaka and colleagues (Procedure **II.2** in Figure 1). They found that halogenated benzenes, such as 1,2,4-trichlorobenzene [25] or iodobenzene [107], which act as good fullerene solvents for stable fullerenes, can also be a source of aryl radicals that functionalize unstable fullerenes and make them soluble in the form of aryl monoadducts. The solvent thus plays a dual role, as a reactant in the addition reaction of unstable fullerenes and simultaneously as a solvent for stable fullerenes and stable derivatives of unstable fullerenes. Conventional HPLC can then be used in further separation steps. This in situ chemical functionalization during the extraction was successful in fishing out a number of unstable metallofullerenes, such as La@$C_{72}$ [108] La@$C_{74}$ [109,110], La@$C_{80}$ [111], and minor isomers of La@$C_{82}$ [112]. Since these EMFs could not be isolated in their pristine form by Procedure **I**, they were dubbed as "missing fullerenes" by the authors. A detailed review of this strategy can be found in Ref. [25]. The success of procedure **II.2** is limited by the reactivity of the applied solvent/reactant to the targeted metallofullerenes. It has not been possible to identify a universal solvent/reactant for all types of metallofullerenes, because the reactivity of metallofullerenes varies so much based on the topology of the $\pi$-electron distribution and the energetics of the frontier orbitals. For example, we have found that the derivatization and isolation of highly interesting dimetallofullerenes with a single-electron metal–metal bond is not feasible via the in situ chemical functionalization strategy.

*4.4. Redox-Extraction and Derivatization of Anions*

The stabilization/separation strategy that appeared to be successful for unstable dimetallofullerenes is denoted in Figure 1 as Procedure **II.3**. In the first step, EMFs are redox-extracted from the soot with hot DMF, resulting in a mixture of EMF anions. As discussed above, the separation of the EMF anions is not straightforward. Instead, EMF anions can be chemically functionalized to produce stable neutral derivatives. For instance, thermal trifluoroalkylation of a dried Y-EMF mixture extracted with DMF gave a mixture of EMF-$CF_3$ derivatives, from which pure Y@$C_{82}(CF_3)_5$ was isolated by HPLC and characterized by NMR [113].

Recently, we have developed two versions of Procedure **II.3**, which involved the functionalization of the metallofullerene anions with the formation of neutral compounds. Hereby, the insoluble metallofullerenes are converted into derivatives that are soluble in fullerene solvents and can be further separated by HPLC to obtain the derivatives of desired metallofullerenes in high purity. In the first method, we successfully applied benzyl bromide $PhCH_2Br$ to react with the metallofullerene anions in DMF [66]. The reaction required mild heating and presumably proceeded as a nucleophilic substitution of bromide with metallofullerene anions, producing benzyl monoadducts of metallofullerenes (see Figure 3a for details). An alternative mechanism, proposed in Ref. [114], involves the electron transfer from the EMF anion to benzyl bromide in the first step, followed by the radical coupling reaction. The benzyl derivatives of metallofullerenes are well soluble in fullerene solvents, allowing their subsequent HPLC separation to yield the high-purity compounds. Using this strategy, we have synthesized and isolated a series of dimetallofullerene derivatives $M_2@I_h$-$C_{80}(CH_2Ph)$ ($M_2$ = $Sc_2$, $Y_2$, $Gd_2$, $Tb_2$, TbY, TbGd, $Dy_2$, $Ho_2$, $Er_2$), among which $Tb_2@C_{80}(CH_2Ph)$ and $Dy_2@C_{80}(CH_2Ph)$ show particularly remarkable single-molecule magnetic properties [20,65–67].

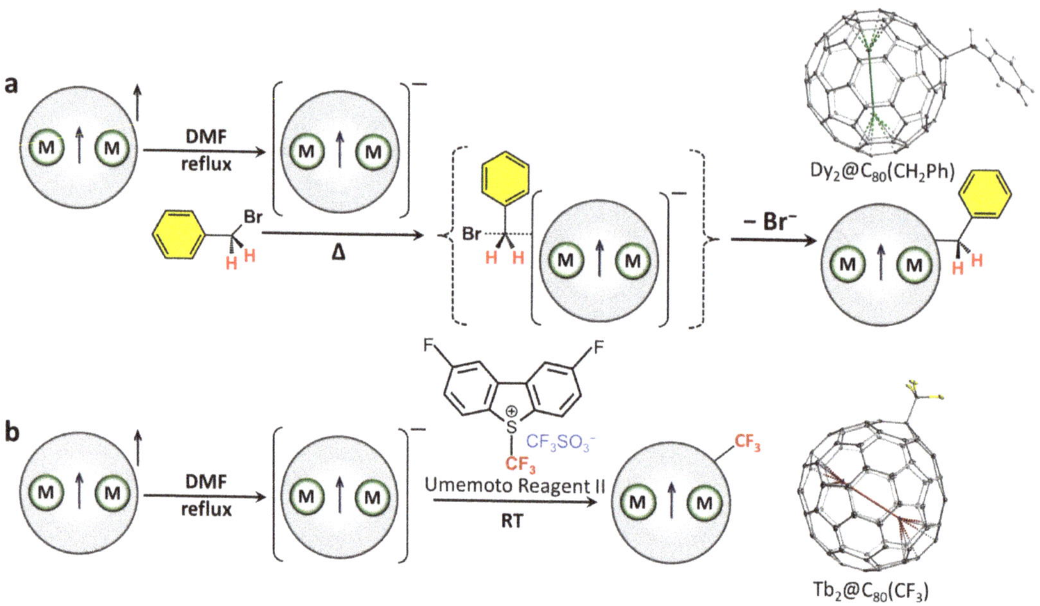


**Figure 3.** Two modifications of procedure **II.3**: (**a**) benzyl bromide route applied to stabilize $M_2@C_{80}$ in the form of $M_2@C_{80}(CH_2Ph)$; (**b**) $CF_3$ functionalization route leading to $M_2@C_{80}(CF_3)$. Also shown are molecular structures of $Dy_2@I_h$-$C_{80}(CH_2Ph)$ and $Tb_2@I_h$-$C_{80}(CF_3)$ determined by single-crystal X-ray diffraction in Refs. [66,115]. Vertical arrows denotes unpaired electron localized on the metal-metal bond.

The unusual magnetic properties are the reason why, until recently, the main focus in the exploitation of the anion functionalization strategy **II.3** has been on the dimetallofullerenes $M_2@I_h$-$C_{80}$. However, $M_2@I_h$-$C_{80}$ with a single-electron metal–metal bond is not the only unstable EMF that can be solubilized in the anionic form during DMF extraction. In fact, the DMF extract contains anions of many monometallofullerenes, and, similarly to $M_2@C_{80}{}^-$, they also undergo mono-benzylation in the reaction with benzyl bromide. This procedure can thus be used to fish out other unstable "missing" EMFs from the fullerene soot. Recently, Yang and colleagues used the benzyl bromide functionalization route to successfully isolate and structurally characterize $Dy@C_{80}(CH_2Ph)$ with the $C_{2v}(5)$ cage isomer [116]. Since $Dy@C_{2v}(5)$-$C_{80}$ is not soluble in fullerene solvents, it had never been properly characterized in the previous studies of Dy-EMFs. This example shows that the benzylation route can be very useful for the stabilization and structural elucidation of various unstable EMFs, and it is very likely that its exploration will help in the discovery of novel elusive fullerene structures.

The versatility of the thermal reaction with benzyl bromide in its application to EMF anions, while being very useful for exploratory studies, becomes a drawback when targeting a specific EMF derivative. The fact that the reaction works well with most of the known metallofullerene anions suggest that it has low selectivity to specific types of metallofullerenes. Furthermore, for EMFs with relatively low symmetry, benzylation does not show particularly high regioselectivity and can yield several isomers of adducts. Maeda et al. showed that the thermal reaction of $La@C_{2v}(9)$-$C_{82}$ anion with benzyl bromide produces two major and two minor isomers of $La@C_{82}(CH_2Ph)$ [114]. Thus, the benzylation of an anionic EMF extract that already contains many EMFs produces an even greater variety of benzyl adducts. The complexity of the product mixture then complicates the HPLC separation. Figure 4 illustrates that the isolation of $Tb_2@C_{80}(CH_2Ph)$ after benzylation of the Tb-EMF anionic mixture is a rather tedious process, requiring three HPLC steps, including recycling HPLC [65].

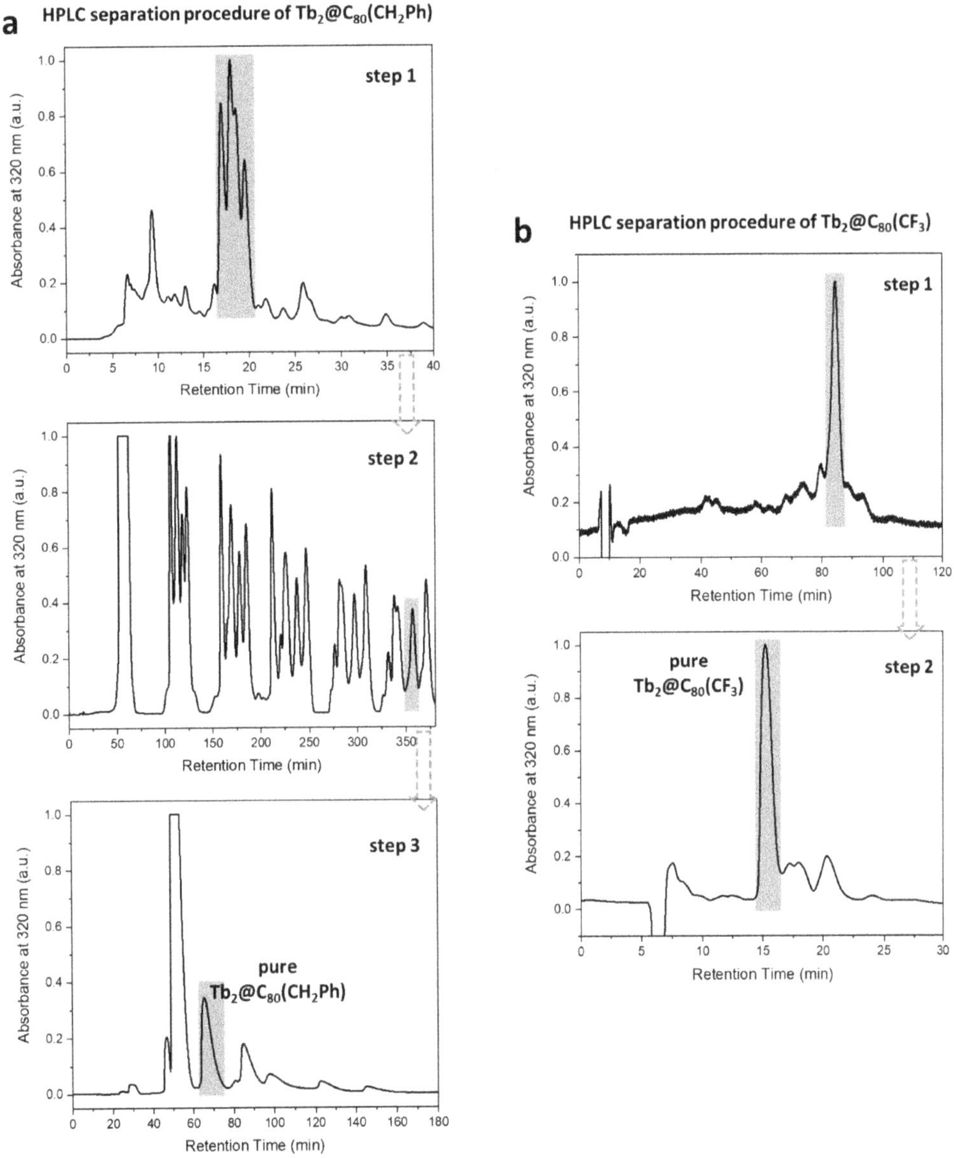


**Figure 4.** HPLC separation procedure of crude reaction products: (**a**) thermal benzylation of the Tb-EMF anionic extract in DMF, the target compound—$Tb_2@C_{80}(CH_2Ph)$ [65]; (**b**) electrophilic trifluoromethylation of Tb-EMF anions in DMF at room temperature, the target compound—$Tb_2@C_{80}(CF_3)$ [115]. Shaded are the fractions containing the target compound and subjected to the next separation step. Note that $Tb_2@C_{80}(CH_2Ph)$ is only a small fraction of the crude product in (**a**), while $Tb_2@C_{80}(CF_3)$ is the main component of the crude product in (**b**). Adapted with permission from Ref. [65]. Copyright 2019, Springer Nature. Adapted with permission from Ref. [115]. Copyright 2021 American Chemical Society.

Another possible source of complication in benzylation is that EMF monoanions do not react with benzyl bromide at room temperature, and the reaction requires heating to proceed. However, we have observed that even slight overheating changes the reaction mechanism from controllable monoaddition to uncontrolled polyaddition of benzyl groups. Presumably, heating facilitates homolytic cleavage of the C–Br bond and formation of benzyl radicals, which then react with fullerene anions by a radical addition mechanism. The latter appeared to be very unselective, and the resulting mixtures of benzyl multiadducts could not be separated into individual compounds. Lowering the reaction temperature requires an increase in the reactivity of EMF anions, which can be achieved by the reduction beyond the monoanionic state. Li et al. showed that benzyl bromide reacts with the Gd@$C_{2v}$(9)-$C_{82}^{3-}$ trianion already at room temperature to give several isomers of Gd@$C_{82}(CH_2Ph)$ [117].

Seeking for a different functionalization reaction for EMF anions that would help to circumvent the drawbacks of the benzyl bromide route, we found that the electrophilic $CF_3$ addition can be particularly useful for the isolation of $M_2$@$I_h$-$C_{80}$ species [115]. Umemoto's Reagent II (Figure 3b), which can be formally described as a source of $CF_3^+$, reacts with anionic EMF extract in DMF instantaneously and at room temperature. DMF is then removed under reduced pressure and the residue is extracted with toluene, in which neutral $CF_3$ derivatives are well soluble. The toluene solution can then be subjected to further analysis and HPLC separation. When the $CF_3^+$ reagent was taken in a large excess, the reaction gave $CF_3$ multiadducts. However, it showed remarkable selectivity toward $[Tb_2@I_h\text{-}C_{80}]^-$, and under an optimized reagent ratio, $Tb_2$@$I_h$-$C_{80}(CF_3)$ was obtained as the main toluene-soluble product. As a result, the time required to isolate the pure monoadduct was dramatically reduced compared to the isolation of $Tb_2$@$I_h$-$C_{80}(CH_2Ph)$, since the HPLC isolation of $Tb_2$@$I_h$-$C_{80}(CF_3)$ required a much smaller number of injections and could be accomplished in two simple steps (see Figure 4, right panel for details). Electrophilic trifluoromethylation thus paves the way for further in-depth studies of the properties and applications of dimetallofullerenes $M_2$@$I_h$-$C_{80}$ with metal–metal bonds. The method is less efficient for EMFs with other carbon cages, as their trifluoromethylation is significantly less selective, and the yield of their monoadducts is considerably lower. Nonetheless, since $CF_3$ adducts crystallize easier than their benzyl analogs, the $CF_3$ route should be more efficient in the structural studies of unstable fullerenes.

In another recent work, we used electrophilic trifluoromethylation to elucidate the stability and molecular structures of Nd dimetallofullerenes [56]. A comparison of the extraction of arc-discharge soot containing Pr and Nd EMFs showed that while $Pr_2$@$C_{80}$ and other Pr dimetallofullerenes are soluble in $CS_2$, Nd dimetallofullerenes could not be extracted by $CS_2$, but were soluble in hot DMF. Thus, the boundary between stable and unstable lanthanide dimetallofullerenes was found to be between Pr and Nd. The anionic Nd-EMF extract in DMF was then reacted with Umemoto's Reagent II and showed similar selectivity toward $Nd_2$@$I_h$-$C_{80}$ as in the earlier study of Tb-EMFs. At the same time, analysis of minor HPLC fractions identified another isomer of $Nd_2$@$C_{80}(CF_3)$. Using single-crystal X-ray diffraction, the molecular structures of two isolated isomers were elucidated as $Nd_2$@$D_{5h}$(6)-$C_{80}(CF_3)$ and $Nd_2$@$I_h$(7)-$C_{80}(CF_3)$. Figure 5 illustrates the transition from unstable to stable forms of $Nd_2$@$C_{80}$ isomers by plotting the evolution of the spin density distribution from unstable $Nd_2$@$C_{80}$ to stable anion $[Nd_2@C_{80}]^-$ and stable neutral $Nd_2$@$C_{80}(CF_3)$. The presence of the single-electron Nd–Nd bond in all three forms is well evidenced by the spin density lobes localized between Nd atoms, while the spin density on the fullerene cage is present only in $Nd_2$@$C_{80}$, but is absent in the anion and $CF_3$ monoadduct.

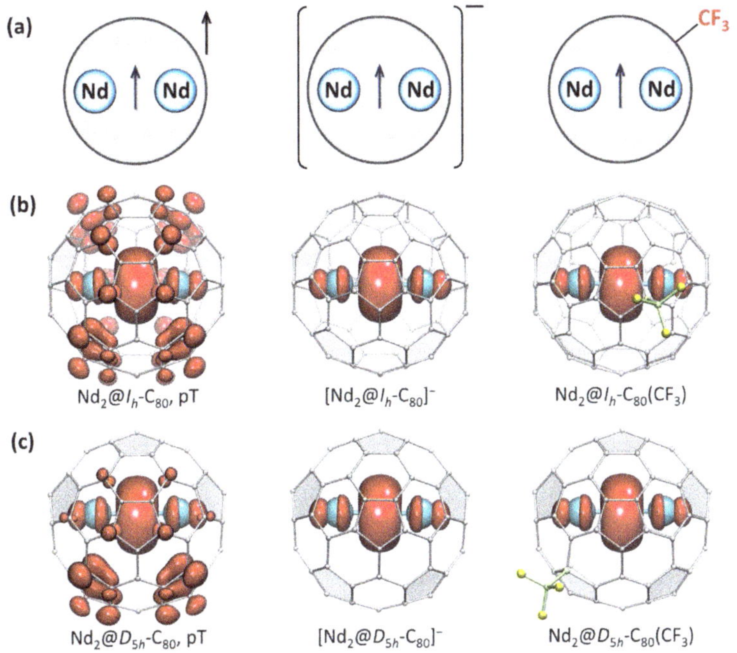


**Figure 5.** (**a**) Schematic description of neutral $Nd_2@C_{80}$ in pseudo-triplet (pT) state (two unpaired valence electrons), $[Nd_2@C_{80}]^-$ anion with unpaired electron on the Nd–Nd bonding orbital, and $Nd_2@C_{80}(CF_3)$ with unpaired electron on the Nd–Nd bonding orbital. (**b**) DFT-computed valence spin density distribution in $Nd_2@C_{80}$, $[Nd_2@C_{80}]^-$ anion and $Nd_2@C_{80}(CF_3)$ with $I_h$-$C_{80}$ cage. (**c**) DFT-computed valence spin density distribution in $Nd_2@C_{80}$, $[Nd_2@C_{80}]^-$ anion and $Nd_2@C_{80}(CF_3)$ with $D_{5h}$-$C_{80}$ cage. Adapted with permission from Ref. [56]. Copyright 2023 The Authors, published by Wiley-VCH GmbH.

## 5. Conclusions

In summary, we overviewed the general strategies to fish out metallofullerenes from the carbon soot, with the emphasis on the recently developed anion functionalization strategy (procedure **II.3**). The most appealing potential of this approach is that it specifically targets the novel structures missing in conventional separation routine (procedure **I**). This potential of the anion functionalization to stabilize and isolate "missing" metallofullerenes is not deeply exploited yet, and it is expected that many more novel metallofullerenes will be discovered with this strategy. The discovery of these novel structures will certainly renew our understandings to the coordination structures and physicochemical properties of metallofullerenes. It is worth noting that the chemical functionalization of metallofullerenes is a burgeoning field with a large variety of reactions already studied towards improved regioselectivity and physicochemical property modification [4,118–124]. In this regard, the $PhCH_2$ and $CF_3$ addition reactions merely exemplify a proof of concept, with the vast possibilities in this direction using the currently available versatile chemical functionalization methods yet to be explored.

**Author Contributions:** Writing, funding acquisition, F.L. and A.A.P. All authors have read and agreed to the published version of the manuscript.

**Funding:** This work was funded by Deutsche Forschungsgemeinschaft (grants LI 3055/3-1, PO 1602/7-1, and PO 1602/11-1).

**Data Availability Statement:** Data are contained within the article.

**Conflicts of Interest:** The authors declare no conflict of interest.

## References


1. Popov, A.A.; Yang, S.; Dunsch, L. Endohedral Fullerenes. *Chem. Rev.* **2013**, *113*, 5989–6113. [CrossRef]
2. Lu, X.; Feng, L.; Akasaka, T.; Nagase, S. Current status and future developments of endohedral metallofullerenes. *Chem. Soc. Rev.* **2012**, *41*, 7723–7760. [CrossRef]
3. Yang, S.; Wei, T.; Jin, F. When metal clusters meet carbon cages: Endohedral clusterfullerenes. *Chem. Soc. Rev.* **2017**, *46*, 5005–5058. [CrossRef]
4. Wang, S.; Zhang, X.; Tan, X.; Li, H.; Dai, S.; Yao, B.; Liu, X.; He, Y.; Jin, F. Recent Progress on the Functionalization of Endohedral Metallofullerenes. *Inorganics* **2023**, *11*, 346. [CrossRef]
5. Li, Y.; Sun, Y.; Kopcha, W.P.; Zhang, J. Advances in Regioselective Functionalization of Endohedral Metallofullerenes. *Chin. J. Chem.* **2023**, *41*, 2025–2034. [CrossRef]
6. Kopcha, W.P.; Biswas, R.; Sun, Y.; Chueng, S.-T.D.; Dorn, H.C.; Zhang, J. Water-soluble endohedral metallofullerenes: New horizons for biomedical applications. *Chem. Commun.* **2023**, *59*, 13551–13561. [CrossRef] [PubMed]
7. Nie, H.; Zhao, C.; Shi, Z.; Jia, C.; Guo, X. Single-Molecule Fullerenes: Current Stage and Perspective. *ACS Mater. Lett.* **2022**, *11*, 1037–1052. [CrossRef]
8. Li, Y.; Kopcha, W.; Rodriguez-Fortea, A.; Zhang, J. Multicomponent Reactions Among Alkyl Isocyanides, sp reactants, and sp2 Carbon Cages. *Synlett* **2022**, *33*, 907–912.
9. Li, W.; Wang, C.; Wang, T. Metallofullertube: From Tubular Endohedral Structures to Properties. *ChemPhysChem* **2022**, *23*, e202200507. [CrossRef]
10. Li, M.; Zhao, R.; Dang, J.; Zhao, X. Theoretical study on the stabilities, electronic structures, and reaction and formation mechanisms of fullerenes and endohedral metallofullerenes. *Coord. Chem. Rev.* **2022**, *471*, 214762. [CrossRef]
11. Shen, W.; Bao, L.; Lu, X. Endohedral Metallofullerenes: An Ideal Platform of Sub-Nano Chemistry. *Chin. J. Chem.* **2022**, *40*, 275–284. [CrossRef]
12. Li, W.; Wang, C.-R.; Wang, T. Molecular structures and magnetic properties of endohedral metallofullerenes. *Chem. Commun.* **2021**, *57*, 10317–10326. [CrossRef] [PubMed]
13. Akhanova, N.Y.; Shchur, D.V.; Pomytkin, A.P.; Zolotarenko, A.D.; Zolotarenko, A.D.; Gavrylyuk, N.A.; Ualkhanova, M.; Bo, W.; Ang, D. Methods for the Synthesis of Endohedral Fullerenes. *J. Nanosci. Nanotechnol.* **2021**, *21*, 2446–2459. [CrossRef] [PubMed]
14. Wang, S.; Chang, Q.; Zhang, G.; Li, F.; Wang, X.; Yang, S.; Troyanov, S.I. Structural Studies of Giant Empty and Endohedral Fullerenes. *Front. Chem.* **2020**, *8*, 607712. [CrossRef] [PubMed]
15. Shen, W.; Hu, S.; Lu, X. Endohedral Metallofullerenes: New Structures and Unseen Phenomena. *Chem. Eur. J.* **2020**, *26*, 5748–5757. [CrossRef] [PubMed]
16. Jalife, S.; Arcudia, J.; Pan, S.; Merino, G. Noble gas endohedral fullerenes. *Chem. Sci.* **2020**, *11*, 6642–6652. [CrossRef]
17. Guan, R.; Chen, M.; Jin, F.; Yang, S. Strain Release of Fused Pentagons on Fullerene Cage via Chemical Functionalization. *Angew. Chem. Int. Ed.* **2020**, *59*, 1048–1073. [CrossRef] [PubMed]
18. Wang, T.; Wang, C. Functional Metallofullerene Materials and Their Applications in Nanomedicine, Magnetics, and Electronics. *Small* **2019**, *15*, e1901522. [CrossRef]
19. Spree, L.; Popov, A.A. Recent advances in single molecule magnetism of dysprosium-metallofullerenes. *Dalton Trans.* **2019**, *48*, 2861–2871. [CrossRef]
20. Liu, F.; Spree, L.; Krylov, D.S.; Velkos, G.; Avdoshenko, S.M.; Popov, A.A. Single-Electron Lanthanide-Lanthanide Bonds Inside Fullerenes toward Robust Redox-Active Molecular Magnets. *Acc. Chem. Res.* **2019**, *52*, 2981–2993. [CrossRef]
21. Jin, P.; Li, Y.; Magagula, S.; Chen, Z. Exohedral functionalization of endohedral metallofullerenes: Interplay between inside and outside. *Coord. Chem. Rev.* **2019**, *388*, 406–439. [CrossRef]
22. Feng, L.; Hao, Y.; Liu, A.; Slanina, Z. Trapping Metallic Oxide Clusters inside Fullerene Cages. *Acc. Chem. Res.* **2019**, *52*, 1802–1811. [CrossRef] [PubMed]
23. Cai, W.; Chen, C.-H.; Chen, N.; Echegoyen, L. Fullerenes as Nanocontainers That Stabilize Unique Actinide Species Inside: Structures, Formation, and Reactivity. *Acc. Chem. Res.* **2019**, *52*, 1824–1833. [CrossRef] [PubMed]
24. Yang, W.; Li, X.; Chen, N. Recent Progress on Endohedral Metallic Fullerenes. *Gen. Chem.* **2018**, *4*, 180004.
25. Yamada, M.; Akasaka, T.; Nagase, S. Salvaging Reactive Fullerenes from Soot by Exohedral Derivatization. *Angew. Chem. Int. Ed.* **2018**, *57*, 13394–13405. [CrossRef]
26. Popov, A.A. Redox-active metal–metal bonds between lanthanides in dimetallofullerenes. *Curr. Opin. Electrochem.* **2018**, *8*, 73–80. [CrossRef] [PubMed]
27. Bao, L.; Peng, P.; Lu, X. Bonding inside and outside Fullerene Cages. *Acc. Chem. Res.* **2018**, *51*, 810–815. [CrossRef]
28. Li, T.; Dorn, H.C. Biomedical Applications of Metal-Encapsulated Fullerene Nanoparticles. *Small* **2017**, *13*, 1603152. [CrossRef] [PubMed]
29. Kako, M.; Nagase, S.; Akasaka, T. Functionalization of Endohedral Metallofullerenes with Reactive Silicon and Germanium Compounds. *Molecules* **2017**, *22*, 1179. [CrossRef]
30. Cerón, M.R.; Maffeis, V.; Stevenson, S.; Echegoyen, L. Endohedral Fullerenes: Synthesis, Isolation, Mono and Bis-Functionalization. *Inorg. Chim. Acta* **2017**, *468*, 16–27. [CrossRef]

31. Abella, L.; Wang, Y.; Rodríguez-Fortea, A.; Chen, N.; Poblet, J.M. Current status of oxide clusterfullerenes. *Inorg. Chim. Acta* **2017**, *468*, 91–104. [CrossRef]
32. Shinohara, H. Another big discovery—Metallofullerenes. *Philos. Trans. R. Soc. Lond. A Math. Phys. Eng. Sci.* **2016**, *374*, 20150325. [CrossRef]
33. Romero, E.L.; Echegoyen, L. Electron spin resonance spectroscopy of empty and endohedral fullerenes. *J. Phys. Org. Chem.* **2016**, *29*, 781–792. [CrossRef]
34. Zhang, Y.; Popov, A.A. Transition-Metal and Rare-Earth-Metal Redox Couples inside Carbon Cages: Fullerenes Acting as Innocent Ligands. *Organometallics* **2014**, *33*, 4537–4549. [CrossRef]
35. Zhang, J.; Dorn, H.C. NMR Studies of the Dynamic Motion of Encapsulated Ions and Clusters in Fullerene Cages: A Wheel Within a Wheel. *Fuller. Nanotub. Carbon Nanostruct.* **2014**, *22*, 35–46. [CrossRef]
36. Wang, T.; Wang, C. Endohedral Metallofullerenes Based on Spherical $I_h$-$C_{80}$ Cage: Molecular Structures and Paramagnetic Properties. *Acc. Chem. Res.* **2014**, *47*, 450–458. [CrossRef] [PubMed]
37. Lu, X.; Bao, L.; Akasaka, T.; Nagase, S. Recent progress in the chemistry of endohedral metallofullerenes. *Chem. Commun.* **2014**, *50*, 14701–14715. [CrossRef] [PubMed]
38. Jin, P.; Tang, C.; Chen, Z. Carbon Atoms Trapped in Cages: Metal Carbide Clusterfullerenes. *Coord. Chem. Rev.* **2014**, *270–271*, 89–111. [CrossRef]
39. Ghiassi, K.B.; Olmstead, M.M.; Balch, A.L. Gadolinium-containing endohedral fullerenes: Structures and function as magnetic resonance imaging (MRI) agents. *Dalton Trans.* **2014**, *43*, 7346–7358. [CrossRef] [PubMed]
40. Garcia-Borras, M.; Osuna, S.; Luis, J.M.; Swart, M.; Sola, M. The role of aromaticity in determining the molecular structure and reactivity of (endohedral metallo)fullerenes. *Chem. Soc. Rev.* **2014**, *43*, 5089–5105. [CrossRef]
41. Cerón, M.R.; Li, F.-F.; Echegoyen, L.A. Endohedral fullerenes: The importance of electronic, size and shape complementarity between the carbon cages and the corresponding encapsulated clusters. *J. Phys. Org. Chem.* **2014**, *27*, 258–264. [CrossRef]
42. Zhang, J.; Stevenson, S.; Dorn, H.C. Trimetallic Nitride Template Endohedral Metallofullerenes: Discovery, Structural Characterization, Reactivity, and Applications. *Acc. Chem. Res.* **2013**, *46*, 1548–1557. [CrossRef] [PubMed]
43. Xie, Y.-P.; Lu, X.; Akasaka, T.; Nagase, S. New features in coordination chemistry: Valuable hints from X-ray analyses of endohedral metallofullerenes. *Polyhedron* **2013**, *52*, 3–9. [CrossRef]
44. Lu, X.; Akasaka, T.; Nagase, S. Carbide Cluster Metallofullerenes: Structure, Properties, and Possible Origin. *Acc. Chem. Res.* **2013**, *46*, 1627–1635. [CrossRef] [PubMed]
45. Cong, H.; Yu, B.; Akasaka, T.; Lu, X. Endohedral metallofullerenes: An unconventional core–shell coordination union. *Coord. Chem. Rev.* **2013**, *257*, 2880–2898. [CrossRef]
46. Yang, S. Synthesis, Separation, and Molecular Structures of Endohedral Fullerenes. *Curr. Org. Chem.* **2012**, *16*, 1079–1094. [CrossRef]
47. Popov, A.A.; Avdoshenko, S.M.; Pendas, A.M.; Dunsch, L. Bonding between strongly repulsive metal atoms: An oxymoron made real in a confined space of endohedral metallofullerenes. *Chem. Commun.* **2012**, *48*, 8031–8050. [CrossRef]
48. Akasaka, T.; Lu, X. Structural and electronic properties of endohedral metallofullerenes. *Chem. Rec.* **2012**, *12*, 256–269. [CrossRef]
49. Yang, S.; Liu, F.; Chen, C.; Jiao, M.; Wei, T. Fullerenes encaging metal clusters-clusterfullerenes. *Chem. Commun.* **2011**, *47*, 11822–11839. [CrossRef]
50. Rodriguez-Fortea, A.; Balch, A.L.; Poblet, J.M. Endohedral metallofullerenes: A unique host-guest association. *Chem. Soc. Rev.* **2011**, *40*, 3551–3563. [CrossRef]
51. Popov, A.A.; Dunsch, L. Electrochemistry In Cavea: Endohedral Redox Reactions of Encaged Species in Fullerenes. *J. Phys. Chem. Lett.* **2011**, *2*, 786–794. [CrossRef]
52. Liu, G.; Wu, Y.; Porfyrakis, K. Synthesis and Chemistry of Endohedral Fullerenes. *Curr. Org. Chem.* **2011**, *15*, 1197–1207. [CrossRef]
53. Kovalenko, V.I.; Khamatgalimov, A.R. Open-shell fullerene $C_{74}$: Phenalenyl-radical substructures. *Chem. Phys. Lett.* **2003**, *377*, 263–268. [CrossRef]
54. Hu, S.; Liu, T.; Shen, W.; Slanina, Z.; Akasaka, T.; Xie, Y.; Uhlik, F.; Huang, W.; Lu, X. Isolation and Structural Characterization of Er@$C_{2v}$(9)-$C_{82}$ and Er@$C_s$(6)-$C_{82}$: Regioselective Dimerization of a Pristine Endohedral Metallofullerene Induced by Cage Symmetry. *Inorg. Chem.* **2019**, *58*, 2177–2182. [CrossRef]
55. Bao, L.; Pan, C.; Slanina, Z.; Uhlik, F.; Akasaka, T.; Lu, X. Isolation and Crystallographic Characterization of the Labile Isomer of Y@$C_{82}$ Cocrystallized with Ni(OEP): Unprecedented Dimerization of Pristine Metallofullerenes. *Angew. Chem. Int. Ed.* **2016**, *55*, 9234–9238. [CrossRef]
56. Yang, W.; Velkos, G.; Rosenkranz, M.; Schiemenz, S.; Liu, F.; Popov, A.A. Nd-Nd Bond in $I_h$ and $D_{5h}$ Cage Isomers of $Nd_2$@$C_{80}$ Stabilized by Electrophilic $CF_3$ Addition. *Adv. Sci.* **2024**, *11*, 2305190. [CrossRef]
57. Wang, Z.; Kitaura, R.; Shinohara, H. Metal-Dependent Stability of Pristine and Functionalized Unconventional Dimetallofullerene $M_2$@$I_h$-$C_{80}$. *J. Phys. Chem. C* **2014**, *118*, 13953–13958. [CrossRef]
58. Fu, W.; Zhang, J.; Fuhrer, T.; Champion, H.; Furukawa, K.; Kato, T.; Mahaney, J.E.; Burke, B.G.; Williams, K.A.; Walker, K.; et al. $Gd_2$@$C_{79}$N: Isolation, Characterization, and Monoadduct Formation of a Very Stable Heterofullerene with a Magnetic Spin State of S = 15/2. *J. Am. Chem. Soc.* **2011**, *133*, 9741–9750. [CrossRef]

59. Zuo, T.; Xu, L.; Beavers, C.M.; Olmstead, M.M.; Fu, W.; Crawford, T.D.; Balch, A.L.; Dorn, H.C. $M_2@C_{79}N$ (M = Y, Tb): Isolation and Characterization of Stable Endohedral Metallofullerenes Exhibiting M...M Bonding Interactions inside Aza[80]fullerene Cages. *J. Am. Chem. Soc.* **2008**, *130*, 12992–12997. [CrossRef]
60. Jin, F.; Xin, J.; Guan, R.; Xie, X.-M.; Chen, M.; Zhang, Q.; Popov, A.A.; Xie, S.-Y.; Yang, S. Stabilizing Three-Center Single-Electron Metal-Metal Bond in a Fullerene Cage. *Chem. Sci.* **2021**, *12*, 6890–6895. [CrossRef]
61. Hu, S.; Zhao, P.; Li, B.; Yu, P.; Yang, L.; Ehara, M.; Jin, P.; Akasaka, T.; Lu, X. Cluster-Geometry-Associated Metal–Metal Bonding in Trimetallic Carbide Clusterfullerenes. *Inorg. Chem.* **2022**, *61*, 11277–11283. [CrossRef]
62. Iiduka, Y.; Wakahara, T.; Nakahodo, T.; Tsuchiya, T.; Sakuraba, A.; Maeda, Y.; Akasaka, T.; Yoza, K.; Horn, E.; Kato, T.; et al. Structural determination of metallofullerene $Sc_3C_{82}$ revisited: A surprising finding. *J. Am. Chem. Soc.* **2005**, *127*, 12500–12501. [CrossRef] [PubMed]
63. Yamada, M.; Kurihara, H.; Suzuki, M.; Saito, M.; Slanina, Z.; Uhlik, F.; Aizawa, T.; Kato, T.; Olmstead, M.M.; Balch, A.L.; et al. Hiding and Recovering Electrons in a Dimetallic Endohedral Fullerene: Air-Stable Products from Radical Additions. *J. Am. Chem. Soc.* **2015**, *137*, 232–238. [CrossRef] [PubMed]
64. Bao, L.; Chen, M.; Pan, C.; Yamaguchi, T.; Kato, T.; Olmstead, M.M.; Balch, A.L.; Akasaka, T.; Lu, X. Crystallographic Evidence for Direct Metal–Metal Bonding in a Stable Open-Shell $La_2@I_h$-$C_{80}$ Derivative. *Angew. Chem. Int. Ed.* **2016**, *55*, 4242–4246. [CrossRef]
65. Liu, F.; Velkos, G.; Krylov, D.S.; Spree, L.; Zalibera, M.; Ray, R.; Samoylova, N.A.; Chen, C.-H.; Rosenkranz, M.; Schiemenz, S.; et al. Air-stable redox-active nanomagnets with lanthanide spins radical-bridged by a metal–metal bond. *Nat. Commun.* **2019**, *10*, 571. [CrossRef]
66. Liu, F.; Krylov, D.S.; Spree, L.; Avdoshenko, S.M.; Samoylova, N.A.; Rosenkranz, M.; Kostanyan, A.; Greber, T.; Wolter, A.U.B.; Büchner, B.; et al. Single molecule magnet with an unpaired electron trapped between two lanthanide ions inside a fullerene. *Nat. Commun.* **2017**, *8*, 16098. [CrossRef]
67. Zaripov, R.B.; Kandrashkin, Y.E.; Salikhov, K.M.; Büchner, B.; Liu, F.; Rosenkranz, M.; Popov, A.A.; Kataev, V. Unusually large hyperfine structure of the electron spin levels in an endohedral dimetallofullerene and its spin coherent properties. *Nanoscale* **2020**, *12*, 20513–20521. [CrossRef]
68. Bubnov, V.P.; Laukhina, E.E.; Kareev, I.E.; Koltover, V.K.; Prokhorova, T.G.; Yagubskii, E.B.; Kozmin, Y.P. Endohedral Metallofullerenes: A Convenient Gram-Scale Preparation. *Chem. Mater.* **2002**, *14*, 1004–1008. [CrossRef]
69. Huang, H.; Yang, S. Toward Efficient Synthesis of Endohedral Metallofullerenes by Arc Discharge of Carbon Rods Containing Encapsulated Rare Earth Carbides and Ultrasonic Soxhlet Extraction. *Chem. Mater.* **2000**, *12*, 2715–2720. [CrossRef]
70. Diener, M.D.; Alford, J.M. Isolation and properties of small-bandgap fullerenes. *Nature* **1998**, *393*, 668–671. [CrossRef]
71. Kubozono, Y.; Noto, T.; Ohta, T.; Maeda, H.; Kashino, S.; Emura, S.; Ukita, S.; Sogabe, T. Extractions of $Ca@C_{60}$ and $Sr@C_{60}$ with Aniline. *Chem. Lett.* **1996**, *25*, 453–454. [CrossRef]
72. Fuchs, D.; Rietschel, H.; Michel, R.H.; Fischer, A.; Weis, P.; Kappes, M.M. Extraction and Chromatographic Elution Behavior of Endohedral Metallofullerenes: Inferences Regarding Effective Dipole Moments. *J. Phys. Chem.* **1996**, *100*, 725–729. [CrossRef]
73. Ding, J.; Yang, S. Efficient N,N-Dimethylformamide Extraction of Endohedral Metallofullerenes for HPLC Purification. *Chem. Mater.* **1996**, *8*, 2824–2827. [CrossRef]
74. Kubozono, Y.; Ohta, T.; Hayashibara, T.; Maeda, H.; Ishida, H.; Kashino, S.; Oshima, K.; Yamazaki, H.; Ukita, S.; Sogabe, T. Preparation and Extraction of $Ca@C_{60}$. *Chem. Lett.* **1995**, *24*, 457–458. [CrossRef]
75. Xiao, J.; Savina, M.R.; Martin, G.B.; Francis, A.H.; Meyerhoff, M.E. Efficient HPLC Purification of Endohedral Metallofullerenes on a Porphyrin-Silica Stationary Phase. *J. Am. Chem. Soc.* **1994**, *116*, 9341–9342. [CrossRef]
76. Anderson, M.R.; Dorn, H.C.; Stevenson, S.A. Making connections between metallofullerenes and fullerenes: Electrochemical investigations. *Carbon* **2000**, *38*, 1663–1670. [CrossRef]
77. Liu, B.B.; Zou, G.T.; Yang, H.B.; Yu, S.; Lu, J.S.; Liu, Z.Y.; Liu, S.Y.; Xu, W.G. Synthesis, extraction and electronic structure of $Ce@C_{2n}$. *J. Phys. Chem. Solids* **1997**, *58*, 1873–1876. [CrossRef]
78. Sun, D.Y.; Liu, Z.Y.; Guo, X.H.; Xu, W.G.; Liu, S.Y. High-yield extraction of endohedral rare-earth fullerenes. *J. Phys. Chem. B* **1997**, *101*, 3927–3930. [CrossRef]
79. Kubozono, Y.; Maeda, H.; Takabayashi, Y.; Hiraoka, K.; Nakai, T.; Kashino, S.; Emura, S.; Ukita, S.; Sogabe, T. Extractions of $Y@C_{60}$, $Ba@C_{60}$, $La@C_{60}$, $Ce@C_{60}$, $Pr@C_{60}$, $Nd@C_{60}$ and $Gd@C_{60}$ with aniline. *J. Am. Chem. Soc.* **1996**, *118*, 6998–6999. [CrossRef]
80. Kareev, I.E.; Shulga, Y.M.; Bubnov, V.P.; Kozlovski, V.I.; Dodonov, A.F.; Martynenko, M.V.; Yagubskii, E.B. Investigation of composition of endometallofullerene extracts. *Fuller. Nanotub. Carbon Nanostruct.* **2004**, *12*, 59–63. [CrossRef]
81. Kareev, I.E.; Bubnov, V.P.; Laukhina, E.E.; Dodonov, A.F.; Kozovski, V.I.; Yagubskii, E.B. Experimental evidence in support of the formation of anionic endohedral metallofullerenes during their extraction with N,N-dimethylformamide. *Fuller. Nanotub. Carbon Nanostruct.* **2004**, *12*, 65–69. [CrossRef]
82. Raebiger, J.W.; Bolskar, R.D. Improved Production and Separation Processes for Gadolinium Metallofullerenes. *J. Phys. Chem. C* **2008**, *112*, 6605–6612. [CrossRef]
83. Bolskar, R.D.; Alford, J.M. Chemical oxidation of endohedral metallofullerenes: Identification and separation of distinct classes. *Chem. Commun.* **2003**, *11*, 1292–1293. [CrossRef] [PubMed]
84. Aoyagi, S.; Nishibori, E.; Sawa, H.; Sugimoto, K.; Takata, M.; Miyata, Y.; Kitaura, R.; Shinohara, H.; Okada, H.; Sakai, T.; et al. A layered ionic crystal of polar $Li@C_{60}$ superatoms. *Nat. Chem.* **2010**, *2*, 678–683. [CrossRef] [PubMed]

85. Okada, H.; Komuro, T.; Sakai, T.; Matsuo, Y.; Ono, Y.; Omote, K.; Yokoo, K.; Kawachi, K.; Kasama, Y.; Ono, S.; et al. Preparation of endohedral fullerene containing lithium ($Li@C_{60}$) and isolation as pure hexafluorophosphate salt ($[Li^+@C_{60}][PF_6^-]$). *RSC Adv.* **2012**, *2*, 10624–10631. [CrossRef]
86. Tsuchiya, T.; Wakahara, T.; Shirakura, S.; Maeda, Y.; Akasaka, T.; Kobayashi, K.; Nagase, S.; Kato, T.; Kadish, K.M. Reduction of endohedral metallofullerenes: A convenient method for isolation. *Chem. Mat.* **2004**, *16*, 4343–4346. [CrossRef]
87. Laukhina, E.E.; Bubnov, V.P.; Estrin, Y.I.; Golod, Y.A.; Khodorkovskii, M.A.; Koltover, V.K.; Yagubskii, E.B. Novel proficient method for isolation of endometallofullerenes from fullerene-containing soots by two-step o-xylene-N,N-dimethylformamide extraction. *J. Mater. Chem.* **1998**, *8*, 893–895. [CrossRef]
88. Yamamoto, K.; Funasaka, H.; Takahashi, T.; Akasaka, T. Isolation of an ESR-Active Metallofullerene of $La@C_{82}$. *J. Phys. Chem.* **1994**, *98*, 2008–2011. [CrossRef]
89. Solodovnikov, S.P.; Tumanskii, B.L.; Bashilov, V.V.; Lebedkin, S.F.; Sokolov, V.I. Spectral study of reactions of $La@C_{82}$ and $Y@C_{82}$ with amino-containing solvents. *Russ. Chem. Bull.* **2001**, *50*, 2242–2244. [CrossRef]
90. Tsuchiya, T.; Wakahara, T.; Lian, Y.F.; Maeda, Y.; Akasaka, T.; Kato, T.; Mizorogi, N.; Nagase, S. Selective extraction and purification of endohedral metallofullerene from carbon soot. *J. Phys. Chem. B* **2006**, *110*, 22517–22520. [CrossRef]
91. Tagmatarchis, N.; Shinohara, H. Production, Separation, Isolation, and Spectroscopic Study of Dysprosium Endohedral Metallofullerenes. *Chem. Mater.* **2000**, *12*, 3222–3226. [CrossRef]
92. Sun, B.; Gu, Z. Solvent-dependent Anion Studies on Enrichment of Metallofullerene. *Chem. Lett.* **2002**, *31*, 1164–1165. [CrossRef]
93. Lu, X.; Li, H.; Sun, B.; Shi, Z.; Gu, Z. Selective reduction and extraction of $Gd@C_{82}$ and $Gd_2@C_{80}$ from soot and the chemical reaction of their anions. *Carbon* **2005**, *43*, 1546–1549. [CrossRef]
94. Shustova, N.B.; Kuvychko, I.V.; Bolskar, R.D.; Seppelt, K.; Strauss, S.H.; Popov, A.A.; Boltalina, O.V. Trifluoromethyl derivatives of insoluble small-HOMO-LUMO-gap hollow higher fullerenes. NMR and DFT structure elucidation of $C_2$-($C_{74}$-$D_{3h}$)$(CF_3)_{12}$, $C_s$-($C_{76}$-$T_d$(2))$(CF_3)_{12}$, $C_2$-($C_{78}$-$D_{3h}$(5))$(CF_3)_{12}$, $C_s$-($C_{80}$-$C_{2v}$(5))$(CF_3)_{12}$, and $C_2$-($C_{82}$-$C_2$(5))$(CF_3)_{12}$. *J. Am. Chem. Soc.* **2006**, *128*, 15793–15798. [CrossRef]
95. Shustova, N.B.; Newell, B.S.; Miller, S.M.; Anderson, O.P.; Bolskar, R.D.; Seppelt, K.; Popov, A.A.; Boltalina, O.V.; Strauss, S.H. Discovering and verifying elusive fullerene cage isomers: Structures of $C_2$-$p^{11}$-($C_{74}$-$D_{3h}$)$(CF_3)_{12}$ and $C_2$-$p^{11}$-($C_{78}$-$D_{3h}$(5))$(CF_3)_{12}$. *Angew. Chem. Int. Ed.* **2007**, *46*, 4111–4114. [CrossRef]
96. Wang, Z.; Nakanishi, Y.; Noda, S.; Niwa, H.; Zhang, J.; Kitaura, R.; Shinohara, H. Missing Small-Bandgap Metallofullerenes: Their Isolation and Electronic Properties. *Angew. Chem. Int. Ed.* **2013**, *52*, 11770–11774. [CrossRef] [PubMed]
97. Nakagawa, A.; Nishino, M.; Niwa, H.; Ishino, K.; Wang, Z.; Omachi, H.; Furukawa, K.; Yamaguchi, T.; Kato, T.; Bandow, S.; et al. Crystalline functionalized endohedral $C_{60}$ metallofullerides. *Nat. Commun.* **2018**, *9*, 3073. [CrossRef]
98. Nakagawa, A.; Aoyagi, S.; Omachi, H.; Ishino, K.; Nishino, M.; Rio, J.; Ewels, C.; Shinohara, H. Isolation and structure determination of missing fullerenes $Gd@C_{74}(CF_3)_n$ through in situ trifluoromethylation. *R. Soc. Open Sci.* **2018**, *5*, 181015. [CrossRef] [PubMed]
99. Wang, Z.; Aoyagi, S.; Omachi, H.; Kitaura, R.; Shinohara, H. Isolation and Structure Determination of a Missing Endohedral Fullerene $La@C_{70}$ through In Situ Trifluoromethylation. *Angew. Chem. Int. Ed.* **2016**, *55*, 199–202. [CrossRef] [PubMed]
100. Xie, S.Y.; Gao, F.; Lu, X.; Huang, R.B.; Wang, C.R.; Zhang, X.; Liu, M.L.; Deng, S.L.; Zheng, L.S. Capturing the labile fullerene[50] as $C_{50}Cl_{10}$. *Science* **2004**, *304*, 699. [CrossRef] [PubMed]
101. Tan, Y.-Z.; Chen, R.-T.; Liao, Z.-J.; Li, J.; Zhu, F.; Lu, X.; Xie, S.-Y.; Li, J.; Huang, R.-B.; Zheng, L.-S. Carbon arc production of heptagon-containing fullerene[68]. *Nat. Commun.* **2011**, *2*, 420. [CrossRef]
102. Zhong, Y.-Y.; Chen, Z.-C.; Du, P.; Cui, C.-H.; Tian, H.-R.; Shi, X.-M.; Deng, S.-L.; Gao, F.; Zhang, Q.; Gao, C.-L.; et al. Double Negatively Curved $C_{70}$ Growth through a Heptagon-Involving Pathway. *Angew. Chem. Int. Ed.* **2019**, *58*, 14095–14099. [CrossRef] [PubMed]
103. Wang, C.R.; Shi, Z.Q.; Wan, L.J.; Lu, X.; Dunsch, L.; Shu, C.Y.; Tang, Y.L.; Shinohara, H. $C_{64}H_4$: Production, isolation, and structural characterizations of a stable unconventional fulleride. *J. Am. Chem. Soc.* **2006**, *128*, 6605–6610. [CrossRef]
104. Tian, H.-R.; Chen, M.-M.; Wang, K.; Chen, Z.-C.; Fu, C.-Y.; Zhang, Q.; Li, S.-H.; Deng, S.-L.; Yao, Y.-R.; Xie, S.-Y.; et al. An Unconventional Hydrofullerene $C_{66}H_4$ with Symmetric Heptagons Retrieved in Low-Pressure Combustion. *J. Am. Chem. Soc.* **2019**, *141*, 6651–6657. [CrossRef]
105. Xie, F.-F.; Chen, Z.-C.; Zhang, M.; Xie, X.-M.; Chen, L.-F.; Tian, H.-R.; Deng, S.-L.; Xie, S.-Y.; Zheng, L.-S. Capturing nonclassical $C_{70}$ with double heptagons in low-pressure combustion. *Chem. Commun.* **2022**, *58*, 9814–9817. [CrossRef] [PubMed]
106. Moreno-Vicente, A.; Mulet-Gas, M.; Dunk, P.W.; Poblet, J.M.; Rodríguez-Fortea, A. Probing the formation of halogenated endohedral metallofullerenes: Predictions confirmed by experiments. *Carbon* **2018**, *129*, 750–757. [CrossRef]
107. Maeda, Y.; Tsuchiya, T.; Kikuchi, T.; Nikawa, H.; Yang, T.; Zhao, X.; Slanina, Z.; Suzuki, M.; Yamada, M.; Lian, Y.; et al. Effective Derivatization and Extraction of Insoluble Missing Lanthanum Metallofullerenes $La@C_{2n}$ (n=36–38) with Iodobenzene. *Carbon* **2016**, *98*, 67–73. [CrossRef]
108. Wakahara, T.; Nikawa, H.; Kikuchi, T.; Nakahodo, T.; Rahman, G.M.A.; Tsuchiya, T.; Maeda, Y.; Akasaka, T.; Yoza, K.; Horn, E.; et al. $La@C_{72}$ having a non-IPR carbon cage. *J. Am. Chem. Soc.* **2006**, *128*, 14228–14229. [CrossRef]
109. Nikawa, H.; Kikuchi, T.; Wakahara, T.; Nakahodo, T.; Tsuchiya, T.; Rahman, G.M.A.; Akasaka, T.; Maeda, Y.; Yoza, K.; Horn, E.; et al. Missing metallofullerene $La@C_{74}$. *J. Am. Chem. Soc.* **2005**, *127*, 9684–9685. [CrossRef]

110. Lu, X.; Nikawa, H.; Kikuchi, T.; Mizorogi, N.; Slanina, Z.; Tsuchiya, T.; Nagase, S.; Akasaka, T. Radical Derivatives of Insoluble La@$C_{74}$: X-ray Structures, Metal Positions, and Isomerization. *Angew. Chem. Int. Ed.* **2011**, *50*, 6356–6359. [CrossRef]
111. Nikawa, H.; Yamada, T.; Cao, B.P.; Mizorogi, N.; Slanina, Z.; Tsuchiya, T.; Akasaka, T.; Yoza, K.; Nagase, S. Missing Metallofullerene with $C_{80}$ Cage. *J. Am. Chem. Soc.* **2009**, *131*, 10950–10954. [CrossRef]
112. Akasaka, T.; Lu, X.; Kuga, H.; Nikawa, H.; Mizorogi, N.; Slanina, Z.; Tsuchiya, T.; Yoza, K.; Nagase, S. Dichlorophenyl Derivatives of La@$C_{3v}$(7)-$C_{82}$: Endohedral Metal Induced Localization of Pyramidalization and Spin on a Triple-Hexagon Junction. *Angew. Chem. Int. Ed.* **2010**, *49*, 9715–9719. [CrossRef]
113. Kareev, I.E.; Lebedkin, S.F.; Bubnov, V.P.; Yagubskii, E.B.; Ioffe, I.N.; Khavrel, P.A.; Kuvychko, I.V.; Strauss, S.H.; Boltalina, O.V. Trifluoromethylated endohedral metallofullerenes: Synthesis and characterization of Y@$C_{82}(CF_3)_5$. *Angew. Chem. Int. Ed.* **2005**, *44*, 1846–1849. [CrossRef]
114. Maeda, Y.; Akita, S.; Suzuki, M.; Yamada, M.; Akasaka, T.; Kobayashi, K.; Nagase, S. Controlling the reactivity of La@$C_{82}$ by reduction: Reaction of the La@$C_{82}$ anion with alkyl halide with high regioselectivity. *Beilstein J. Org. Chem.* **2023**, *19*, 1858–1866. [CrossRef]
115. Wang, Y.; Velkos, G.; Israel, N.J.; Rosenkranz, M.; Büchner, B.; Liu, F.; Popov, A.A. Electrophilic Trifluoromethylation of Dimetallofullerene Anions en Route to Air-Stable Single-Molecule Magnets with High Blocking Temperature of Magnetization. *J. Am. Chem. Soc.* **2021**, *143*, 18139–18149. [CrossRef] [PubMed]
116. Han, X.; Xin, J.; Yao, Y.; Liang, Z.; Qiu, Y.; Chen, M.; Yang, S. Capturing the Long-Sought Dy@$C_{2v}$(5)-$C_{80}$ via Benzyl Radical Stabilization. *Nanomaterials* **2022**, *12*, 3291. [CrossRef] [PubMed]
117. Zhou, X.; Yao, Y.-R.; Hu, Y.; Yang, L.; Yang, S.; Zhang, Y.; Zhang, Q.; Peng, P.; Jin, P.; Li, F.-F. Reactivity of Open-Shell Metallofullerene Anions: Synthesis, Crystal Structures, and Electrochemical Properties of Benzylated Gd@$C_{2v}$-$C_{82}$. *Inorganics* **2023**, *11*, 349. [CrossRef]
118. Yang, S.; Ioffe, I.N.; Troyanov, S.I. Chlorination-Promoted Skeletal Transformations of Fullerenes. *Acc. Chem. Res.* **2019**, *52*, 1783–1792. [CrossRef] [PubMed]
119. Lebedeva, M.A.; Chamberlain, T.W.; Khlobystov, A.N. Harnessing the Synergistic and Complementary Properties of Fullerene and Transition-Metal Compounds for Nanomaterial Applications. *Chem. Rev.* **2015**, *115*, 11301–11351. [CrossRef]
120. Boltalina, O.V.; Popov, A.A.; Kuvychko, I.V.; Shustova, N.B.; Strauss, S.H. Perfluoroalkylfullerenes. *Chem. Rev.* **2015**, *115*, 1051–1105. [CrossRef] [PubMed]
121. Yamada, M.; Akasaka, T.; Nagase, S. Carbene Additions to Fullerenes. *Chem. Rev.* **2013**, *113*, 7209–7264. [CrossRef] [PubMed]
122. Tzirakis, M.D.; Orfanopoulos, M. Radical Reactions of Fullerenes: From Synthetic Organic Chemistry to Materials Science and Biology. *Chem. Rev.* **2013**, *113*, 5262–5321. [CrossRef]
123. Lu, X.; Akasaka, T.; Nagase, S. Chemistry of endohedral metallofullerenes: The role of metals. *Chem. Commun.* **2011**, *47*, 5942–5957. [CrossRef] [PubMed]
124. Yamada, M.; Akasaka, T.; Nagase, S. Endohedral Metal Atoms in Pristine and Functionalized Fullerene Cages. *Acc. Chem. Res.* **2010**, *43*, 92–102. [CrossRef] [PubMed]

inorganics

MDPI

*Review*

# Process Parameter Optimisation for Endohedral Metallofullerene Synthesis via the Arc-Discharge Method

**Sapna Sinha [1,*], Karifa Sanfo [2], Panagiotis Dallas [3], Sujay Kumar [4] and Kyriakos Porfyrakis [2,*]**

[1] McGovern Institute and K. Lisa Yang Center for Bionics, Massachusetts Institute for of Technology, Cambridge, MA 02139, USA
[2] Faculty of Engineering and Science, University of Greenwich, Central Avenue, Chatham Maritime, Kent ME4 4TB, UK
[3] Institute of Nanoscience & Nanotechnology, NCSR "Demokritos", Patr. Gregoriou E' & 27 Neapoleos Str., Agia Paraskevi, 15341 Athens, Greece; p.dallas@inn.demokritos.gr
[4] Department of Physics, Babasaheb Bhimrao Ambedkar Bihar University, Muzaffarpur 842001, Bihar, India
* Correspondence: ssapna@mit.edu (S.S.); k.porfyrakis@greenwich.ac.uk (K.P.)

**Abstract:** Fullerenes have a unique structure, capable of both encapsulating other molecules and reacting with those on the exterior surface. Fullerene derivatives have also been found to have enormous potential to address the challenges of the renewable energy sector and current environmental issues, such as in the production of n-type materials in bulk heterojunction solar cells, as antimicrobial agents, in photocatalytic water treatment processes, and in sensor technologies. Endohedral metallofullerenes, in particular, can possess unpaired electron spins, driven by the enclosed metal atom or cluster, which yield valuable magnetic properties. These properties have significant potential for applications in molecular magnets, spin probes, quantum computing, and devices such as quantum information processing,, atomic clocks, and molecular magnets. However, the intrinsically low yield of endohedral fullerenes remains a huge obstacle, impeding not only their industrial utilization but also the synthesis and characterization essential for exploring novel applications. The low yield and difficulty in separation of different types of endohedral fullerenes results in the usage of a large amount of solvents and energy, which is detrimental to the environment. In this paper, we analyse the methodologies proposed by various researchers and identify the critical synthesis parameters that play a role in increasing the yields of fullerenes.

**Keywords:** endohedral metallofullerenes; EMFs; arc discharge; bottom-up synthesis; in situ doping

## 1. Introduction

Carbon, the building block of life and the fourth most abundant element in the universe, is one of the most indispensable materials known to humankind. In addition to the two naturally occurring allotropes, graphite and diamond, carbon is also capable of taking several remarkable physical forms on the nanoscale. One such distinct allotrope in the family of carbon nanomaterials is the fullerene. Its discovery [1] sparked a great deal of curiosity and interest amongst the scientific community due to its unique closed, cage-like structure, composed solely from $sp^2$ hybridized carbon [2,3]. Within the first few years after the discovery of buckyball $C_{60}$, the term 'fullerene' achieved a uniform acceptance as the general name for these carbon cages. Fullerenes can have different numbers of carbon atoms, shapes, and symmetry; hence, the largely accepted notation system [4] is to subscript the number of carbon atoms to state the class of the fullerene. The structure of $C_{60}$, the most abundant buckyball, has been verified to consist of 12 five-membered rings and 20 six-membered rings [2]. Most of the fullerenes obey the isolated pentagon rule (IPR), where pentagons are exclusively surrounded by hexagons, which increases their stability. However, several fullerenes, especially a number of endohedral fullerenes (see next section), do not follow the IPR rule [5]. Instead they follow the maximum pentagon separation rule, where they exhibit maximum pentagon separation for maximum stability [6].

Fullerenes are known to have different isomers, with $C_{60}$ having 1812 theoretically different isomers [7,8]. These isomers can have different symmetry, form, and energetic stability. An example of the different structures of the six most stable isomers of $C_{80}$ is shown in Figure 1, with $D_{5d}$ and $D_2$ being the most stable and abundantly found isomers [9]. Fullerene isomers are also known to have different oxidation potential and chemical reactivity. For instance, the $D_3$ isomer of $C_{78}$ is easier to oxidize than the $C_{2v}$ isomer of the same fullerene by about 0.25 V [10]. Another very intriguing feature of the fullerenes is that they have a hollow space, which is large enough to encapsulate other atoms, ions, molecules, and various kinds of clusters, giving rise to a new class of hybrid materials called endohedral fullerenes.

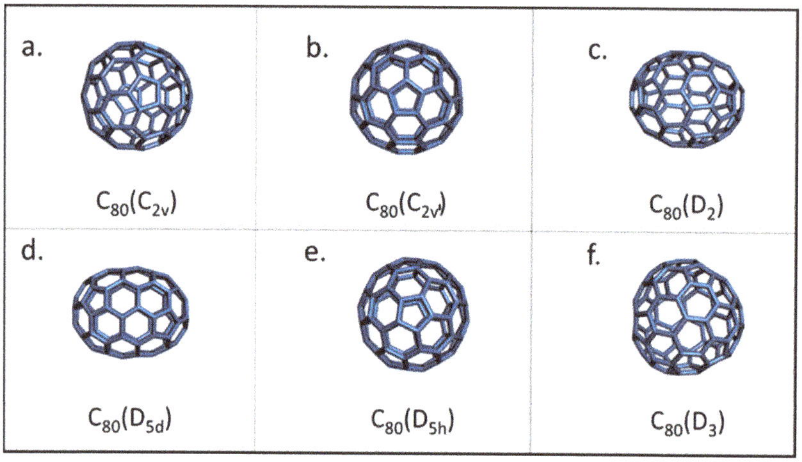


**Figure 1.** (**a**–**f**) Schematic representation of the different isomers of the $C_{80}$ cage.

## 2. Structure of Endohedral Fullerenes

The ability of different sizes of fullerene cages to incorporate different types of metals and clusters can lead to the formation of materials with novel solid-state properties. The robust nature of the carbon cage plays a very important role in isolating the atoms from the exterior environment, while the electronic properties of the cage can be tuned by the selection of the atoms inside. La@$C_{82}$ was among the first discovered endohedral fullerene, where the symbol @ denotes the inclusion of a La atom inside a $C_{82}$ cage. Since then, fullerenes encapsulating metallic [11], carbide [12,13], nitride [14], and many methano- [15] and oxide- [16] clusters have been synthesized and isolated over the years. In fact, fullerenes containing single atoms of elements that otherwise are not stable, such as noble gases [17–19] and non-metals [20] such as nitrogen [21], have also been synthesized. Transition metals and actinides were amongst the challenging elements to encapsulate within fullerenes, primarily due to their reactivity and the difficulties associated with their incorporation into the graphite rod for arc-discharge synthesis. However, over the past decade, significant progress has been made in also encapsulating various transition metals within fullerenes. [22] In these cases, while describing the diversity of the fullerene cages, prefixes and cage symmetries are also used. For instance, $Sc_4(\mu_3\text{-}O)_3@I_h(7)\text{-}C_{80}$ represents the endohedral cluster group, cage structure, and point-group symmetry, following the nomenclature of Fowler and Manolopoulos [4].

In 1991, researchers succeeded in encapsulating metal atoms, which opened a complete new set of possibilities in the field of material development, generally referred to as endohedral metallofullerenes [23]. Among the endohedral fullerenes, the trimetallic nitride-containing cages of the type $M_3N@C_{2n}$ (where M = Sc, Gd; 2n = 68–96), shown in Figure 2, have been one of the most studied materials since their discovery in 1999 [14] due to their high yields during synthesis and their appropriate HOMO-LUMO levels, desired for applications in solar cells. However, despite the comparable high yields, empty fullerenes still constituted as major product during the synthesis process. The yields of $M_3N@C_{2n}$ were further improved by Dunsch and Yang [24], who introduced ammonia during fullerene synthesis and produced an endohedral fullerene, $Sc_3N@I_h$-$C_{80}$, as the majority product for the first time. This family of fullerenes has an especially high yield because of the charge transfer from the encaged cluster, which results in a stable ion pair and stabilizes the higher fullerene cages of various isomeric structures. In addition, the large HOMO-LUMO gaps give further kinetic stability to the hybrid molecule.

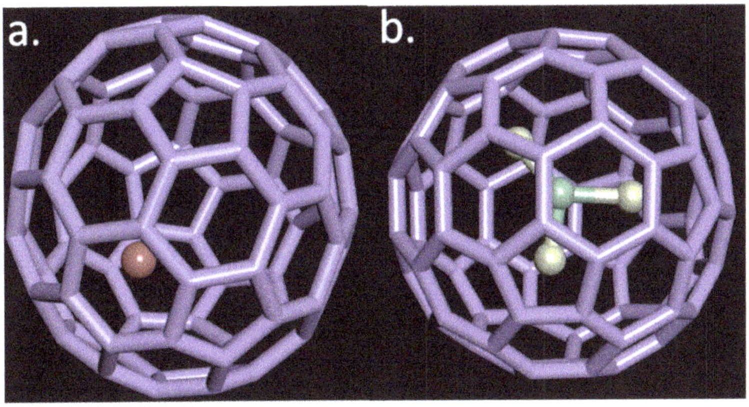


**Figure 2.** A representative figure of two endohedral metallofullerenes, (**a**) monometallic in a $C_{82}$ cage and (**b**) trimetallic nitride encapsulated in a $C_{80}$ fullerene cage. The red and the yellow balls correspond to the metal ion encapsulated in the carbon cage and the green to nitrogen.

### 2.1. *Endohedral Metallofullerenes: EMFs*

In the years following the discovery of $C_{60}$, the study of empty cage fullerenes heavily dominated the field of nanomaterials. The focus on empty cage buckyballs often resulted in the blind assumption that endohedral metallofullerenes (denoted as EMFs) possessed the same structural or chemical properties as the corresponding empty cages. However, over the years, after more studies were conducted on individual EMFs, it became established that they exhibit novel structural, chemical, and electronic properties, which were completely different to those of their parent empty cages. For instance, carbon cages of EMFs have different isomeric structures as compared to their parent fullerene cage, and they also show different solubility in individual solvents [25]. As for their electronic and chemical behaviour, their chemical reactions with radicals [26] and many other cyclo-addition [27] reactions show significantly different patterns and redox potentials [28]. For instance, the reduction potentials of $La@C_{82}$ and $La_2@C_{80}$ are 0.70 V and 0.81 V lower than that of $C_{60}$ [29]. The reason for such a huge difference is largely attributed to the covalent interaction between the enclosed species and the carbon cage as well as the charge transfer phenomenon from the metallic species to the fullerene cage.

The most studied metallofullerenes involve rare earth metals, such as La, Sc, Dy, and Tm [23,30–35], and main group metals such as Li and Ca [36]. There is an extensive array of EMFs that have one, two, three, or even four metal atoms trapped inside them [37]. Alternatively, they can also contain $M_3N$ [14] or $M_2C_2$ units [12], where M refers to metals. The key to understanding the electronic properties of these hybrid materials is through studying the degree of charge transfer from the metal species to the exterior cage, which alters its electronic state. While the number of electrons transferred from a majority of metal atoms to the cage is known and verified using X-ray diffraction, absorption, and other spectroscopic methods [31–35], we observed that contradictory results still exist in the literature regarding the effect of the various parameters on the arc synthesis and the formation mechanism of these fullerenes. For instance, Dunk et al. claimed that the formation mechanism is governed by a charge transfer between the metal and the cage [38]. Through detailed mechanistic studies, they revealed that the $C_2$ insertion during the bottom-up synthesis becomes more difficult as the oxidation state of the metal increases. This charge transfer and the nature of the cluster inside the cage also dictates a series of properties, such as electrochemical behaviour and light emission. In fact, Ito et al. studied $Er_2@C_{82}$ and $Er_2C_2@C_{82}$, which showed nearly identical cage properties with a 6-negatively charged $C_{82}{}^{6-}$, using fluorescence spectroscopy [39]. Most notably, this cage charge contradicted an earlier publication by Wang et al. claiming that the $Er_2@C_{82}$ and $Er_2C_2@C_{82}$ compounds have a 4- negatively charged cage, i.e., $C_{82}{}^{4-}$ [40]. Furthermore, Ito et al. showed that the characteristic emission of the erbium ions at 1520 nm was influenced by the symmetry of the cage, an observation later echoed by Plant et al. [41] Until as recently as 2020, previously unknown factors, such as enantiomers and enthalpy, were discovered to play vital and evolving roles in the synthesis mechanism for fullerenes $C_{2n}$ ($50 \leq 2n \leq 70$) [42]. After a critical analysis of research articles [43,44], it can be concluded that the complete mechanism of their formation, either through a bottom-up or top-down approach, and the charge transfer itself remains elusive and highly controversial [45]. In fact, the insights derived from experimental results have also been contradictory to theoretical results in the field. For instance, it has been suggested that factors such as the $C_2$ concentration and temperature may dictate the more favourable direction of the reversible $C_2$ injection/ejection reaction during the synthetic process [46]. However, theoretical studies have highlighted that the $C_2$ injection is more energetically favourable than its $C_2$ ejection counterpart, contradicting the experimental data on the importance of temperature and concentration [47,48]. The missing key links and low synthesis yield of fullerenes make it difficult to investigate the fullerene formation mechanism. Utilizing both experimental and theoretical frameworks is therefore necessary to answer some of the fundamental questions of this field.

As briefly mentioned in the previous section, the isomeric structures of the carbon cages encapsulating metals differ from their empty counterparts [34]. Further study of the positions and movements of the encapsulated atoms is also important in determining the chemical and physical properties of EMFs, including their magnetic behaviour, electron paramagnetic resonance (EPR), and electrochemical properties [49]. The structure of the EMFs themselves had long been speculated, but the first crucial evidence came through in 1995 when Shinohara et al. experimentally confirmed that the metal atom is indeed encapsulated within the fullerene cage [31]. The synchrotron X-ray powder diffraction patterns of $Y@C_{82}$ and the maximum entropy method (MEM) revealed that the yttrium atom was displaced from the centre and instead tightly bound to the cage [50]. The study also confirmed strong dipole–dipole and charge transfer interaction between the different molecules [49]. MEM/Rietveld analysis has long been used to determine the structures of metallofullerenes [31,50]. However, the results (structure and position of metals inside the carbon cage) obtained by this method have been in question in recent years. This started when Wang et al. [12] discovered that the scandium carbide endohedral fullerene, $Sc_3C_2@C_{80}$, was, in fact, what had long been considered to be a trimetallofullerene $Sc_3@C_{82}$ [51]. These incoherences emphasize the importance of $^{13}C$ NMR and synchrotron X-ray structural analysis in determining the structures and encapsulated cluster inside the fullerene cage and suggest the unreliability of the decades of results obtained exclusively using MEM/Rietveld analysis when determining the structures of metallofullerenes [52–54]. Consequently, the structures and positions of metals inside a fullerene have been re-examined in the past few years [55–58]. Furthermore, it was also observed that these metal clusters had motions and trajectories inside the cage [59], which also changed positions depending on the temperature [60,61].

### 2.2. Environmental Application

A massive number of publications in fullerene-related research show the possibility of using fullerenes for a variety of environmental and energy applications. The characteristics of fullerenes that give them an advantage over other materials include their robust carbon structure, absorption of light in the visible range, electron-accepting capabilities, and more recently, superconductivity [62]. All these properties make fullerenes especially sought-after materials for energy applications, especially organic solar cells [63]. They have become, and remain, an important part of high-performance organic solar cells.

### 2.3. Synthesis

Over the past few decades, EMFs have been proposed for a number of applications, including energy harvesting in organic solar cells [64] and MRI contrast agents [65–67]. EMFs have been synthesized through a range of different methodologies, such as laser ablation of graphite [23,68], arc discharge [30,33], ion bombardment [36,69], resistive heating [70], and electron initiation inside carbon nanotubes (CNTs) within aberration-corrected high-resolution transmission electron spectroscopy (AC-HRTEM) (CNT acts as a nanoreactor container to facilitate the formation of endohedral fullerenes from metals and the amorphous carbon atoms trapped inside. The process is energetically enabled under the presence of an electron beam inside an AC-HRTEM.) [71]. Some of these popular methods are summarized in Table 1 below.

**Table 1.** Advantages and disadvantages of different synthetic approaches.

| EMF Synthesis Method | Advantages | Disadvantages |
| --- | --- | --- |
| Arc discharge [72] | • Low setup and maintenance costs<br>• Fewer structural defects<br>• Access to the broadest catalogue of EMF | • Difficult to scale-up<br>• Requires extensive purification<br>• Low yield (1–30%) |
| Laser ablation [72] | • Higher purity of products<br>• Useful for mechanistic studies | • High cost for reactor building and maintenance |
| Molecular surgery [72,73] | • Low energy impact<br>• Simple HPLC purification | • Several synthetic steps required to open/close the fullerenes cage<br>• Harmful synthetic solvents<br>• Limited to non-metal and alkali-metal EMFs<br>• Low yield (~0.1 to 1%) |
| Ion implantation [72] | • Lower pressure required (5–10 mbar)<br>• Low-temperature process, where ions are accelerated towards the carbon cages<br>• Simple HPLC purification possible | • Limited to non-metal and alkali-metal EMFs<br>• Very low yield ($10^{-5}$ to $10^{-4}$%) |

The Kratschmer–Huffman arc method has been widely used and accepted as the most efficient and cost-effective technique to produce EMFs and, hence, is the focus of this review. Figure 3a shows a schematic diagram of a typical standard direct current (D.C.) arc-discharge apparatus, and Figure 3b shows the actual image of one in the Porfyrakis laboratory in the Department of Materials at the University of Oxford.

While the Kratschmer–Huffman method immediately became the most popular choice of EMF synthesis, it was not exempt from certain limitations. Most notably, only 60–70% of the graphite electrode was transferred to the carbon condensate (CC) containing the fullerenes and EMF, while the remaining 30–40% was lost as graphite build-up on the cathode. This disadvantage has now been eliminated by utilizing a high frequency arc discharge to supply the arc. The symmetrical setup of the electrodes allowed reaching up to a 100% conversion of the electrode material into the fullerenes and EMF containing CC [74]. Additionally, high-frequency arc discharge can also be used for further mechanistic investigation of fullerene synthesis.

A metal oxide/graphite composite is used as the anode, whereas graphite is used as the cathode, and both of them are arced in the D.C. mode in the presence of an inert gas (usually helium). Despite three decades of extensive research and various proposed pathways, a well-established formation mechanism for the synthesis of fullerenes and EMFs is still missing. Their ambiguous reaction pathways, alongside the violent conditions involving the plasma formation, arc, and chemical ionization that are required for the synthesis, are a major drawback in studying the synthesis mechanisms of these unique carbon nanostructures. Furthermore, a few other drawbacks hinder the scaling-up of their synthesis and consequently the applications of endohedral metallofullerenes. The overall yield of the arc-discharge synthesis of EMFs is low, with the ratio of EMF-to-empty-cage typically at around 1% [49]. As a result, a mixture of empty cage and endohedral fullerenes is typically produced, thus limiting the applications of the existing chromatographic methodologies for their fast and efficient separation. This results in the use of excessive amounts of solvents to efficiently separate them, which is neither ecological nor environmentally friendly. Therefore, it becomes important to review the synthesis of fullerenes. Shinohara published the first detailed scientific review of EMF research in 2000 [75], followed by an extensive update on endohedral fullerenes by Popov et al. in 2012 [76]. Many other groups have addressed the synthesis and properties of endohedral fullerenes [76–79], including nitride fullerenes [24], carbide cluster fullerenes [80], and oxide cluster fullerenes. Herein, we focus primarily on critically discussing the key findings in the field for optimizing of synthesis parameters of endohedral metallofullerenes and the new chemically activated method, with a particular emphasis on the post-2000 science along with comparisons to the early methodologies. The yield of the EMF has been observed to be highly dependent on the pressure of He gas, the arc current, the arc gap, and the composition of anodes, among other factors [51,81]. Liu [82] and Bandow [83] have argued that other parameters, such as the presence of catalysts (Cu, Fe, etc.), the back-burning of anodes, and in situ activation during arc discharge, also have an effect on the EMF yield. However, the latter two parameters would not increase the yield by a huge fraction because they do not modify the arc conditions drastically or change/influence the synthesis chemistry, e.g., lower the activation energy of the formation of the endohedral metallofullerenes. However, in order to maximize the efficiency of the method, a balance between the time spent during the synthesis and increase in yield needs to be established. Predominant aspects of overcoming the above challenge are anode composition, ionization potential, catalysts, and helium pressure in synthesis [74].

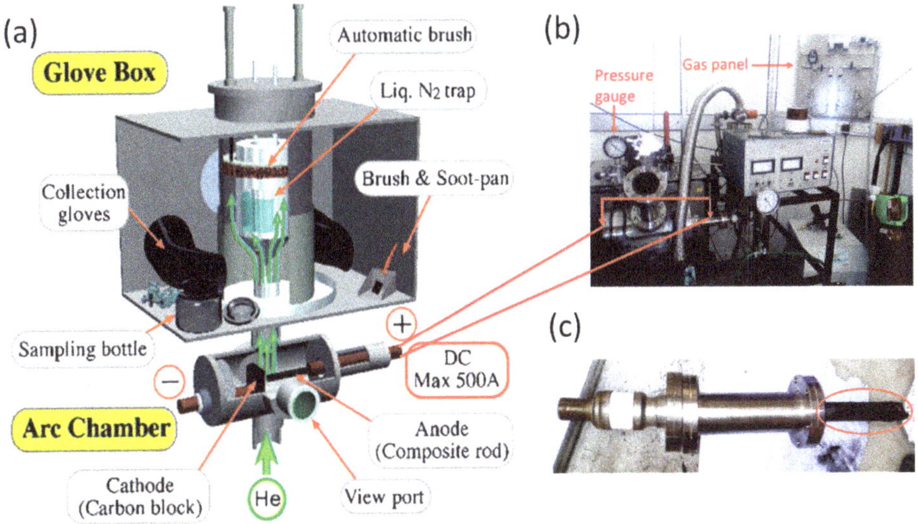


**Figure 3.** (**a**) A schematic diagram of a typical arc-discharge apparatus used for synthesizing fullerenes (Reprinted/adapted with permission from Ref. [84]. Copyright (c) 2007 The Japan Society of Applied Physics). (**b**) An image of an actual arc-discharge apparatus from the Porfyrakis laboratory at the University of Oxford. (**c**) Image of a metal oxide/graphite composite rod used during an arc-discharge process.

In recent years, the Kratschmer–Huffman arc design has also been modified mechanically in several ways to make it more environmentally friendly. For instance, the use of demineralized coal electrodes, minimizing the use of current and using direct current instead of high alternating current are a few ways in which the process has been made more efficient and less energy intensive [85]. Recently, Kyesmen et al. showed scaled-up production yield of fullerenes by employing such methods and including efficient energy usage by using resistive heating of one of the electrodes along its length [86]. Such green methodologies should be further explored for future research purposes and industrial-scale production of fullerenes to ensure an environment friendly approach towards using nanomaterials.

## 3. Reaction Parameters and Pathways

In this section, we will delve into the critical parameters that are essential for achieving the highest yield of EMFs, with a special focus on encapsulation in the $C_{82}$ cage.

### 3.1. Anode Composition

Metal oxide (e.g., $La_2O_3$, $Y_2O_3$, etc.) mixed with graphite has generally been used as the anode material during the arc-discharge process. However, there has been no general consensus on the ratio of metal/carbon used in the metal. Inakuma et al. also showed that the abundance of different types of EMF present in the soot depends heavily on the metal/carbon mixing ratio in the composite rods [51,81]. The M/C ratio has varied widely over the years, also depending on the metals that are incorporated inside the graphite, as broadly summarized in Table 2. Furthermore, it has also been found that anodes made out of metal-carbide rods actually produce EMFs in a bigger yield than metal-oxide rods, which has been experimentally proved by several authors [82,83]. Using La as the example metal, the authors showed that the presence of $LaC_2$ in supersaturated carbon vapor plays an important role in generating the metallofullerenes. However, this conclusion is in contrast with the work published later by Lian et al. [87], who studied Ni-La alloy anode rods for the preparation of lanthanum-containing metallofullerenes (La@$C_{82}$ (I), La@$C_{82}$ (II), and $La_2$@$C_{80}$). It was found that $LaNi_2$ is a much better dopant than conventional $La_2O_3$ in the anode rod, and the ratio of La:C is very important for increasing the yield. The absence of oxygen in the $LaNi_2$ dopant can partly be the reason for such observations as well as the possible catalytic effect of Nickel in the formation of fullerenes. The authors also found that the use of $LaNi_2$ instead of $La_2O_3$ enhanced the amount of raw soot produced from arc discharge. The authors attributed the results to reduction in the amount of formation of cathode deposits and debris. The optimum conditions increased the yield of La-EMF by almost 30% compared to conventional substrates. Comparing the existing literature in this field, as summarized in Table 2, it seems rather hard to deduce which single metal/carbon ratio in the anode would produce the best yields for different types of metallofullerenes. It surely is a huge determinant of the yield of a particular type of EMF, but more research, both experimental and theoretical, needs to be undertaken in this direction to test different combinations of metals for anodes and also to explore the chemistry involved.

### 3.2. *Ionization Potential*

The ionization potential (I.P.) of the EMFs has been known to play a very important role in EMF extraction and chemical reactivity, but it only recently came to light that the first ionization potential of the metal itself plays an important role in EMF synthesis. Ross et al. deduced that the I.P. of the metal atom not only influences the efficiency of the synthesis but may partly determine the metallofullerene formation [88]. Their group noted that when graphite anode rods were impregnated with Y/La atoms and with Sc/Y atomic composition in equal ratio, there was a greater abundance of La- and Y-containing metallofullerenes than Y or Sc, respectively. The results suggested that metals with lower I.P. (La < Y < Sc) tend to react more efficiently, i.e., the growth reaction of metal ions and fullerene precursor species plays an important role in the arc-discharge process. This is consistent with other experimental studies that show that lanthanides are easier to encapsulate inside a fullerene cage compared to other alkali earth metals or transition metals [89]. Comparing the yields of the synthesis of Pt-EMF [90,91], Fe-EMFs [92], and Sc-EMFs ($Pt_{I.P.} = 8.96 > Fe_{I.P.} = 7.90 > Sc_{I.P.} = 6.56$) [93], Pt-EMFs have the lowest yield, whereas the Sc-EMFs have the highest yield, even though the covalent radii of the metals are not very different (Pt = 1.17 A; Fe = 1.3 A; Sc = 1.44 A). Since the overall yield of EMFs is generally low, and their synthesis is a very time-consuming process, it is important that the choice of the metal atom, which can potentially form EMFs, is based on their ionization potential and chemical nature of bonding with the carbon clusters.

### 3.3. *Effect of Catalyst*

The catalytic growth of carbon nanomaterials has long been known from the synthesis of carbon nanotubes [94], and more recently, graphene [95]. However, the use of catalysts for the synthesis of fullerenes was not developed until 2008, when Stevenson et al. [16] doped the graphite rods with copper nitrate in a modified arc chamber. They observed that the presence of metallic copper increased the overall yield of endohedral fullerenes. It still remains unclear if it also facilitates the formation of endohedral mono-metallofullerenes and if the ratio of metallofullerenes to other empty cages in the carbon soot is, in fact, higher than previously reported. Other groups have noted that using transition metal alloys (e.g., Ni) in the anode composition [87] increases the yield of the EMFs. But again, results published in the literature make it difficult to judge whether the transition metal plays an active role as a catalyst or if it just reduces the detrimental effects that oxygen has, whist using metal oxide composite rods during the arc discharge. We deduce that catalysts have not been studied in depth largely because of the difficulty in producing homogeneous composite anode rods containing different catalysts such that they can be actually used systematically during the arc operation. The development of new methods for efficient dispersion of catalytic particles inside the electrodes during their fabrication could help solve these problems. Studying the effect of metal clusters or metal alloys as additives in the arc discharge, could be worthwhile not only for increasing the yield of EMFs but also for elucidating the formation pathways for fullerenes and carbon nanotubes in the arc reactor chamber.

### 3.4. *Effect of Helium Pressure*

The presence of helium in the arc-discharge chamber is well known to be beneficial for the formation of endohedral fullerenes [96]. It is still not clear what the exact mechanism behind it is, but it is believed that He, being a small and inert species, facilitates collisions between carbon atoms without taking part in reactions. Therefore, it enhances the possibility of carbon–carbon bond formation. There is also a general consensus that varying the pressure of helium leads to different yields of the EMFs. A literature survey (Table 2) in the field shows that using lower pressures of helium (50–100 mBar) has been the standard arc-discharge parameter until recently, when researchers started using high He pressures (up to 500–700 mBar) for EMF synthesis. In 2017, Qian et al. [97] published the first theoretical study on how the self-assembly of fullerene during the arc-discharge process is influenced by the presence of helium. They also proposed how the concentration of helium can change the distribution of different fullerene formation in the arc. Similar results were shown for synthesizing fullerenes in a high-frequency arc-discharge system, as reported by Churilov et al. [74]. The authors experimented with a range of helium pressures (30–150 kPa) and their effect on the formation of different types of Y-EMFs, and concluded that 120 kPa results in the maximum yield for the EMF Y@$C_{82}$. Moreover, empty cages were found to increase with the helium pressure but not the EMFs. The effect of helium pressure in both the traditional arc discharge as well as high-frequency arc discharge suggests a difference in the mechanism of formation of the empty cages and the EMFs. Further studies in this area can help optimize the exact synthetic parameters for the desired type of fullerene. Another helium pressure-dependent study was reported by Cherviakova et al. [98], where they carried out a series of experiments by keeping all the variables constant and only changing the He pressure in the chamber from 30 to 360 mBar. The experiment was carried out for various EMFs, and they found that although 120 mBar helium pressure yields the optimal condition for synthesis of different M@$C_{82}$, there is a huge variation in the obtained optimal yield. For instance, although both Sc@$C_{82}$ and Gd@$C_{82}$ showed the best synthetic yields at 120 kPa He pressure, the yields were very different, with Sc@$C_{82}$ having 9.6 wt% ratio of all fullerenes, as compared to Gd@$C_{82}$, which had only 5.4 wt%. However, there is a massive variation in homemade arc-discharge apparatus as well as differences in the exact operational parameters used for experiments carried out at different places. Therefore, there is a need for a systematic study on a single apparatus, for synthesizing the range of EMFs and determining if the pressure conditions vary depending on the entrapped metals. Moreover, a wider pressure range should also be tested so that the exact reaction pathway behind the formation of different types of EMFs can be properly investigated.

**Table 2.** Summary of the parameters used over the years to synthesize different types of endohedral metallofullerenes.

| Year | Anode | Fullerene | He Pressure (mBar) | Current | Extraction Solvent | References |
|---|---|---|---|---|---|---|
| 1992 | $Sc_2O_3$ | Sc-EMF | 66.6 | - | Toluene and Pyridine and $CS_2$ 1 | [35] |
| 1992 | $Sc_2O_3$ | Sc-EMF | 266.6 | - | Toluene | [32] |
| 1992 | $La_2O_3$ | La-EMF | 266.6 | - | Toluene | [33] |
| 1993 | $Sc_2O_3$ M/C = 2.3/100 atomic ratio | Sc-EMF | 66 | 220 A | $CS_2$ | [81] |
| 1993 | 8 wt% $La_2O_3$ | La@$C_{82}$ | 133.3 | 2.3 A/mm$^2$ | Toluene | [83] |
| 1994 | $Sc_2O_3$ M/C = 1:10 | | 66–133 | 500 A | $CS_2$ | [51] |

**Table 2.** *Cont.*

| Year | Anode | Fullerene | He Pressure (mBar) | Current | Extraction Solvent | References |
|---|---|---|---|---|---|---|
| 1995 | $La_2O_3$ (La/C = 2:100) | La@$C_{2n}$ | 133 | 200 A | Toluene | [82] |
| 1995 | $Gd_2O_3$ (1–2% atomic%) | Gd-EMF | 60 | 250 A | TCB [2] | [99] |
| 1995 | - | Y@$C_{82}$ | 66 | 500 A | Toluene | [31] |
| 1996 | $CeO_2$ | Ce@$C_{82}$ | 66 | | DMF | [100] |
| 1996 | Tm/C = 1:50 | Tm@$C_{82}$ | 200 | 175 A | Toluene | [34] |
| 1996 | $Pr_6O_{11}$ (Pr/C = 1:100) | Pr@$C_{82}$, $Pr_2$@$C_{80}$ | 66 | | DMF [3] | [101] |
| 1998 | - | Sc@$C_{82}$ | 66 | 300–400 A | Toluene | [93] |
| 1999 | $Sc_2O_3$ Sc/C = 2.8/100 | $Sc_3$@$C_{82}$ | 66 | 500 A | $CS_2$ | [11] |
| 2002 | M/C = 1/100 (M = La,Y) | La@$C_{2n}$ [4], Y@$C_{2n}$ | 160 | 90 A | o-xylene+ DMF/DMA/DMSO | [102] |
| 2002 | Tm/C composite rod | Tm@$C_{82}$ | 240 | 600 A | TCB | [103] |
| 2002 | | Gd@$C_{82}$ | 960 | 50 A | DMF | [104] |
| 2003 | $Gd_2O_3$ | Gd@$C_{60}$ | 133 | - | Dichlorobenzene | [105] |
| 2004 | $MNi_2$ | All EMF | 960 | 50 A | DMF (~6–7%) $CS_2$ (~1.5%) Pyridine (~1.5%) Aniline (~1%) [5] | [106] |
| 2005 | CeO2 (Ce/C = 2/100) | $Ce_2$@$C_{80}$ | 66.6 | 70 A | TEA/acetone [6] | [107] |
| 2006 | $Sc_2O_3$ Sc/C = 2/100 | Sc- EMF | 66 | 150 A, 40 V | TCB | [108] |
| 2010 | $Gd_2O_3$ | Gd@$C_{82}$ | 200 | 90 A | DMF liq./liq. extraction | [109] |
| 2014 | $Gd_2O_3$ | Gd@$C_{82}$, Gd@$C_{84}$ | 506 | 110–120 A | o-xylene and DMF | [110] |
| 2017 | $Gd_2O_3$ (M:C = 1:1) mass ratio | Gd@$C_{82}$ | 1200 | 50–400 A | $C_5H_5N$ | [98] |
| 2017 | $Sc_2O_3$ (M:C = 1:1) mass ratio | Sc@$C_{82}$ | 1200 | 50–400 A | $C_5H_5N$ | [98] |
| 2017 | $Er_2O_3$ (M:C = 1:1) mass ratio | Er@$C_{82}$ | 1200 | 50–400 A | $CS_2$ | [98] |
| 2017 | TiO (M:C = 1:1) mass ratio | Ti@$C_{82}$ | 600 | 50–400 A | $CS_2$ | [98] |
| 2017 | $Y_2O_3O$ (M:C = 1:1) mass ratio | Y@$C_{82}$, $Y_2$@$C_{82}$ | 600 | 50–400 A | $CS_2$ | [98] |
| 2017 | $Y_2O_3O$ (M:C = 1:1) mass ratio | Y@$C_{82}$ | 1200 | 50–400 A | $C_5H_5N$ | [98] |
| 2023 | $Tb_4O_7$ (M:C = 1:15) molar ratio | Tb@$C_{82}$ | 400 mbar | 100 A | of $CS_2$/toluene | [111] |

[1] Carbon Disulfide; [2] 1,2,4-Trichlorobenzene; [3] N,N-Dimethylformamide; [4] $68 \geq 2n \geq 96$; [5] Extraction yield by weight percent of (Tb, Y, La, Gd)-EMF to soot; [6] Triethylamine.

### *3.5. Chemically Activated Arc Discharge*

#### 3.5.1. In Situ Doping

Another way to address the low yield and the unwanted mixture of various types of fullerenes is by changing the reaction conditions through in situ doping of the graphite/metal electrodes, which are to be vaporized during the arc discharge. These methodologies require the incorporation of reactive gases and/or solid compounds that alter the conditions and influence the reaction equilibrium towards the formation of higher cages and, ideally, EMFs. One of the most typical examples is the use of reactive gases, e.g., $NH_3$ [112], whose incorporation leads to the scaled-up synthesis of trimetallic nitride fullerenes, for instance, making the $Sc_3N@I_h$-$C_{80}$ the third most abundant fullerene, after $C_{60}$ and $C_{70}$. Another alternative is the doping of the graphite electrodes themselves with solid organic compounds. Core drilled graphite rods have been repeatedly doped with active inorganic or organic compounds. These compounds could be calcium cyanamide [113], urea and, most importantly, inorganic nitrate compounds. In that direction, Stevenson et al. pioneered the CAPTEAR method (the name stands for "Chemically Adjusting Plasma Temperature, Energy and Reactivity") (Figure 4). They core-drilled and filled the graphite/$Sc_2O_3$ electrodes with copper nitrate powder and in a later study with metallic Cu, in order to study the reactive species generated during the synthesis procedure [114,115]. The decomposition of copper nitrate led to an exothermic reaction and the generation of reactive nitrogen gases, oxygen, and vaporized copper. A detailed analysis of the decomposition products can be seen in the references herein. These factors influenced the temperature, energy, conductivity of the arc, and reactivity of the plasma. Stevenson et al. observed that this particular arc discharge reaction leads to the exclusive and highly selective formation of TNT-type metallofullerenes ($Sc_3N@I_h$-$C_{80}$) and that the empty cage fullerenes were formed with a low yield. In their pioneering 2007 publication, they observed that this preferential formation of EMFs, based on HPLC calculated yield, was about 80% wt as compared to standard yield of 5–30 wt%. They observed that from the 13 mg of extracted fullerenes, 12 mg were identified as $Sc_3N@I_h$-$C_{80}$.

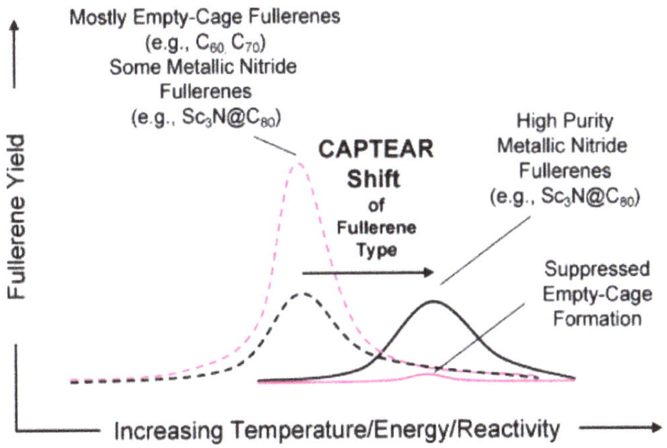


**Figure 4.** Parameter optimization of the CAPTEAR method to synthesize the desired endohedral fullerene. (Reprinted/adapted with permission from Ref. [114]. Copyright © 2007, American Chemical Society).

Expanding this method further, we recently developed a similar methodology for the study of the synthesis of gadolinium metallofullerenes and carbon nanotubes. We constructed phase diagrams regarding the effect of pressure and doping levels with copper nitrate on the overall yields. Figure 5 shows a phase diagram (reproduced from [116]) summarizing the ratio between the various fullerene families, depending on different reaction conditions. Another interesting aspect of this method is that when high doping with copper nitrate is applied, the yield of $C_{60}$ and $C_{70}$ appears to be inversed. During a conventional arc discharge reaction, the ratio of $C_{60}$ to $C_{70}$ is 3:1; however, under high levels of chemical doping, $C_{70}$ appears to have a higher yield compared to $C_{60}$. The results can be attributed to the difference in thermodynamic stability of these two fullerenes.

The thermodynamic parameters for the formation of $C_{60}$ were determined long after its initial discovery, mainly due to the early difficulties in producing sufficient amounts of the material and the low purity. Interestingly, despite the fact that the fullerene is a strained molecule and hence has much higher energy in free form as compared to diamond and graphite, it can still be formed in sufficient quantities. Higher cages than $C_{60}$ have a different structure and symmetry, with $C_{70}$, the second most abundant fullerene to exceed $C_{60}$, has a binding energy per carbon atom of more than 0.02 eV greater than $C_{60}$, whose binding energy is 7.40 eV/C [117]. Despite the thermodynamic favour of $C_{70}$, $C_{60}$ is kinetically favoured. According to previous reports, the heat of formation for $C_{70}$ is 2555 kJ/mol, and 2327 kJ/mol for $C_{60}$ [118]. The comparison between $C_{60}$ and $C_{70}$ is shown in Figure 6. Similarly, fullerenes with more carbon atoms and EMFs themselves, which exhibit different isomers with different strains, will give a more complicated picture for determining the thermodynamic parameter of their synthesis [119].

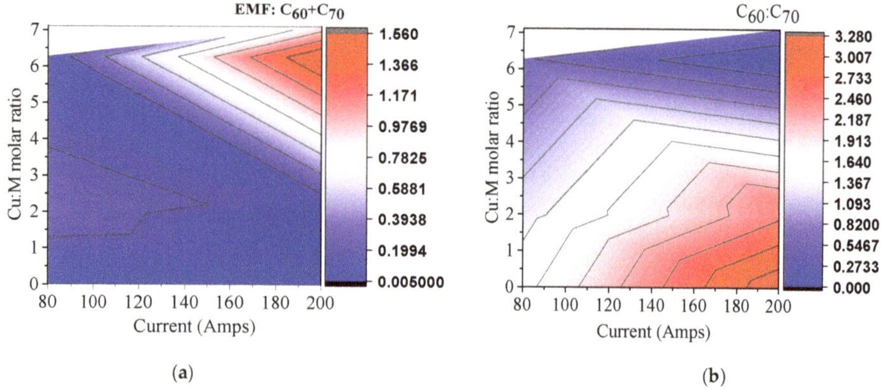


**Figure 5.** (**a**) The ratio of the higher fullerenes (>$C_{88}$) and EMF to the abundant $C_{60}/C_{70}$ vs. the current and Cu-to-rare-earth-metal molar ratio. (**b**) The ratio of $C_{60}$ to $C_{70}$ vs. the current and Cu-to-rare-earth-metal molar ratio (Reprinted/adapted with permission from [116]. Copyright 2006, Royal Society of Chemistry).

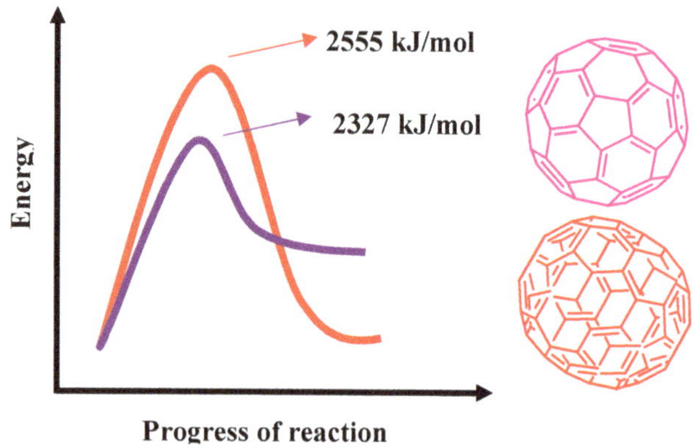


**Figure 6.** Kinetic vs. thermodynamic behaviour of the two most abundant fullerenes $C_{60}$ (in purple) and $C_{70}$ (in red). Reproduced from [118].

3.5.2. Effect of Reactive Gases: Methane and Ammonia

Chemically activated arc discharge presents a novel route not only for the control of the reaction pathway and hence the overall yield of the fullerenes, but also for the synthesis of novel EMFs with exciting properties, simply by carefully tailoring and studying the reaction conditions. Chen et al. [120] obtained different yields of dysprosium sulphite EMFs, a class of materials with interesting magnetic properties, by changing the levels of methane pressure used during synthesis. The dysprosium disulphide EMFs exhibit unique magnetic properties, with three energy barriers at 10.5, 48, and 1232 K. In general, a variety of magnetic behaviour can be observed in EMFs depending on the type of the incarcerated cluster. For example, the carbide $Er_2C_2@C_{82}$ behaves as a paramagnet, while the monometallic $Er@C_{82}$ shows an antiferromagnetic behaviour. An important dimetallic EMF, $ErSc_2N@C_{80}$, has even shown a long life of photoisomerization, which could be measured for 12 h at 20 K without any degradation. The synthesis of these EMFs is highly important, as they hold promising applications as molecular memory elements [121] and materials for spintronics and telecommunications device applications. EMFs comprising erbium and scandium are of special interest, as erbium has important optical properties (1.52 μm emission at the telecommunications window), and scandium atoms have large nuclear spin ($I = 2/7$), which is suitable for spin probe and atomic clock applications.

For the efficient synthesis of dysprosium sulphide EMFs, methane has been a valuable addition during the arc-discharge synthesis of metallofullerenes as a reactive gas, similar to the use of ammonia during the synthesis of fullerenes containing trimetallic nitride [122,123]. Popov demonstrated that the simple addition of 20 mbar $CH_4$ led to significant reduction in the formation of empty cage fullerenes during the synthesis of dysprosium sulphide fullerenes. The reaction mechanism in both methane- and ammonia-mediated synthesis is dominated by the hydrogen atom present in the gases. Hydrogen suppresses the growth of empty cage fullerenes and enables the more selective synthesis of mixed-metal and trimetallic nitride EMFs. On the other hand, other additives may have detrimental effects on the empty cage/EMF ratio. For example, during the graphite/metal oxide arc evaporation in the presence of $SO_2$, the empty fullerenes are the main products.

Furthermore, the nitrogen-containing atmosphere may lead not only to carbon cages but to a different class of fullerenes, the azafullerenes, where a nitrogen replaces a carbon atom and inherits other interesting properties. One of these azafullerenes, $Gd_2@C_{79}N$, exhibits a long spin relaxation time and a high magnetic spin state and is suitable for SQUID and pulsed EPR studies. In the case of dimetallic azafullerenes, gadolinium appears to have a higher yield compared to yttrium [124]. The gadolinium atoms couple in a ferromagnetic manner.

The formations of carbon nanotubes and metal carbides are two other reactions that take place during the arc vaporization and could be used as another observation to support either the bottom-up or the top-down mechanism for fullerene synthesis. Regarding the proposed mechanism of the charge transfer between the metal and the cage that stabilizes the EMF, see the work by Shinohara and Kroto [38].

In another example, Shinohara managed to isolate the missing low band gap endohedral metallofullerenes simply by adding Teflon tubes inside the arc reactor, in a short distance from the centre of the arc. During the arc-discharge process, the Teflon decomposed to $CH_2$ and $CF_3$ reactive groups that functionalize, in situ, the surface of the fullerenes. The resulting perfluorinated EMFs were more stable than their pristine counterparts, which are notoriously difficult to isolate with any common solvent [125]. Examples of the isolated fullerenes are $Y@C_{70}(CF_3)_3$, $Y@C_{72}(CF_3)$, $Y@C_{74}(CF_3)$ (I), $Y@C_{74}(CF_3)$ (II), and $Y@C_{74}(CF_3)_3$, as identified through MALDI-TOF analysis. The authors performed optical and DFT studies on these following analysis of other missing, small band gap EMFs. By further expanding the gallery of doping agents and mapping the temperature and pressure conditions, for example, by incorporating thermal and light sensors in the reactor and identifying the ideal molar ratio among the compounds, it is expected that the hindrances regarding the fullerene applications will be overcome. For example, in the case of nitrogen-based doping compounds, the reactivity in a CAPTEAR reaction appears to be based on the specific nitrogen anion, further demonstrating the complexity of the reaction [126].

## 4. Solvent Extraction

Solvent extraction is the next most important step in extraction of the EMFs from the carbon soot. Across much of the literature, the separation of fullerenes and/or EMFs from the soot is typically achieved by the Soxhlet extraction (SE) method [127]. Hence, the yield of extraction is dependent on the type of soot, the number of extraction cycles, and the solvents circulated through the sample. As discussed in the introduction, the metal inside the EMFs is not located in the centre of the carbon cage and is shifted towards one of the ends; therefore, this imparts a dipole moment to the EMF. Thus, the extraction of EMFs from the soot also requires a solvent of higher dipole moment, such as DMF, aniline, or

DMSO, instead of toluene or other solvents of low dipole moments that have traditionally been employed for extracting empty cages [110]. One of the common ways of extracting the maximum amount of fullerenes is by involving a two-step extraction method, where first a low-polarity solvent is used to extract the empty cages (>60%), followed by a high-polarity solvent for maximum extraction of the EMFs [110]. It is also important to note that different solvents work for different types of EMF. For instance, under constant parameters, Chervyakova et al. [98] showed that $CS_2$ could efficiently extract 11 wt% of Y@$C_{82}$ out of all fullerenes, compared to 5.0 wt % and 2.3 wt% yield of Sc@$C_{82}$ and Gd@$C_{82}$, respectively. Moreover, although $CS_2$ and $C_5H_5N$ are both considered a good solvent for extracting EMFs, it was observed that the overall yield of M@$C_{82}$ could vary widely between the types of solvents used under different synthetic parameters. For instance, at 360kPa He pressure, $CS_2$ extraction yielded about 12 times more Sc@$C_{82}$ than the use of $C_5H_5N$. Whereas at 120kPa, $CS_2$ extraction resulted in a threefold decrease in the Sc@$C_{82}$ yield as compared to using $C_5H_5N$. The authors have not commented on these observations, but perhaps a detailed study in this direction could help understand the relationship between synthesis processes and the resultant complex reactions of fullerenes as well as the corresponding methods to extract them efficiently using a particular solvent.

It has also been reported that porous glass, used for holding the soot and the solvent, increases the area of contact of the soot with the solvents and thus leads to a two- to threefold increase in the efficiency of the extraction process [110].

The latest methods and devices for the extraction of fullerenes have seen a shifted focus, aiming to address the limitations of the Soxhlet extraction method instead of optimising its parameters. In fact, the typical solvent extraction can be time-consuming, with the extraction requiring up to 48 h and the need for constant monitoring of stability parameters. Additionally, an efficient SE is still wasteful and utilizes environmentally harmful solvents, as listed in the previous paragraph [128,129]. Recently, Churilov et al. [130] devised a new method of extraction for empty cage fullerenes, called the mechanical extraction method, where a rotating shaft is used to rotate and grind the fullerenes, thus breaking the van der Waals bonds between the fullerenes and insoluble carbon particles from the soot. This also helps in separating the soluble fullerenes from the insoluble fullerenes in the solvent, thus efficiently separating the two in the extraction process. The combination of the mechanical influence with the filtration led to a massive reduction in the time taken for extraction (~15 min) compared to the traditional Soxhlet extraction method, whilst maintaining a similar resultant weight percentage of fullerene-to-soot-extraction ratio. The same group went on to demonstrate the efficiency of the method with higher cage empty fullerenes and EMFs [131]. They showed that the mechanical extraction method could extract 0.4 wt% higher weight fullerenes, which consisted of about 25% EMFs. Although the authors showed this methodology to work well with both EMFs and empty cages, we believe that it is worth using the mechanical extraction method and Soxhlet extraction to efficiently extract empty cage fullerenes and subsequently EMFs, respectively. Moreover, improvements can be further made in the mechanical extraction process by introducing heating in the system. This could lead to increased breaking of the fullerene-carbon particle bond and their increased solubility in the solvent.

## 5. Conclusions

In conclusion, we summarized some of the important parameters needed to optimize the synthesis of endohedral metallofullerenes. It is important to increase the yield of these nanomaterials in order to carry out in-depth investigation of their synthesis mechanism as well as to ensure a sustainable use for future electronic and opto-electronic devices. Challenges still exist in scaling-up the synthesis of individual fullerene molecules and optimizing separation of individual fullerenes from the obtained soot. While the conventional arc-discharge method, based on the vaporization of graphite composites, seems to have some limitations, the doping of the rods, despite being a more tedious and uncontrollable process, can be further expanded and provide new insights and pathways. New doping agents and optimized reaction conditions can be developed for further studying the role of different parameters during synthesis. Our ongoing research has a final aim of mapping the temperature and pressure conditions during the arc-discharge process by utilizing tailor-made arc-discharge reactors and carefully controlling and recording the reaction parameters and conditions. Incorporation of thermocouples inside the reactor to map the temperature during the synthesis will help in understanding the reaction conditions that are best suited for optimum synthesis yield of the endohedral fullerenes of interest [132]. This mapping is important not only for the scaled-up production of EMFs but also for the synthesis of new clusters, presenting either new properties or combining optical, optoelectronic, electrochemical, and magnetic properties. Examples of such materials are EMFs that incorporate erbium with gadolinium or dysprosium in different ratios, with the former providing the optical properties and the latter providing the magnetic behaviour. New advances, such as the construction of atomic clocks with N@$C_{60}$ [133], can provide new directions for the applications of endohedral fullerenes. In particular, EMFs of both well-known (such as Y@$C_{82}$, Sc@$C_{82}$, and La@$C_{82}$) or newly developed materials with different electron and nuclear spin numbers (such as phosphorous-containing endohedral fullerenes) could potentially be used for atomic clock applications. The incorporation of machine learning (ML) and artificial intelligence (AI) approaches for the synthesis of EMFs could revolutionize the industrial-scale production of these molecules by defining optimal process parameters and complementing traditional techniques such as the Design of Experiments (DoE) approach [134], favoured by engineers for many decades. This is a direction in which we plan to endeavour in the near future.

Finally, a combination of optical and paramagnetic properties [135–137] can also lead to new carbon nanotube–EMF peapods or EMF–graphene hybrid nanomaterials [138] with novel properties. Even in organic photovoltaics, fullerenes have recently been incorporated in ternary systems [139]. The incorporation of ternary systems including EMFs and non-fullerene acceptors could provide further avenues for higher-efficiency solar cells. Thus, the science and technology of EMFs will be a viable research topic for years to come.

**Author Contributions:** Conceptualization, S.S. and K.P.; methodology, S.S.; software, S.S.; writing—original draft preparation, S.S., P.D. and K.P.; writing—review and editing, S.S., K.S., S.K., P.D. and K.P. All authors have read and agreed to the published version of the manuscript.

**Funding:** This research was funded by EPSRC grant number EP/K030108/1.

**Acknowledgments:** S.S. would like to thank the Schmidt Science Fellows for financial support. K.P. would like to acknowledge EPSRC for grant support (EP/K030108/1).

**Conflicts of Interest:** The authors declare no conflict of interest.

## References

1. Kroto, H.W.; Heath, J.R.; O Brien, S.C.; Curl, R.F.; Smalley, R.E. $C_{60}$: Buckminsterfullerene. *Nature* **1985**, *318*, 162–163. [CrossRef]
2. Hawkins, J.M.; Meyer, A.; Lewis, T.A.; Loren, S.; Hollander, F.J. Crystal Structure of Osmylated $C_{60}$: Confirmation of the Soccer Ball Framework. *Science* **1991**, *252*, 312–313. [CrossRef]
3. Wudl, F. The chemical properties of buckminsterfullerene ($C_{60}$) and the birth and infancy of fulleroids. *Accounts Chem. Res.* **1992**, *25*, 157–161. [CrossRef]
4. Akasaka, T.; Lu, X. Structural and electronic properties of endohedral metallofullerenes. *Chem. Rec.* **2012**, *12*, 256–269. [CrossRef]
5. Kobayashi, K.; Nagase, S. A stable unconventional structure of $Sc_2$@$C_{66}$ found by density functional calculations. *Chem. Phys. Lett.* **2002**, *362*, 373–379. [CrossRef]
6. Rodríguez-Fortea, A.; Alegret, N.; Balch, A.L.; Poblet, J.M. The maximum pentagon separation rule provides a guideline for the structures of endohedral metallofullerenes. *Nat. Chem.* **2010**, *2*, 55–961. [CrossRef] [PubMed]
7. Schwerdtfeger, P.; Wirz, L.N.; Avery, J. The topology of fullerenes. *WIREs Comput. Mol. Sci.* **2015**, *5*, 96–145. [CrossRef] [PubMed]
8. Sure, R.; Hansen, A.; Schwerdtfeger, P.; Grimme, S. Comprehensive theoretical study of all 1812 $C_{60}$ isomers. *Phys. Chem. Chem. Phys.* **2017**, *19*, 14296–14305. [CrossRef] [PubMed]
9. Rios, C.; Salcedo, R. Computational Termochemistry Study of the $C_{80}$ Isomers and Their Endo Lanthanum Complexes by Applying Homodesmotic and Isodesmic Reactions. *Molecules* **2012**, *17*, 14588–14601. [CrossRef] [PubMed]

10. Yang, Y.; Arias, F.; Echegoyen, L.; Chibante, L.P.F.; Flanagan, S.; Robertson, A.; Wilson, L.J. Reversible Fullerene Electrochemistry: Correlation with the HOMO-LUMO Energy Difference for $C_{60}$, $C_{70}$, $C_{76}$, $C_{78}$, and $C_{84}$. *J. Am. Chem. Soc.* **1995**, *117*, 7801–7804. [CrossRef]
11. Takata, M.; Nishibori, E.; Sakata, M.; Inakuma, M.; Yamamoto, E.; Shinohara, H. Triangle Scandium Cluster Imprisoned in a Fullerene Cage. *Phys. Rev. Lett.* **1999**, *83*, 2214–2217. [CrossRef]
12. Wang, C.R.; Kai, T.; Tomiyama, T.; Yoshida, T.; Kobayashi, Y.; Nishibori, E.; Shinohara, H. A Scandium Carbide Endohedral Metallofullerene: $(Sc_2C_2)@C_{84}$. *Angew. Chem. Int. Ed.* **2001**, *40*, 397–399. [CrossRef]
13. Wang, T.-S.; Chen, N.; Xiang, J.-F.; Li, B.; Wu, J.-Y.; Xu, W.; Jiang, L.; Tan, K.; Shu, C.-Y.; Lu, X.; et al. Russian-Doll-Type Metal Carbide Endofullerene: Synthesis, Isolation, and Characterization of $Sc_4C_2@C_{80}$. *J. Am. Chem. Soc.* **2009**, *131*, 16646–16647. [CrossRef]
14. Stevenson, S.; Rice, G.; Glass, T.; Harich, K.; Cromer, F.; Jordan, M.R.; Craft, J.; Hadju, E.; Bible, R.; Olmstead, M.M.; et al. Small-bandgap endohedral metallofullerenes in high yield and purity. *Nature* **1999**, *401*, 55–57. [CrossRef]
15. Krause, M.; Ziegs, F.; Popov, A.A.; Dunsch, L. Entrapped Bonded Hydrogen in a Fullerene: The Five-Atom Cluster $Sc_3CH$ in $C_{80}$. *ChemPhysChem* **2007**, *8*, 537–540. [CrossRef]
16. Stevenson, S.; Mackey, M.A.; Stuart, M.A.; Phillips, J.P.; Easterling, M.L.; Chancellor, C.J.; Olmstead, M.M.; Balch, A.L. A Distorted Tetrahedral Metal Oxide Cluster inside an Icosahedral Carbon Cage. Synthesis, Isolation, and Structural Characterization of $Sc_4(\mu_3\text{-}O)_2@I_h\text{-}C_{80}$. *J. Am. Chem. Soc.* **2008**, *130*, 11844–11845. [CrossRef]
17. Syamala, M.S.; Cross, R.J.; Saunders, M. $^{129}Xe$ NMR Spectrum of Xenon Inside $C_{60}$. *J. Am. Chem. Soc.* **2002**, *124*, 6216–6219. [CrossRef]
18. Saunders, M.; Jiménez-Vázquez, H.A.; Cross, R.J.; Poreda, R.J. Stable Compounds of Helium and Neon: $He@C_{60}$ and $Ne@C_{60}$. *Science* **1993**, *259*, 1428–1430. [CrossRef] [PubMed]
19. Saunders, M.; Jimenez-Vazquez, H.A.; Cross, R.J.; Mroczkowski, S.; Gross, M.L.; Giblin, D.E.; Poreda, R.J. Incorporation of helium, neon, argon, krypton, and xenon into fullerenes using high pressure. *J. Am. Chem. Soc.* **1994**, *116*, 2193–2194. [CrossRef]
20. Larsson, J.A.; Greer, J.C.; Harneit, W.; Weidinger, A. Phosphorous trapped within buckminsterfullerene. *J. Chem. Phys.* **2002**, *116*, 7849–7854. [CrossRef]
21. Mauser, H.; Hirsch, A.; van Eikema Hommes, N.J.; Clark, T.; Pietzak, B.; Weidinger, A.; Dunsch, L. Stabilization of Atomic Nitrogen Inside C60. *Angew. Chem. Int. Ed. Engl.* **1997**, *36*, 2835–2838. [CrossRef]
22. Cai, W.; Zhang, M.; Echegoyen, L.; Lu, X. Recent Advances in Endohedral Metallofullerenes. *Fundam. Res.* 2023, *in press*. [CrossRef]
23. Chai, Y.; Guo, T.; Jin, C.; Haufler, R.E.; Chibante, L.P.F.; Fure, J.; Wang, L.; Alford, J.M.; Smalley, R.E. Fullerenes with metals inside. *J. Phys. Chem.* **1991**, *95*, 7564–7568. [CrossRef]
24. Dunsch, L.; Yang, S. Metal Nitride Cluster Fullerenes: Their Current State and Future Prospects. *Small* **2007**, *3*, 1298–1320. [CrossRef]
25. Tsuchiya, T.; Wakahara, T.; Lian, Y.; Maeda, Y.; Akasaka, T.; Kato, T.; Mizorogi, N.; Nagase, S. Selective Extraction and Purification of Endohedral Metallofullerene from Carbon Soot. *J. Phys. Chem. B* **2006**, *110*, 22517–22520. [CrossRef]
26. Sato, S.; Maeda, Y.; Guo, J.D.; Yamada, M.; Mizorogi, N.; Nagase, S.; Akasaka, T. Mechanistic Study of the Diels–Alder Reaction of Paramagnetic Endohedral Metallofullerene: Reaction of $La@C_{82}$ with 1,2,3,4,5-Pentamethylcyclopentadiene. *J. Am. Chem. Soc.* **2013**, *135*, 5582–5587. [CrossRef] [PubMed]
27. Bickelhaupt, F.M.; Solà, M.; Fernández, I. Understanding the Reactivity of Endohedral Metallofullerenes: $C_{78}$ versus $Sc_3N@C_{78}$. *Chem. A Eur. J.* **2015**, *21*, 5760–5768. [CrossRef]
28. Yutaka, M.; Akasaka, T.; Nagase, S. *Advances in Carbon Nanomaterials;* Jenny Stanford Publishing: Sigapore, 2012; Volume 269.
29. Suzuki, T.; Maruyama, Y.; Kato, T.; Kikuchi, K.; Achiba, Y. Electrochemical properties of $La@C_{82}$. *J. Am. Chem. Soc.* **1993**, *115*, 11006–11007. [CrossRef]
30. Kikuchi, K.; Suzuki, S.; Nakao, Y.; Nakahara, N.; Wakabayashi, T.; Shiromaru, H.; Saito, K.; Ikemoto, I.; Achiba, Y. Isolation and characterization of the metallofullerene $LaC_{82}$. *Chem. Phys. Lett.* **1993**, *216*, 67–71. [CrossRef]
31. Takata, M.; Umeda, B.; Nishibori, E.; Sakata, M.; Saitot, Y.; Ohno, M.; Shinohara, H. Confirmation by X-ray diffraction of the endohedral nature of the metallofullerene $Y@C_{82}$. *Nature* **1995**, *377*, 46–49. [CrossRef]
32. Yannoni, C.S.; Hoinkis, M.; de Vries, M.S.; Bethune, D.S.; Salem, J.R.; Crowder, M.S.; Johnson, R.D. Scandium Clusters in Fullerene Cages. *Science* **1992**, *256*, 1191–1192. [CrossRef]
33. Johnson, R.D.; de Vries, M.S.; Salem, J.; Bethune, D.S.; Yannoni, C.S. Electron paramagnetic resonance studies of lanthanum-containing $C_{82}$. *Nature* **1992**, *355*, 239–240. [CrossRef]
34. Kirbach, U.; Dunsch, L. The Existence of Stable $Tm@C_{82}$ Isomers. *Angew. Chem. Int. Ed. Engl.* **1996**, *35*, 2380–2383. [CrossRef]
35. Shinohara, H.; Sato, H.; Ohkohchi, M.; Ando, Y.; Kodama, T.; Shida, T.; Kato, T.; Saito, Y. Encapsulation of a scandium trimer in $C_{82}$. *Nature* **1992**, *357*, 52–54. [CrossRef]
36. Campbell, E.E.; Fanti, M.; Hertel, I.V.; Mitzner, R.; Zerbetto, F. The hyperpolarisability of an endohedral fullerene: $Li@C_{60}$. *Chem. Phys. Lett.* **1998**, *288*, 131–137. [CrossRef]
37. Deng, Q.; Junghans, K.; Popov, A.A. Carbide clusterfullerenes with odd number of carbon atoms: Molecular and electronic structures of $Sc_4C@C_{80}$, $Sc_4C@C_{82}$, and $Sc_4C_3@C_{80}$. *Theor. Chem. Accounts* **2015**, *134*, 10. [CrossRef]

38. Dunk, P.W.; Mulet-Gas, M.; Nakanishi, Y.; Kaiser, N.K.; Rodríguez-Fortea, A.; Shinohara, H.; Poblet, J.M.; Marshall, A.G.; Kroto, H.W. Bottom-up formation of endohedral mono-metallofullerenes is directed by charge transfer. *Nat. Commun.* **2014**, *5*, 5844. [CrossRef] [PubMed]
39. Ito, Y.; Okazaki, T.; Okubo, S.; Akachi, M.; Ohno, Y.; Mizutani, T.; Shinohara, H. Enhanced 1520 nm Photoluminescence from $Er^{3+}$ Ions in Di-erbium-carbide Metallofullerenes $(Er_2C_2)@C_{82}$ (Isomers I, II, and III). *ACS Nano* **2007**, *1*, 456–462. [CrossRef] [PubMed]
40. Wang, J.; Zhao, Y.; Lee, P.-H.; Irle, S. $Er^{3+}$ Photoluminescence in Er2@$C_{82}$ and Er2C2@$C_{82}$ Metallofullerenes Elucidated by Density Functional Theory. *Inorg. Chem.* **2017**, *56*, 6576–6583. [CrossRef] [PubMed]
41. Plant, S.R.; Dantelle, G.; Ito, Y.; Ng, T.C.; Ardavan, A.; Shinohara, H.; Porfyrakis, K. Acuminated fluorescence of $Er^{3+}$ centres in endohedral fullerenes through the incarceration of a carbide cluster. *Chem. Phys. Lett.* **2009**, *476*, 41–45. [CrossRef]
42. Li, M.-Y.; Zhao, Y.-X.; Han, Y.-B.; Yuan, K.; Nagase, S.; Ehara, M.; Zhao, X. Theoretical Investigation of the Key Roles in Fullerene-Formation Mechanisms: Enantiomer and Enthalpy. *ACS Appl. Nano Mater.* **2020**, *3*, 547–554. [CrossRef]
43. Yang, S.; Zalibera, M.; Rapta, P.; Dunsch, L. Charge-Induced Reversible Rearrangement of Endohedral Fullerenes: Electrochemistry of Tridysprosium Nitride Clusterfullerenes $Dy_3N@C_2n$ (2n =78, 80). *Chem. A Eur. J.* **2006**, *12*, 7848–7855. [CrossRef]
44. Popov, A.A.; Avdoshenko, S.M.; Cuniberti, G.; Dunsch, L. Dimerization of Radical-Anions: Nitride Clusterfullerenes versus Empty Fullerenes. *J. Phys. Chem. Lett.* **2011**, *2*, 1592–1600. [CrossRef]
45. Zhang, J.; Bowles, F.L.; Bearden, D.W.; Ray, W.K.; Fuhrer, T.; Ye, Y.; Dixon, C.; Harich, K.; Helm, R.F.; Olmstead, M.M.; et al. A missing link in the transformation from asymmetric to symmetric metallofullerene cages implies a top-down fullerene formation mechanism. *Nat. Chem.* **2013**, *5*, 880–885. [CrossRef]
46. Crichton, R.; Zhang, J. *Handbook of Fullerene Science and Technology*; Springer Nature: Singapore, 2022.
47. Mulet-Gas, M.; Abella, L.; Dunk, P.W.; Rodríguez-Fortea, A.; Kroto, H.W.; Poblet, J.M. Small endohedral metallofullerenes: Exploration of the structure and growth mechanism in the $Ti@C_{2n}$ (2n = 26–50) family. *Chem. Sci.* **2014**, *6*, 675–686. [CrossRef] [PubMed]
48. Yao, Y.R.; Chen, Z.C.; Chen, L.; Zheng, S.Y.; Yang, S.; Deng, S.L.; Zheng, L.S. Two Metastable Endohedral Metallofullerenes $Sc_2C_2@C_1(39656)\text{-}C_{82}$ and $Sc_2C_2@C_1(51383)\text{-}C_{84}$:Direct-$C_2$-Insertion Products from Their Most Stable Precursors. *J. Am. Chem. Soc.* **2023**, *145*, 16778–16786. [CrossRef]
49. Yamada, M.; Akasaka, T.; Nagase, S. Endohedral Metal Atoms in Pristine and Functionalized Fullerene Cages. *Acc. Chem. Res.* **2009**, *43*, 92–102. [CrossRef] [PubMed]
50. Takata, M. The MEM/Rietveld method with nano-applications—Accurate charge-density studies of nano-structured materials by synchrotron-radiation powder diffraction. *Acta Crystallogr. A* **2008**, *64*, 232–245. [CrossRef]
51. Shinohara, H.; Inakuma, M.; Hayashi, N.; Sato, H.; Saito, Y.; Kato, T.; Bandow, S. Spectroscopic Properties of Isolated $Sc_3@C_{82}$ Metallofullerene. *J. Phys. Chem.* **1994**, *98*, 8597–8599. [CrossRef]
52. Senapati, L.; Schrier, J.; Whaley, K.B. Reply to Comment on "Electronic Transport, Structure, and Energetics of Endohedral $Gd@C_{82}$ Metallofullerenes". *Nano Lett.* **2005**, *5*, 2341. [CrossRef]
53. Kobayashi, K.; Nagase, S. Theoretical study of structures and dynamic properties of $Sc_3@C_{82}$. *Chem. Phys. Lett.* **1999**, *313*, 45–51. [CrossRef]
54. Yamada, M.; Wakahara, T.; Lian, Y.; Tsuchiya, T.; Akasaka, T.; Waelchli, M.; Mizorogi, N.; Nagase, S.; Kadish, K.M. Analysis of Lanthanide-Induced NMR Shifts of the $Ce@C_{82}$ Anion. *J. Am. Chem. Soc.* **2006**, *128*, 1400–1401. [CrossRef] [PubMed]
55. Guo, Y.-J.; Zheng, H.; Yang, T.; Nagase, S.; Zhao, X. Theoretical Insight into the Ambiguous Endohedral Metallofullerene $Er_3C_{74}$: Covalent Interactions among Three Lanthanide Atoms. *Inorg. Chem.* **2015**, *54*, 8066–8076. [CrossRef]
56. Lu, X.; Akasaka, T.; Nagase, S. Carbide Cluster Metallofullerenes: Structure, Properties, and Possible Origin. *Accounts Chem. Res.* **2013**, *46*, 1627–1635. [CrossRef] [PubMed]
57. Popov, A.A.; Zhang, L.; Dunsch, L. A Pseudoatom in a Cage: Trimetallofullerene $Y_3@C_{80}$ Mimics $Y_3N@C_{80}$ with Nitrogen Substituted by a Pseudoatom. *ACS Nano* **2010**, *4*, 795–802. [CrossRef] [PubMed]
58. Zhao, P.; Zhao, X.; Ehara, M. Theoretical Insight into $Sc_2C_{76}$: Carbide Clusterfullerene $Sc_2C_2@C_{74}$ versus Dimetallofullerene $Sc_2@C_{76}$. *Inorg. Chem.* **2017**, *56*, 10195–10203. [CrossRef]
59. Jin, P.; Hao, C.; Li, S.; Mi, W.; Sun, Z.; Zhang, J.; Hou, Q. Theoretical Study on the Motion of a La Atom Inside a $C_{82}$ Cage. *J. Phys. Chem. A* **2006**, *111*, 167–169. [CrossRef] [PubMed]
60. Liu, F.; Velkos, G.; Krylov, D.S.; Spree, L.; Zalibera, M.; Ray, R.; Samoylova, N.A.; Chen, C.-H.; Rosenkranz, M.; Schiemenz, S.; et al. Air-stable redox-active nanomagnets with lanthanide spins radical-bridged by a metal–metal bond. *Nat. Commun.* **2019**, *10*, 571. [CrossRef]
61. Hao, Y.; Wang, Y.; Dubrovin, V.; Avdoshenko, S.M.; Popov, A.A.; Liu, F. Caught in Phase Transition: Snapshot of the Metallofullerene $Sc_3N@C_{70}$ Rotation in the Crystal. *J. Am. Chem. Soc.* **2021**, *143*, 612–616. [CrossRef]
62. Speller, E.M. The significance of fullerene electron acceptors in organic solar cell photo-oxidation. *Mater. Sci. Technol.* **2017**, *33*, 924–933. [CrossRef]
63. Ganesamoorthy, R.; Sathiyan, G.; Sakthivel, P. Review: Fullerene based acceptors for efficient bulk heterojunction organic solar cell applications. *Sol. Energy Mater. Sol. Cells* **2017**, *161*, 102–148. [CrossRef]
64. Ross, R.B.; Cardona, C.M.; Guldi, D.M.; Sankaranarayanan, S.G.; Reese, M.O.; Kopidakis, N.; Peet, J.; Walker, B.; Bazan, G.C.; Van Keuren, E.; et al. Endohedral fullerenes for organic photovoltaic devices. *Nat. Mater.* **2009**, *8*, 208–212. [CrossRef]

65. Li, T.; Xiao, L.; Yang, J.; Ding, M.; Zhou, Z.; LaConte, L.; Jin, L.; Dorn, H.C.; Li, X. Trimetallic Nitride Endohedral Fullerenes Carboxyl-$Gd_3N@C_{80}$: A New Theranostic Agent for Combating Oxidative Stress and Resolving Inflammation. *ACS Appl. Mater. Interfaces* **2017**, *9*, 17681–17687. [CrossRef]
66. Lu, J.; Sabirianov, R.F.; Mei, W.N.; Gao, Y.; Duan, C.G.; Zeng, X. Structural and Magnetic Properties of $Gd_3N@C_{80}$. *J. Phys. Chem. B* **2006**, *110*, 23637–23640. [CrossRef]
67. Zhang, J.; Ye, Y.; Chen, Y.; Pregot, C.; Li, T.; Balasubramaniam, S.; Dorn, H.C. $Gd_3N@C_{84}(OH)x$: A New Egg-Shaped Metallofullerene Magnetic Resonance Imaging Contrast Agent. *J. Am. Chem. Soc.* **2014**, *136*, 2630–2636. [CrossRef]
68. Weaver, J.; Chai, Y.; Kroll, G.; Jin, C.; Ohno, T.; Haufler, R.; Guo, T.; Alford, J.; Conceicao, J.; Chibante, L.; et al. XPS probes of carbon-caged metals. *Chem. Phys. Lett.* **1992**, *190*, 460–464. [CrossRef]
69. Murphy, T.A.; Pawlik, T.; Weidinger, A.; Höhne, M.; Alcala, R.; Spaeth, J.M. Observation of Atomlike Nitrogen in Nitrogen-Implanted Solid $C_{60}$. *Phys. Rev. Lett.* **1996**, *77*, 1075–1078. [CrossRef]
70. Funasaka, H.; Yamamoto, K.; Sakurai, K.; Ishiguro, T.; Sugiyama, K.; Takahashi, T.; Kishimoto, Y. Preparation of Fullerene Derivatives by Resistive Heating with Graphite Crucible. *Full- Sci. Technol.* **1993**, *1*, 437–448. [CrossRef]
71. Sinitsa, A.S.; Chamberlain, T.W.; Zoberbier, T.; Lebedeva, I.V.; Popov, A.M.; Knizhnik, A.A.; McSweeney, R.L.; Biskupek, J.; Kaiser, U.; Khlobystov, A.N. Formation of Nickel Clusters Wrapped in Carbon Cages: Toward New Endohedral Metallofullerene Synthesis. *Nano Lett.* **2017**, *17*, 1082–1089. [CrossRef] [PubMed]
72. Popov, A.A. Synthesis and Molecular Structures of Endohedral Fullerenes. In *Nanostructure Science and Technology*; Lockwood, D., Ed.; Springer: Berlin/Heidelberg, Germany, 2017; pp. 1–34.
73. Bloodworth, S.; Whitby, R.J. Synthesis of endohedral fullerenes by molecular surgery. *Commun. Chem.* **2022**, *5*, 121. [CrossRef] [PubMed]
74. Churilov, G.; Popov, A.; Vnukova, N.; Dudnik, A.; Samoylova, N.; Glushenko, G. Controlled synthesis of fullerenes and endohedral metallofullerenes in high frequency arc discharge. *Full-Nanotub. Carbon Nanostructures* **2016**, *24*, 675–678. [CrossRef]
75. Shinohara, H. Endohedral metallofullerenes. *Rep. Prog. Phys.* **2000**, *63*, 843–892. [CrossRef]
76. Chen, N.; Ortiz, A.L.; Echegoyen, L. *Fullerenes: Principles and Applications*, 2nd ed; The Royal Society of Chemistry: London, UK, 2011; Chapter 2, pp. 12–65.
77. Yang, S. Synthesis, Separation, and Molecular Structures of Endohedral Fullerenes. *Curr. Org. Chem.* **2012**, *16*, 1079–1094. [CrossRef]
78. Lu, X.; Feng, L.; Akasaka, T.; Nagase, S. Current status and future developments of endohedral metallofullerenes. *Chem. Soc. Rev.* **2012**, *41*, 7723–7760. [CrossRef]
79. Lu, X.; Bao, L.; Akasaka, T.; Nagase, S. Recent progress in the chemistry of endohedral metallofullerenes. *Chem. Commun.* **2014**, *50*, 14701–14715. [CrossRef]
80. Jin, P.; Tang, C.; Chen, Z. Carbon atoms trapped in cages: Metal carbide clusterfullerenes. *Coord. Chem. Rev.* **2014**, *270–271*, 89–111. [CrossRef]
81. Shinohara, H.; Yamaguchi, H.; Hayashi, N.; Sato, H.; Ohkohchi, M.; Ando, Y.; Saito, Y. Isolation and spectroscopic properties of scandium fullerenes ($Sc_2@C_{74}$, $Sc_2@C_{82}$, and $Sc_2@C_{84}$). *J. Phys. Chem.* **1993**, *97*, 4259–4261. [CrossRef]
82. Liu, B.-B.; Xu, W.-G.; Liu, Z.-Y.; Yang, H.-B.; Li, M.-H.; Liu, S.-Y.; Zou, G.-T. High yield synthesis and extraction of La@C2n. *Solid State Commun.* **1996**, *97*, 407–410. [CrossRef]
83. Bandow, S.; Shinohara, H.; Saito, Y.; Ohkohchi, M.; Ando, Y. High yield synthesis of lanthanofullerenes via lanthanum carbide. *J. Phys. Chem.* **1993**, *97*, 6101–6103. [CrossRef]
84. Kitaura, R.; Shinohara, H. Endohedral Metallofullerenes and Nano-Peapods. *Jpn. J. Appl. Phys.* **2007**, *46*, 881. [CrossRef]
85. Lu, W.; Baek, J.B.; Dai, L. (Eds.) *Carbon Nanomaterials for Advanced Energy Systems*; Wiley: Hoboken, NJ, USA, 2015.
86. Kyesmen, P.I.; Onoja, A.; Amah, A.N. Fullerenes synthesis by combined resistive heating and arc discharge techniques. *SpringerPlus* **2016**, *5*, 1323. [CrossRef] [PubMed]
87. Lian, Y.; Yang, S.; Yang, S. Revisiting the Preparation of $La@C_{82}$ (I and II) and $La_2$ $@C_{80}$: Efficient Production of the "Minor" Isomer $La@C_{82}$(II). *J. Phys. Chem. B* **2002**, *106*, 3112–3117. [CrossRef]
88. Ross, M.M.; Nelson, H.H.; Callahan, J.H.; McElvany, S.W. Production and characterization of metallofullerenes. *J. Phys. Chem.* **1992**, *96*, 5231–5234. [CrossRef]
89. Fedorov, A.; Novikov, P.; Churilov, G. Influence of electron concentration and temperature on endohedral metallofullerene $Me@C_{84}$ formation in a carbon plasma. *Chem. Phys.* **2003**, *293*, 253–261. [CrossRef]
90. Balch, A.L.; Olmstead, M.M. Reactions of Transition Metal Complexes with Fullerenes ($C_{60}$, $C_{70}$, etc.) and Related Materials. *Chem. Rev.* **1998**, *98*, 2123–2166. [CrossRef] [PubMed]
91. Mu, L.; Yang, S.; Feng, R.; Kong, X. Encapsulation of Platinum in Fullerenes: Is That Possible? *Inorg. Chem.* **2017**, *56*, 6035–6038. [CrossRef]
92. Pradeep, T.; Kulkarni, G.U.; Kannan, K.R.; Row, T.N.G.; Rao, C.N.R. A novel iron fullerene ($FeC_{60}$) adduct in the solid state. *J. Am. Chem. Soc.* **1992**, *114*, 2272–2273. [CrossRef]
93. Nishibori, E.; Takata, M.; Sakata, M.; Inakuma, M.; Shinohara, H. Determination of the cage structure of $Sc@C_{82}$ by synchrotron powder diffraction. *Chem. Phys. Lett.* **1998**, *298*, 79–84. [CrossRef]

94. Kiang, C.-H.; VAN Loosdrecht, P.; Beyers, R.; Salem, J.; Bethune, D.; Iii, W.G.; Dorn, H.; Burbank, P.; Stevenson, S. novel structures from arc-vaporized carbon and metals: Single-layer carbon nanotubes and metallofullerenes. *Surf. Rev. Lett.* **1996**, *3*, 765–769. [CrossRef]
95. Novoselov, K.S.; Geim, A.K.; Morozov, S.V.; Jiang, D.; Zhang, Y.; Dubonos, S.V.; Grigorieva, I.V.; Firsov, A.A. Electric field effect in atomically thin carbon films. *Science* **2004**, *306*, 666–669. [CrossRef]
96. Vekselman, V.; Feurer, M.; Huang, T.; Stratton, B.; Raitses, Y. Complex structure of the carbon arc discharge for synthesis of nanotubes. *Plasma Sources Sci. Technol.* **2017**, *26*, 065019. [CrossRef]
97. Qian, H.-J.; Wang, Y.; Morokuma, K. Quantum mechanical simulation reveals the role of cold helium atoms and the coexistence of bottom-up and top-down formation mechanisms of buckminsterfullerene from carbon vapor. *Carbon* **2017**, *114*, 635–641. [CrossRef]
98. Chervyakova, D.I.; Churilov, G.N.; Dudnik, A.I.; Glushenko, G.A.; Kovaleva, E.A.; Kuzubov, A.A.; Vnukova, N.G. The Technology and Setup for High-Throughput Synthesis of Endohedral Metal Fullerenes. In *Proceedings of the Scientific-Practical Conference 'Research and Development—2016*; Springer International Publishing: Berlin/Heidelberg, Germany, 2018; pp. 481–489. [CrossRef]
99. Funasaka, H.; Sakurai, K.; Oda, Y.; Yamamoto, K.; Takahashi, T. Magnetic properties of $Gd@C_{82}$ metallofullerene. *Chem. Phys. Lett.* **1995**, *232*, 273–277. [CrossRef]
100. Ding, J.; Yang, S. Efficient N, N -Dimethylformamide Extraction of Endohedral Metallofullerenes for HPLC Purification. *Chem. Mater.* **1996**, *8*, 2824–2827. [CrossRef]
101. Ding, J.; Yang, S. Isolation and Characterization of $Pr@C_{82}$ and $Pr_2@C_{80}$. *J. Am. Chem. Soc.* **1996**, *118*, 11254–11257. [CrossRef]
102. Bubnov, V.P.; Laukhina, E.E.; Kareev, I.E.; Koltover, V.K.; Prokhorova, T.G.; Yagubskii, E.B.; Kozmin, Y.P. Endohedral Metallofullerenes: A Convenient Gram-Scale Preparation. *Chem. Mater.* **2002**, *14*, 1004–1008. [CrossRef]
103. Kodama, T.; Ozawa, N.; Miyake, Y.; Sakaguchi, K.; Nishikawa, H.; Ikemoto, I.; Achiba, Y. Structural Study of Three Isomers of $Tm@C_{82}$ by $^{13}C$ NMR Spectroscopy. *J. Am. Chem. Soc.* **2002**, *124*, 1452–1455. [CrossRef]
104. Sun, B.; Feng, L.; Shi, Z.; Gu, Z. Improved extraction of metallofullerenes with DMF at high temperature. *Carbon* **2002**, *40*, 1591–1595. [CrossRef]
105. Bolskar, R.D.; Benedetto, A.F.; Husebo, L.O.; Price, R.E.; Jackson, E.F.; Wallace, S.; Alford, J.M. First Soluble $M@C_{60}$ Derivatives Provide Enhanced Access to Metallofullerenes and Permit in Vivo Evaluation of $Gd@C_{60}[C(COOH)_2]_{10}$ as a MRI Contrast Agent. *J. Am. Chem. Soc.* **2003**, *125*, 5471–5478. [CrossRef] [PubMed]
106. Lian, Y.; Shi, Z.; Zhou, X.; Gu, Z. Different Extraction Behaviors between Divalent and Trivalent Endohedral Metallofullerenes. *Chem. Mater.* **2004**, *16*, 1704–1714. [CrossRef]
107. Kodama, T.; Higashi, K.; Ichikawa, T.; Suzuki, S.; Nishikawa, H.; Ikemoto, I.; Kikuchi, K.; Achiba, Y. Selective Extraction of Endohedral Metallofullerenes Using a Mixed Solvent of Triethylamine and Acetone. *Chem. Lett.* **2005**, *34*, 464–465. [CrossRef]
108. Iiduka, Y.; Wakahara, T.; Nakajima, K.; Tsuchiya, T.; Nakahodo, T.; Maeda, Y.; Akasaka, T.; Mizorogi, N.; Nagase, S. $^{13}C$ NMR spectroscopic study of scandium dimetallofullerene, $Sc_2@C_{84}$ vs. $Sc_2C_2@C_{82}$. *Chem. Commun.* **2006**, *19*, 2057–2059. [CrossRef]
109. Grushko, Y.S.; Kozlov, V.S.; Sedov, V.P.; Kolesnik, S.G.; Lebedev, V.T.; Shilin, V.A.; Khodorkovsky, M.A.; Artamonova, T.O.; Shakhmin, A.L.; Shamanin, V.V.; et al. MRI-Contrasting System Based on Water-Soluble Fullerene/Gd-Metallofullerene Mixture. *Full-Nanotub. Carbon Nanostructures* **2010**, *18*, 417–421. [CrossRef]
110. Kozlov, V.S.; Suyasova, M.V.; Lebedev, V.T. Synthesis, extraction, and chromatographic purification of higher empty fullerenes and endohedral gadolinium metallofullerenes. *Russ. J. Appl. Chem.* **2014**, *87*, 121–127. [CrossRef]
111. Dong, W.; Zhou, Q.; Shen, W.; Yang, L.; Jin, P.; Lu, X.; Lian, Y. The Various Packing Structures of $Tb@C_{82}$ (I, II) Isomers in Their Cocrystals with Ni(OEP). *Nanomaterials* **2023**, *13*, 994. [CrossRef]
112. Olmstead, M.M.; de Bettencourt-Dias, A.; Duchamp, J.C.; Stevenson, S.; Marciu, D.; Dorn, H.C.; Balch, A.L. Isolation and Structural Characterization of the Endohedral Fullerene $Sc_3N@C_{78}$. *Angew. Chem. Int. Ed.* **2001**, *40*, 1223–1225. [CrossRef]
113. Yang, S.; Zhang, L.; Zhang, W.; Dunsch, L. A Facile Route to Metal Nitride Clusterfullerenes by Using Guanidinium Salts: A Selective Organic Solid as the Nitrogen Source. *Chem. A Eur. J.* **2010**, *16*, 12398–12405. [CrossRef]
114. Stevenson, S.; Thompson, M.C.; Coumbe, H.L.; Mackey, M.A.; Coumbe, C.E.; Phillips, J.P. Chemically Adjusting Plasma Temperature, Energy, and Reactivity (CAPTEAR) Method Using NO x and Combustion for Selective Synthesis of $Sc_3N@C_{80}$ Metallic Nitride Fullerenes. *J. Am. Chem. Soc.* **2007**, *129*, 16257–16262. [CrossRef]
115. Stevenson, S.; Mackey, M.A.; Thompson, M.C.; Coumbe, H.L.; Madasu, P.K.; Coumbe, C.E.; Phillips, J.P. Effect of copper metal on the yield of $Sc_3N@C_{80}$ metallofullerenes. *Chem. Commun.* **2007**, *41*, 4263–4265. [CrossRef] [PubMed]
116. Dallas, P.; Meysami, S.S.; Grobert, N.; Porfyrakis, K. Classification of carbon nanostructure families occurring in a chemically activated arc discharge reaction. *RSC Adv.* **2016**, *6*, 24912–24920. [CrossRef]
117. Hagelberg, F. *Magnetism in Carbon Nanostructures*; Cambridge University Press: Cambridge, UK, 2017; ISBN 9781107069848.
118. Beckhaus, H.D.; Verevkin, S.; Rüchardt, C.; Diederich, F.; Thilgen, C.; ter Meer, H.U.; Müller, W. $C_{70}$ Is More Stable than $C_{60}$: Experimental Determination of the Heat of Formation of $C_{70}$. *Angew. Chem. Int. Ed. Engl.* **1994**, *33*, 996–998. [CrossRef]
119. Slanina, Z.; FranÇlois, J.P.; Kolb, M.; Bakowies, D.; Thiel, W. Calculated Relative Stabilities of $C_{84}$. *Fuller. Sci. Technol.* **1993**, *1*, 221–230. [CrossRef]
120. Chen, Y.; Willis, M.C. Cu(I)—Catalyzed sulfonylative Suzuki–Miyaura cross-coupling. *Chem. Sci.* **2017**, *8*, 3249–3253. [CrossRef]

121. Morton, J.J.; Tiwari, A.; Dantelle, G.; Porfyrakis, K.; Ardavan, A.; Briggs GA, D. Switchable ErSc2N Rotor within a C80 Fullerene Cage: An Electron Paramagnetic Resonance and Photoluminescence Excitation Study. *Phys. Rev. Lett.* **2008**, *101*, 013002. [CrossRef]
122. Junghans, K.; Schlesier, C.; Kostanyan, A.; Samoylova, N.A.; Deng, Q.; Rosenkranz, M.; Popov, A.A. Methane as a Selectivity Booster in the Arc-Discharge Synthesis of Endohedral Fullerenes: Selective Synthesis of the Single-Molecule Magnet Dy2 TiC@C80 and Its Congener $Dy_2TiC_2@C_{80}$. *Angew. Chem. Int. Ed.* **2015**, *54*, 13411–13415. [CrossRef]
123. Junghans, K.; Rosenkranz, M.; Popov, A.A. $Sc_3CH@C_{80}$: Selective $^{13}C$ enrichment of the central carbon atom. *Chem. Commun.* **2016**, *52*, 6561–6564. [CrossRef]
124. Zhang, J.; Shibata, A.; Ito, M.; Shuto, S.; Ito, Y.; Mannervik, B.; Abe, H.; Morgenstern, R. Synthesis and Characterization of a Series of Highly Fluorogenic Substrates for Glutathione Transferases, a General Strategy. *J. Am. Chem. Soc.* **2011**, *133*, 14109–14119. [CrossRef]
125. Wang, Z.; Nakanishi, Y.; Noda, S.; Niwa, H.; Zhang, J.; Kitaura, R.; Shinohara, H. Missing Small-Bandgap Metallofullerenes: Their Isolation and Electronic Properties. *Angew. Chem. Int. Ed.* **2013**, *52*, 11770–11774. [CrossRef] [PubMed]
126. Liu, F.; Guan, J.; Wei, T.; Wang, S.; Jiao, M.; Yang, S. A Series of Inorganic Solid Nitrogen Sources for the Synthesis of Metal Nitride Clusterfullerenes: The Dependence of Production Yield on the Oxidation State of Nitrogen and Counter Ion. *Inorg. Chem.* **2013**, *52*, 3814–3822. [CrossRef] [PubMed]
127. Jensen, W.B. The Origin of the Soxhlet Extractor. *J. Chem. Educ.* **2007**, *84*, 1913. [CrossRef]
128. Bezmel'nitsyn, V.N.; Eletskii, A.V.; Okun', M. Fullerenes in solutions. *Phys. Uspekhi* **1998**, *41*, 1091–1114. [CrossRef]
129. Raebiger, J.W.; Alford, J.M.; Bolskar, R.D.; Diener, M.D. Chemical redox recovery of giant, small-gap and other fullerenes. *Carbon* **2011**, *49*, 37–46. [CrossRef]
130. Churilov, G.N.; Elesina, V.I.; Dudnik, A.I.; Vnukova, N.G. Ultrafast method of fullerenes extraction from carbon condensate. *Full-Nanotub. Carbon Nanostructures* **2019**, *27*, 225–232. [CrossRef]
131. Elesina, V.I.; Churilov, G.N.; Vnukova, N.G.; Dudnik, A.I.; Osipova, I.V. Filtration process combined with mechanical action, as a method for efficient extraction of endohedral metallofullerenes from carbon soot. *Full-Nanotub. Carbon Nanostructures* **2019**, *27*, 803–807. [CrossRef]
132. Rincón-García, L.; Ismael, A.K.; Evangeli, C.; Grace, I.; Rubio-Bollinger, G.; Porfyrakis, K.; Agraït, N.; Lambert, C.J. Molecular design and control of fullerene-based bi-thermoelectric materials. *Nat. Mater.* **2015**, *15*, 289–293. [CrossRef] [PubMed]
133. Harding, R.T.; Zhou, S.; Zhou, J.; Lindvall, T.; Myers, W.K.; Ardavan, A.; Laird, E.A. Spin Resonance Clock Transition of the Endohedral Fullerene $^{15}N@C_{60}$. *Phys. Rev. Lett.* **2017**, *119*, 140801. [CrossRef] [PubMed]
134. Cao, B.; Adutwum, L.A.; Oliynyk, A.O.; Luber, E.J.; Olsen, B.C.; Mar, A.; Buriak, J.M. How To Optimize Materials and Devices via Design of Experiments and Machine Learning: Demonstration Using Organic Photovoltaics. *ACS Nano* **2018**, *12*, 7434–7444. [CrossRef]
135. Hao, J.; Li, F.; Li, H.; Chen, X.; Zhang, Y.; Chen, Z.; Hao, C. Dynamic motion of an Lu pair inside a C76(Td) cage. *RSC Adv.* **2015**, *5*, 34383–34389. [CrossRef]
136. Roukala, J.; Straka, M.; Taubert, S.; Vaara, J.; Lantto, P. Ratcheting rotation or speedy spinning: EPR and dynamics of $Sc_3C_2@C_{80}$. *Chem. Commun.* **2017**, *53*, 8992–8995. [CrossRef]
137. Krisilov, A.; Nechaev, I.; Kotova, A.; Shikhaliev, K.; Chernov, V.; Zon, B. The role of the encapsulated atom in the vibrational spectra of $La@C_{60}$–$Lu@C_{60}$ lanthanide endofullerenes. *Comput. Theor. Chem.* **2015**, *1054*, 100–108. [CrossRef]
138. Sinha, S.; Sheng, Y.; Griffiths, I.; Young, N.P.; Zhou, S.; Kirkland, A.I.; Porfyrakis, K.; Warner, J.H. *In Situ* Atomic-Level Studies of Gd Atom Release and Migration on Graphene from a Metallofullerene Precursor. *ACS Nano* **2018**, *12*, 10439–10451. [CrossRef]
139. Yu, H.; Wang, Y.; Zou, X.; Yin, J.; Shi, X.; Li, Y.; Zhao, H.; Wang, L.; Ng, H.M.; Zou, B.; et al. Improved photovoltaic performance and robustness of all-polymer solar cells enabled by a polyfullerene guest acceptor. *Nat. Commun.* **2023**, *14*, 2323. [CrossRef] [PubMed]

inorganics

Review

# Recent Progress on the Functionalization of Endohedral Metallofullerenes

Song Wang [1,2,*], Xianming Zhang [1], Xi Tan [2], Hongzhen Li [2], Songxin Dai [2], Bin Yao [2], Xingyan Liu [2], Youzhou He [2] and Fei Jin [3,*]

1 Engineering Research Center for Waste Oil Recovery Technology and Equipment, Ministry of Education, Chongqing Technology and Business University, Chongqing 400067, China; zxm215@126.com
2 Department of Materials Science and Engineering, College of Environment and Resources, Chongqing Technology and Business University, Chongqing 400067, China; 17843952097@163.com (X.T.); 2020412242@email.ctbu.edu.cn (H.L.); 19122079474@163.com (S.D.); yaobinpku@126.com (B.Y.); liuxingyan@ctbu.edu.cn (X.L.); yzhectbu@163.com (Y.H.)
3 Magnetism Key Laboratory of Zhejiang Province, China Jiliang University, Hangzhou 310018, China
* Correspondence: wangsong@ctbu.edu.cn (S.W.); jinfei@cjlu.edu.cn (F.J.)

**Abstract:** Functionalization of endohedral metallofullerenes (EMFs) plays an important role in exploring the reactivity of EMFs and stabilizing missing EMFs, thus conferring tunable properties and turning EMFs into applicable materials. In this review, we present exhaustive progress on the functionalization of EMFs since 2019. Classic functionalization reactions include Prato reactions, Bingel–Hirsch reactions, radical addition reactions, carbene addition reactions, and so on are summarized. And new complicated multi-component reactions and other creative reactions are presented as well. We also discuss the structural features of derivatives of EMFs and the corresponding reaction mechanisms to understand the reactivity and regioselectivity of EMFs. In the end, we make conclusions and put forward an outlook on the prospect of the functionalization of EMFs.

**Keywords:** endohedral metallofullerenes; functionalization; reactivity; regioselectivity; derivatives; DFT calculation

## 1. Introduction

Endohedral metallofullerenes (EMFs) with the metallic atom(s) or cluster encapsulated inside the carbon cages exhibit unique molecular structure and physicochemical properties, conferring EMFs great potential application in biomedicine, optoelectronic, and spin devices [1,2].

Until now, hundreds of EMFs have been reported and it is found that metals from IA, IIA, IIIB, IVB, and VB groups can be entrapped in the carbon cage [1–3], especially Sc, Y, and lanthanides, and form various EMFs including conventional EMFs and cluster EMFs. In the past five years, Th- or U-based actinide EMFs become new family members, showing novel molecular structures [4–6].

Based on the entrapped metallic species, EMFs are usually divided into conventional EMFs including mono-metallofullerenes (e.g., $La@C_{82}$) [7], di-metallofullerenes (e.g., $La_2@C_{80}$) [7], tri-metallofullerenes (e.g., $Sm_3@C_{80}$) [8] and cluster metallofullerenes [2]. In 1999, $Sc_3N@C_{80}$, the first cluster metallofullerenes, was reported and also represented trimetallic nitride template (TNT) EMFs [9]. Later on, various cluster metallofullerenes such as carbides (e.g., $Sc_2C_2@C_{84}$ [10], $TiLu_2C@C_{80}$) [11], methano (e.g., $Sc_3CH@C_{80}$) [12], carbonitride (e.g., $Sc_3CN@C_{80}$) [13], oxide (e.g., $Sc_4O_2@C_{80}$) [14], sulfide (e.g., $Sc_2S@C_{82}$) [15] and cyano-clusters (e.g., $YCN@C_{82}$) [16] metallofullerenes have been successively reported. In particular, the embedded metallic species donates electrons (up to 6e) to the carbon cages [17,18], which significantly stabilizes EMFs and changes their electronic structures compared to that of pristine empty fullerenes.

EMFs show distinctive chemical properties compared to empty fullerenes due to the doping of embedded metallic species [1–3,19–21]. In 1995, Akasaka first reported the chemical reaction of La@$C_{82}$ [22], which starts to tune the properties of EMFs. Up to now, various kinds of methods have been developed for the functionalization of EMFs including photochemical reactions [23,24], Diels–Alder reactions [25,26], Prato reactions [27,28], Bingel–Hirsch reactions [29,30], radical addition reactions [31,32] and so on [1–3]. The inherent properties of EMFs are investigated via functionalization, which is essential access to the application of EMFs. In addition, the functionalization of EMFs is beneficial to acquiring high-quality crystals of EMFs and characterizing the structures of EMFs by forming derivatives that reduce the symmetry of EMFs [33,34]. Hence, the functionalization of EMFs is a popular direction and lots of reviews on the functionalization of EMFs have been published [1–3,19,35].

In this review, we focus on recent progress in the functionalization of EMFs from 2019. Classic functionalization reactions of EMFs with variable endohedral specials, different cage sizes, and isomers are summarized. And new complicated multi-component reactions and other creative reactions are also presented. The structural features of derivatives of EMFs and the corresponding reaction mechanisms are mainly discussed to understand the reactivity and regioselectivity of EMFs. Finally, we make conclusions and put forward an outlook.

## 2. Functionalization Reactions

The functionalization of EMFs since 2019 was summarized in Table 1.

### 2.1. Silylation and Germylation

The first exohedral functionalization of EMFs started from the organosilicon reaction of La@$C_{2v}$(9)-$C_{82}$ in 1995 [22]. Later on, the digermirane reaction of the La@$C_{2v}$(9)-$C_{82}$ was also introduced and showed high reactivity [36]. So far, silylation and germylation reactions of mono-EMFs [37], di-EMFs [38,39], and nitride cluster-fullerenes (NCFs) [40,41] have been realized and exhaustively summarized in Kako's and Jin's reviews [35,42].

Very recently, a novel method for the exclusive separation of $I_h$ and $D_{5h}$ isomers of $Sc_3N$@$C_{80}$ was proposed based on the different reactivity of photoreactions of $Sc_3N$@$I_h$-$C_{80}$ and $Sc_3N$@$D_{5h}$-$C_{80}$ with disilirane, silirane, and digermirane [43]. Specifically, under the same condition, $Sc_3N$@$I_h$-$C_{80}$ reacted readily with them to afford the corresponding 1:1 adducts, whereas $Sc_3N$@$D_{5h}$-$C_{80}$ was recovered without any adducts [43]. According to these results, the separation solution was put forward and involved three steps: selective derivatization of $Sc_3N$@$I_h$-$C_{80}$ with disilirane, silirane, and digermirane, facile high performance liquid chromatography (HPLC) separation of pristine $Sc_3N$@$I_h$-$C_{80}$ and $Sc_3N$@$D_{5h}$-$C_{80}$ derivatives, and thermolysis of corresponding derivatives to recovery pristine $Sc_3N$@$I_h$-$C_{80}$ [43]. Furthermore, the reaction mechanism has been studied by laser flash photolysis experiments [43]. Decay of the transient absorption of $^3Sc_3N$@$I_h$-$C_{80}^*$ was observed to be enhanced in the presence of disilirane, indicating the quenching process; however, the transient absorption of $^3Sc_3N$@$D_{5h}$-$C_{80}^*$ was much less intensive, which could not confirm the quenching of $^3Sc_3N$@$D_{5h}$-$C_{80}^*$ by disilirane [43]. Finally, based on the time-dependent density functional theory (TD-DFT) calculations, the poor electron acceptor property of $^3Sc_3N$@$D_{5h}$-$C_{80}^*$ might decrease the photochemical reactivity toward disilirane, silirane, and digermirane compared to $^3Sc_3N$@$I_h$-$C_{80}^*$ [43].

### 2.2. Prato Reaction (1,3 Dipolar Cycloaddition)

The Prato reaction via 1,3-dipolar cycloaddition of azomethine ylides with alkene is one of the most useful methods of fullerene functionalization due to its high selectivity and feasibility [1,2], which means easily connecting a wide range of addends and functional group to fullerenes. Prato reactions of mono-EMFs [44], di-EMFs [27,45], nitride EMFs [28,46], and carbide EMFs [47] have been reviewed in Dunsch's, Yang's, and Jin's reviews [1,2,35]. Herein, recent progress is summarized and some results are inspiring.

The pyrazole and the ring-fused pyrrole reaction of di-EMF $Y_2@C_{3v}(8)$-$C_{82}$ affords highly regioselective and quantitative mono-adduct [48]. Only one [6,6]-adduct out of the twenty-five different types of nonequivalent C-C bonds of $Y_2@C_{3v}(8)$-$C_{82}$ was got and its molecular structure was identified by the single crystal X-ray diffraction [48]. Theoretical results suggested that both the anisotropic distribution of p-electron density on the $C_{3v}(8)$-$C_{82}$ cage and the local strain of the cage carbon atoms lead to the formation of the highly regioselective and quantitative mono-adduct [48]. Additionally, electrochemical studies revealed that the reversibility of the reductive processes of $Y_2@C_{3v}(8)$-$C_{82}$ was significantly changed by exohedral functionalization, whereas the oxidative process was less influenced [48]. The reason for this phenomenon is that the reduction processes depend on the carbon cage, but the oxidation is mainly influenced by the endohedral $Y_2$ cluster [48].

In 2022, Chen et al. [49] reported the 1,3-dipolar cycloaddition photoreaction of $Sc_3N@D_{3h}$-$C_{78}$ with carbonyl ylide affords two isomeric mono-adducts **1a** and **1b** under light irradiation (see Figure 1a). The single crystal X-ray diffraction confirmed that both mono-adducts **1a** and **1b** have the same addition sites at a closed [6,6] bond and possess the same *cis*-conformation, namely, in which the hydrogen atoms of two methines situated at the same side of the tetrahydrofuran ring with parallel orientation [49]. However, two isomeric mono-adducts **3a** and **3b** were got when $Sc_3N@D_{3h}$-$C_{78}$ was replaced by $C_{60}$ in a similar reaction under the same condition [49]. Surprisingly, the crystal structure of **3b** shows that the addition pattern is located at the closed [6,6]-bond and exhibits a *trans*-conformation, in which the hydrogen atoms of the two methines within the tetrahydrofuran ring located at the opposite sides (see Figure 1b) [49]. In contrast, **3a** has the *cis*-conformation based on the $^1H$ NMR spectrum which has been previously reported by Wang and Yang [50,51]. Furthermore, the DFT calculation results indicated that the synergistic contributions of thermodynamics of adducts, the most reactive C-C bond, and the cis-dipole intermediate from *trans* **2**, result in the high regioselectivity of *cis*-conformations of **1a** and **1b** [49].

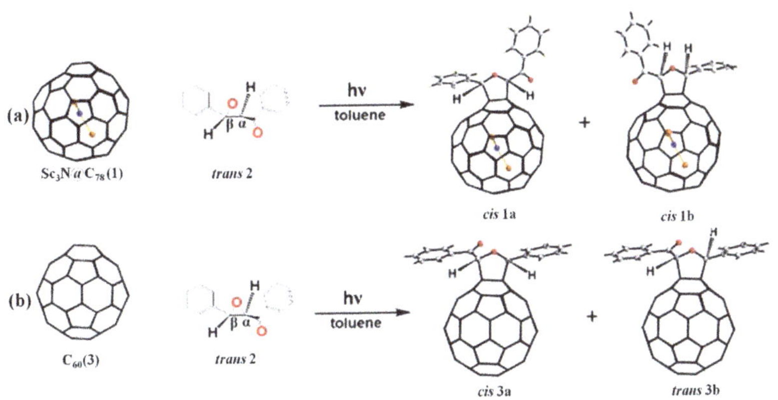


**Figure 1.** The 1,3-dipolar cycloaddition of *trans*-phenyl-(3-phenyl-oxiranyl)-methanone (*trans* **2**) with (**a**) $Sc_3N@D_{3h}$-$C_{78}$ and (**b**) $C_{60}$. Atom key: Sc (brown-yellow), N (blue), O (red). Reproduced from Ref. [49] with permission. Copyright 2021 John Wiley and Sons.

Though the mono-adducts of Prato reaction of $M_3N@C_{80}$ (M = M = Sc, Y, Gd, Dy, Ho, Er, Lu) have been widely reported [1,2], the corresponding bis-adducts are rare. Aroua et al. [52] synthesized the bis-adducts of the Prato reaction of $Gd_3N@I_h$-$C_{80}$ and $Y_3N@I_h$-$C_{80}$ in 2015. Later on, Cerón et al. [53] obtained 1,3-dipolar bis-additions of $Sc_3N@I_h$-$C_{80}$ and $Lu_3N@I_h$-$C_{80}$ using the tether-controlled multi-functionalization method, but the unambiguous crystal structure characterization of these reported Prato bis-adducts are missing. Fortunately, Semivrazhskaya et al. [54] reported one precise crystal structure of two bis-ethylpyrrolidinoadducts of $Gd_3N@I_h$-$C_{80}$ (**4**) obtained by regioselective 1,3-dipolar cycloadditions; see Figure 2a. The crystal structure of the minor-bis-adduct (**7**) shows a $C_2$-symmetric carbon cage with [6,6][6,6]-addition sites, as shown in Figure 2b and a strictly planar $Gd_3N$ cluster, which is much less strained compared to a pyramidalized cluster of the pristine $Gd_3N@I_h$-$C_{80}$. Indeed, two interior Gd atoms coordinated to the $C_{80}$ cage at two $sp^3$ addition sites, releasing the strain of the endohedral $Gd_3N$ cluster in the minor-bis-adduct. However, the structure of major-bis-adduct was presumably asymmetric [6,6][6,6]-addition sites, because its visible - near infrared (vis-NIR) spectra are almost identical to that of the major bis-adduct of $Y_3N@C_{80}$ with an asymmetric [6,6][6,6]-structure. Moreover, it is experimentally observed that the symmetrical minor-bis-adduct of $Gd_3N@I_h$-$C_{80}$ isomerized to the asymmetrical major-adduct and based on the linear transit calculations, a pathway of the isomerization is through the formation of the [6,6][6,6]-bis-adduct, followed by [5,6][6,6]-bis-adduct [54].

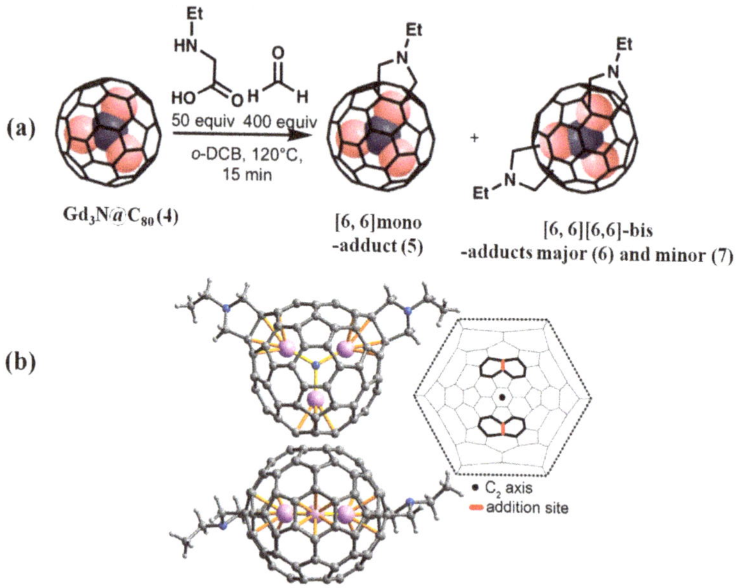


**Figure 2.** (**a**) Prato reactions of $Gd_3N@I_h$-$C_{80}$ (**4**) afford the mono-adducts (**5**) and bis-adducts (**6**) and (**7**). (**b**) X-ray crystal structure of minor-bis- $Gd_3N@C_{80}[C_4H_9N]_2$ **7** (side and top views) and the red thick bonds in the Schlegel diagram represent the additional sites. The black dot in the center of the Schlegel diagram indicates the $C_2$ axis. Reproduced from Ref. [54] with permission. Copyright 2019 American Chemical Society.

It is hard to get the Prato multi-additions of EMFs because of their huge isomers theoretically. Unexpectedly, four tris- and one tetra-isomers for both $Y_3N@I_h$-$C_{80}$ and $Gd_3N@I_h$-$C_{80}$ were obtained in a regioselective manner when $M_3N@I_h$-$C_{80}$(M = Y, Gd) reacted with an excess of N-ethylglycine and formaldehyde [55]. Three of four tris-adducts of $Y_3N@I_h$-$C_{80}$ are [6,6][6,6][6,6] isomers and the remaining one is [6,6][6,6][5,6]-isomer confirmed by the NMR whereas the tetra-adduct is all [6,6] isomer which is similar to that of $Gd_3N@I_h$-$C_{80}$ [55]. And, mutual interconversions among all [6,6] tris-adducts of both $Y_3N@I_h$-$C_{80}$ and $Gd_3N@I_h$-$C_{80}$ are observed at room temperature. Density functional theory (DFT) calculations show that the most stable structures corresponded to adding at the most strained bonds by estimating the relative stabilities of tris- and tetra-adducts formed upon Prato functionalization of the most pyramidalized regions of the fullerene structure [55]. Electron resonance (ESR) measurements of pristine, bis-, and tris-adducts of $Gd_3N@I_h$-$C_{80}$ show that the rotation of the endohedral cluster slowed down as the addition numbers to $C_{80}$ cage increased, which indicates the accommodating of Gd atoms of the relatively large

$Gd_3N$ cluster inner space at the $sp^3$ addition sites [55]. Therefore, the strain release of the $Gd_3N@I_h$-$C_{80}$ leads to the high regioselectivity of the Prato multi-addition reaction.

### 2.3. Bingel–Hirsch Reaction

Bingel–Hirsch reaction is one of the most used methods to synthesize derivations of EMFs [1,2]. The first Bingel–Hirsch reaction of $Gd@C_{60}$ affords multi-adducts $Gd@C_{60}[C(COOH)_2]n$ (n = 1–10) [56]. Surprisingly, the reaction of $La@C_{2v}(9)$-$C_{82}$ obtains four singly bonded adducts as well as one conventional cyclopropane adduct [29,57]. Recently, a highly regioselective Bingel–Hirsch reaction of $Y@C_s(6)$-$C_{82}$ induced by the metal encapsulation was reported by Shen et al. [58]. Three mono-adduct out of 44 possible isomers for the $C_s(6)$-$C_{82}$ cage have been thoroughly isolated, showing high regioselectivity [59]. The single-crystal X-ray diffraction crystallography analysis confirmed that the bromomalonate group was singly bonded to the [5,6,6]-cage carbon of $Y@C_s(6)$-$C_{82}$ in one isomer [59]. Meanwhile, further experimental and theoretical results indicate that three mono-adduct isomers may mutually be regioisomers with bromomalonate singly bonded to different cage carbon atoms having high spin density values caused by the encapsulation of a $Y^{3+}$ ion into the low symmetric $C_s(6)$-$C_{82}$ cage [58].

Despite differences in activity, NCFs such as $M_3N@I_h(7)$-$C_{80}$ (M = Sc, Y, Gd, Lu) [59] and $Gd_3N@C_{2n}$ (2n = 82, 84) [60], typically afford the methanofullerene adduct (via [2 + 1] cycloaddition). The Bingel–Hirsch reaction of $Y_3N@I_h(7)$-$C_{80}$ yields an unexpected [6,6]-open mono-adduct confirmed by a crystallographic study [30]. In contrast, the Bingel–Hirsch reaction of $TiY_2N@I_h(7)$-$C_{80}$, a $Y^{3+}$ ion replaced by a $Ti^{3+}$ ion compared to the $Y_3N@I_h(7)$-$C_{80}$, affords a singly bonded adduct, indicating that change the endohedral atom or clusters can manipulate the regioselectivity of EMFs as well as the addition pattern [61].

Notably, a new route to synthesize the Bingel–Hirsch derivative of EMFs via the cationic metallofullerenes was reported by Hu et al. recently, as shown in Figure 3 [62]. $M_3N@I_h(7)$-$C_{80}$ (M = Sc or Lu) cations were generated by both electrochemical and chemical oxidation and then the cations successfully underwent the typical Bingel–Hirsch reaction [62]. For $Sc_3N@I_h(7)$-$C_{80}$, both [5,6]-open (**10**) and [6,6]-open (**11**) adducts have been obtained whereas the former has never been synthesized via the neutral NCFs [62]. However, only a [6,6]-open adduct (**12**) was generated for $Lu_3N@I_h(7)$-$C_{80}$ [62]. DFT calculations implied that the cationic $M_3N@I_h(7)$-$C_{80}$ was much more reactive than the neutral compound for the Bingel–Hirsch reaction [62]. Furthermore, a new unusual mechanism for the Bingel–Hirsch reaction of the cationic NCFs is proposed, involving an outer-sphere single-electron transfer (SET) process [62]. Namely, the diethyl bromomalonate anion not only acts as a nucleophile, the same role in common Bingel–Hirsch reaction, but also as an electron donor, a new role to stable intermediate $[M_3N@C_{80}(C_2H_5COO)_2CBr]^*$ [62]. Additionally, the energy profiles for the Bingel–Hirsch additions on $M_3N@C_{80}{}^+$ cations for M = Sc and Lu were drawn out to explain the distinguished regioselectivity of $M_3N@C_{80}{}^+$ cations, in which M = Sc (with the [6,6] and [5,6] products) and M = Lu (with only the [6,6] product) [62].

### 2.4. Carbene Addition

The carbene reactions are vital for the functionalization of EMFs as well as for determining the molecular structure of EMFs [3]. Up to now, photochemical carbene reactions of mono-EMFs [24], di-EMFs [63], carbide clusterfullerenes [64], and NCFs [65] have been reported and open fulleroid derivatives were usually achieved. Especially, the N-heterocyclic carbene (NHC) reaction of $Sc_3N@I_h$-$C_{80}$ afforded an abnormal carbene product with C5 as its active center singly connected to the inert [6,6,6] carbon atom of the $C_{80}$ cage [66]. In contrast, normal NHC products of $M_3N@I_h$-$C_{80}$ (M = Sc, Lu) were acquired by the introduction of a little oxygen in the same reaction [67]. Recent results of NHC reaction with $Lu_3N@I_h(7)$-$C_{80}$, $Lu_2@C_{3v}(8)$-$C_{82}$, and $Lu_2@C_{2v}(9)$-$C_{82}$ indicated that the high regioselectivity and preferential formation for mono-adducts mainly originated from the electronic effect including molecular orbitals and electrostatic interactions of the fullerene cages in addition to the steric clash between the NHC and EMFs [68].

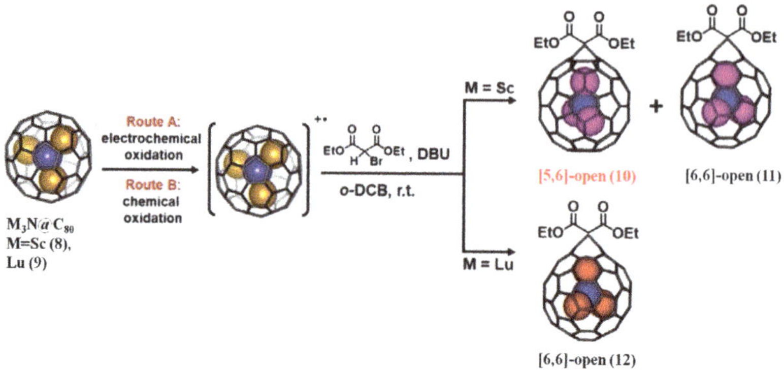


**Figure 3.** Bingel–Hirsch reaction of $M_3N@I_h$-$C_{80}$ (M = Sc or Lu) radical mono-cation under conventional conditions. Reproduced from Ref. [62] with permission. Copyright 2020 John Wiley and Sons.

Recently, the carbene reaction was extended to the actinide EMFs, Th@$C_{3v}$(8)-$C_{82}$ and U@$C_{2v}$(9)-$C_{82}$, and the former afforded three mono-adducts, named Th@$C_{3v}$(8)-$C_{82}$Ad(I, II, III) (Ad = adamantylidene), while four mono-adducts were acquired for the latter, named U@$C_{2v}$(9)-$C_{82}$Ad(I, II, III, IV), presenting remarkably higher reactivity than lanthanide EMFs [69], as shown in Figure 4a. Both Th@$C_{3v}$(8)-$C_{82}$Ad(I) and U@$C_{2v}$(9)-$C_{82}$Ad(I) are [6,6]-open cage structures, which are unambiguously confirmed by single crystal X-ray crystallography [69]. Moreover, isomerization of Th@$C_{3v}$(8)-$C_{82}$Ad(II) and Th@$C_{3v}$(8)-$C_{82}$Ad(III) and U@$C_{2v}$(9)-$C_{82}$Ad(II) and U@$C_{2v}$(9)-$C_{82}$Ad(III) was observed at room temperature; however, Th@$C_{3v}$(8)-$C_{82}$Ad(I) and U@$C_{2v}$(9)-$C_{82}$Ad(I) showed high stability under the same condition [69]. DFT calculations suggested that carbon atoms with the largest negative charge density and POAV(p-orbital axis vector) values are the best sites toward the Ad addition [69]. Furthermore, compared to the lanthanide analogs, the unusually high reactivity of Th@$C_{3v}$(8)-$C_{82}$ and U@$C_{2v}$(9)-$C_{82}$ steamed from much closer metal–cage distance, increased metal-to-cage charge transfer, and strong metal–cage interactions, which is due to the significant contribution of extending Th-5f and U-5f orbitals to the occupied molecular orbitals [69].

Very recently, the first carbene reaction of $Dy_2TiC@I_h$-$C_{80}$, the $\mu_3$-carbido cluster-fullerene ($\mu_3$-CCF) bearing central $\mu_3$-C and Ti(IV) atoms forming a Ti = C double bond, with 2-adamantane-2,3-[3H]-diazirine($AdN_2$) was reported [70], see Figure 4b. Noteworthily, the photochemical carbene reaction of $Dy_2TiC@I_h$-$C_{80}$ with $AdN_2$ affords only one mono-adduct with a [6,6]-open addition pattern, which is unambiguously confirmed by single-crystal X-ray diffraction [70]. Meanwhile, the Ad moiety selectively attacks the [6,6]-bond which is adjacent to the $Ti^{4+}$ ion instead of the two $Dy^{3+}$ ions, thus the encapsulated Ti atom within $Dy_2TiC@I_h$-$C_{80}$-Ad is fully ordered while the two Dy atoms are still disordered [70]. Theoretical calculations indicate that the [6,6]-open adduct is thermodynamic and the Ti(IV) ion plays a decisive role in the high regioselectivity [70]. In contrast, a different type of adduct with the additional sites adjacent to the $Y^{3+}$ ion instead of the $Ti^{3+}$ ion is predicted to be got when a similar reaction of $Y_2TiN@I_h$-$C_{80}$ containing a Ti(III) ion with $AdN_2$ is carried out [70]. Hence, the non-rare-earth metal Ti bearing a high oxidation state within the $\mu_3$-CCF determines the peculiarity of the chemical properties of the $\mu_3$-CCF [70].

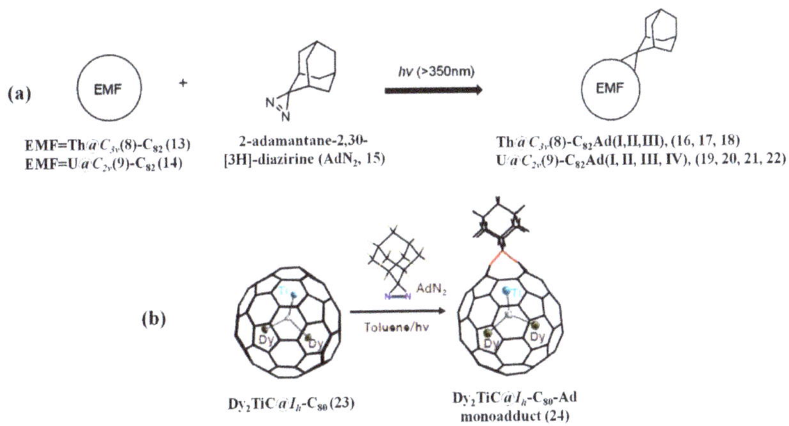


**Figure 4.** (**a**) Reaction of $AdN_2$ ($AdN_2$ = 2-adamantane-2,30-[3H]-diazirine) (**15**) with Th@$C_{3v}$(8)-$C_{82}$ (**13**) or U@$C_{2v}$(9)-$C_{82}$ (**14**). Reproduced from Ref. [69] with permission from the Royal Society of Chemistry. (**b**) Synthetic route of $Dy_2$TiC@$I_h$-$C_{80}$-Ad (**24**) through the photochemical reaction of $Dy_2$TiC@$I_h$-$C_{80}$ (**23**) with $AdN_2$ (**15**). Reproduced from Ref. [70] with permission from the Chinese Chemical Society (CCS), Peking University (PKU), and the Royal Society of Chemistry.

### *2.5. Radical Addition Reaction*

Radical addition reactions of EMFs can generate a wide range of derivatives and especially capture the unstable EMFs to form stable derivatives with a closed-shell electronic configuration. Radical groups such as benzyl or trifluoromethyl were reacted with EMFs including mono-EMFs [71], di-EMFs [31], NFC [72], CCF [32], and cyano-clusters [73] EMFs.

Liu et al. reported that an array of air-stable dimetallofullerene $Ln_2$@$C_{80}$($CH_2$Ph) ($Ln_2$ = $Y_2$, $Gd_2$, $Tb_2$, $Dy_2$, $Ho_2$, $Er_2$, TbY, TbGd) (**25**) was acquired by reacting metallofullerene anions with benzyl-bromide in DMF (N,N-Dimethylformamide) solution [74]. Single-crystal X-ray diffraction of $Ln_2$@$C_{80}$($CH_2$Ph) shows that the benzyl is attached to the cage via a single bond, see Figure 5a [74]. Moreover, there is a covalent lanthanide–lanthanide bond in $Ln_2$@$C_{80}$($CH_2$Ph) and a single electron residing on the metal–metal bonding orbital [74]. Thanks to very strong exchange interactions between 4f moments and the residing single electron, $Tb_2$@$C_{80}$($CH_2$Ph) is a robust nanomagnet and shows a gigantic coercivity of 8.2 Tesla at 5 K and a high 100 s blocking temperature of magnetization of 25.2 K [74].

Recently, a long-sought dysprosium-based EMF Dy@$C_{2v}$(5)-$C_{80}$ was captured in the form of Dy@$C_{2v}$(5)-$C_{80}$($CH_2$Ph)(Ph =-$C_6H_5$) (**26**) from carbon soot with various fullerenes [75]. On the basis of the single crystal X-ray diffraction, the carbon cage is confirmed as a rare $C_{2v}$(5)-$C_{80}$, which is the first case of an EMF composed of a mono-metal rare earth ion encapsulated within this $C_{80}$ cage, as shown in Figure 5b [75]. And, the benzyl group is grafted onto the [5,6,6]-carbon atom via a single bond, which verifies that Dy@$C_{2v}$(5)-$C_{80}$($CH_2$Ph) has an open-shell electron configuration and its electron configuration is $Dy^{3+}$@$C_{2v}$(5)-$C_{80}^{3-}$ [75]. Meanwhile, the encapsulated $Dy^{3+}$ ion is sited underneath the [6,6]-bond and deviated from the symmetry plan of $C_{2v}$(5)-$C_{80}$ [75]. Furthermore, based on theoretical calculations, the bandgap is enlarged from 1.11 eV to 1.86 eV and the LUMO energy level is obviously elevated to about 0.87 eV due to a benzyl radical addition to Dy@$C_{2v}$(5)-$C_{80}$, thus synergistically stabilizing the missing Dy@$C_{2v}$(5)-$C_{80}$ [75].

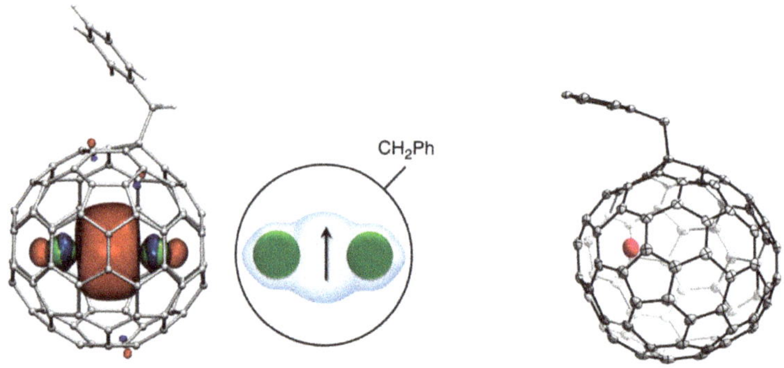


**(a)** $Ln_2@I_h$**-**$C_{80}(CH_2Ph)$ **(25)** **(b)** $Dy@C_{2v}$**(5)-**$C_{80}(CH_2Ph$ **(26)**

**Figure 5.** (**a**) Molecular structure of $Ln_2@C_{80}(CH_2Ph)$. Single-occupied Ln–Ln bonding molecular orbital (left; carbons are gray, hydrogens are white, lanthanides are green), and schematic depiction of the molecule (right; the arrow represents an unpaired electron residing on the Ln–Ln bonding orbital). (**b**) Drawing of $Dy@C_{2v}(5)$-$C_{80}(CH_2Ph)$.

Xu et al. reported that using 1,2,4-trichlorobenzene or iodobenzene to extract the Er-based fullerene soot, $Er@C_{72}(C_6H_3Cl_2)$ and $Er@C_{76}(C_6H_5)$ were obtained by free radical addition [76]. In addition, a high-temperature trifluoromethylated reaction of $Er@C_{2n}$ and $AgCOOCF_3$ afforded several adducts, $Er@C_{76}(CF_3)_5$, $Er@C_{82}(CF_3)_3$, and $Er@C_{82}(CF_3)_5$(I, II, III) [76]. Unexpectedly, all derivatives of $Er@C_{2n}$(2n = 72, 76, 82) exhibit detectable characteristic NIR photoluminescence at around 1520 nm, which stems from the emission of $Er^{3+}$, but pristine $Er@C_{2n}$(2n = 76, 82) show no photoluminescence [76]. Based on the UV-vis-NIR absorption spectrum, HOMO-LUMO gaps of the derivatives are remarkably enlarged compared to that of pristine $Er@C_{2n}$ [76]. And, odd-number addition groups result in a closed-shell electronic structure for the derivatives [76]. Hence, the enlarged HOMO-LUMO gap and the closed-shell electronic structure turn on the photoluminescence activity of $Er^{3+}$ [76]. In addition, the photoluminescence mechanism study suggests that the photoluminescence of the $Er^{3+}$ ion comes from the energy transfer from the fullerene cage to the $Er^{3+}$ ion; however, the photoluminescence of pristine $Er@C_{2n}$ was quenched by energy transfer from the first excited state of $Er^{3+}$ to the fullerene cage [76].

Li et al. reported a novel radical reaction of $Y@C_{2v}(9)$-$C_{82}$ with N-arylbezamidine catalyzed by silver carbonate [77]. The reaction is highly regioselective and affords only one mono-adduct with an imidazoline group. The theoretical calculation reveals the addition group is attached to a specific [5,6]-bond near the Y atom [77].

### *2.6. Electrophilic Trifluoromethylation*

The lack of selectivity of the radical addition reactions results in multi-adducts and a complicated separation process. Unprecedentedly, the high selectivity of electrophilic trifluoromethylation of metallofullerene anions was reported very recently [78]. Specifically, reacting metallofullerene anions extracted by DMF with Umemoto reagent II affords $M_2@C_{80}(CF_3)$ (M = Tb, Y) mono-adducts as the major product, indicating higher selectivity of electrophilic trifluoromethylation than that of benzyl bromide reaction (see Figure 6a) [78]. The single-crystal X-ray diffraction analysis shows that the $CF_3$ group is attached to the pentagon/hexagon/hexagon junction ([5,6,6] position) of the $I_h$-$C_{80}$ cage via a single bond, as shown in Figure 6b [78]. Similarly, a single electron remains between two Tb ions, forming single-electron metal–metal bonds with the formal metal oxidation state of $Tb^{2.5+}$ [78]. As expected, $Tb_2@C_{80}(CF_3)$ is also a robust single-molecule magnet based on magnetic characterizations, which is comparable to the benzyl mono-adduct $Tb_2@C_{80}(CH_2Ph)$ [78]. Therefore, electrophilic trifluoromethylation is a very excellent approach to stabilize metallofullerene anions $M_2@C_{80}$, which is a simple, fast room-temperature reaction with high selectivity and needs less time for required HPLC separation compared to the benzylation reaction of $M_2@C_{80}$ [78].

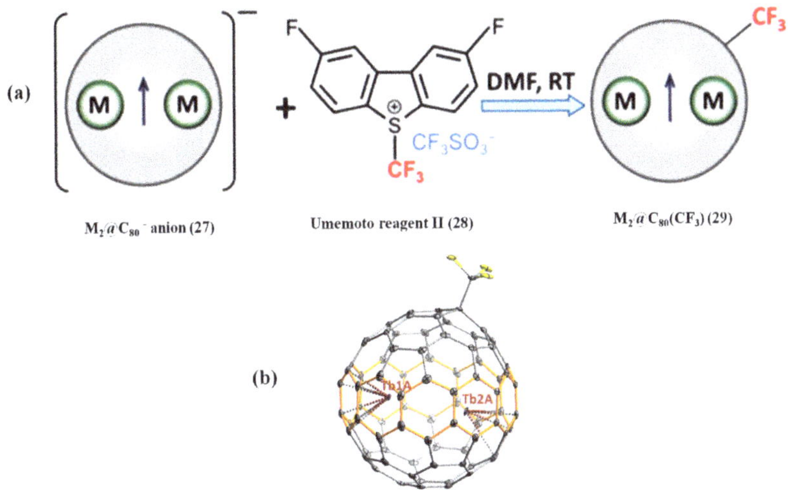


**Figure 6.** (**a**) Scheme of the reaction between $M_2@C_{80}^-$ anion and Umemoto reagent II with the formation of $M_2@C_{80}(CF_3)$. (**b**) $Tb_2@C_{80}(CF_3)$ molecule from the single crystal showing the main Tb sites Tb1A and Tb2A. Reproduced from Ref. [78] with permission. Copyright 2021 American Chemical Society.

### 2.7. Coordination Reaction

The organometallic complexes of EMFs usually bear the $\eta^2$ coordination fashion [79], and the $\eta^1$-coordinated complexes of EMFs with only one metal–cage bond are very desirable. Until 2019, Xie et al. [80] reported a coordination reaction of $Re_2(CO)_{10}$ (**31**) to paramagnetic Y@$C_{2v}$(9)-$C_{82}$ (**30**) by an efficient radical coupling, and an unprecedented η1-coordinated complex: $Re(CO)_5$-$\eta^1$-Y@$C_{2v}$(9)-$C_{82}$ (**32**) was obtained, as shown in Figure 7a. Crystallographic results confirmed that the $Re(CO)_5$ moiety coordinates to a [5,6,6]-carbon atom of the $C_{82}$ cage via a single Re-C σ bond [80]. Furthermore, Vis-NIR and ESR results verified that the $Re(CO)_5$ moiety transfers one electron to the EMF, resulting in a closed-shell electronic structure similar to anionic (Y@$C_{82}$)- species with high stability [80]. In contrast, replacing with diamagnetic $Sc_3N@I_h(9)$-$C_{82}$ and $Lu_2@C_{2v}(9)$-$C_{82}$ in the same reaction, no product is yielded, further confirming a radical coupling process [80].

Bao et al. [81] reported that reacting mono-metallofullerenes, Eu@$C_2$(5)-$C_{82}$ (**33**) and Eu@$C_2$(13)-$C_{84}$) (**34**), with a tungsten complex $W(CO)_4(Ph_2PC_2H_4PPh_2)$ (**35**) afford the highly regioselective product; alternatively, only one product was acquired for each reaction. Single crystal X-ray crystallography shows that in both products (**36**, **37**), the tungsten moiety coordinates to a [6,6]-bond of the cage with an $\eta^2$-fashion, as shown in Figure 7b [81]. And, after the coordination of the tungsten moiety, the motion of the internal Eu ion in both adducts is restricted to a certain extent, which is due to changing the electrostatic potentials inside the cage supported by theoretical calculations [81]. Moreover, the observed high regioselectivity is explained by electron transfer from the endohedral Eu to the cage, which significantly changed the LUMO distribution on $C_2$(5)-$C_{82}$/$C_2$(13)-$C_{84}$ based on theoretical calculations [81]. Therefore, the interplay between the endohedral and the exohedral metallic moiety is achieved in a single molecule system via intramolecular charge transfer [81].

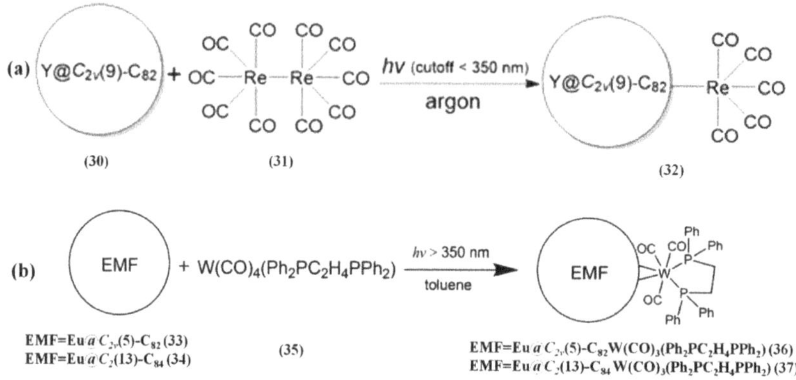


**Figure 7.** (**a**) Reaction scheme of Y@$C_{2v}$(9)-$C_{82}$ with $Re_2(CO)_{10}$. Reproduced from Ref. [80] with permission from the Royal Society of Chemistry. (**b**) Reaction of $W(CO)_4(Ph_2PC_2H_4PPh_2)$ with Eu@$C_2$(5)-$C_{82}$ or Eu@$C_2$(13)-$C_{84}$. Reproduced from Ref. [81] with permission. Copyright 2019 American Chemical Society.

### 2.8. Methoxide Reaction

Unexpected bis-methoxyl adducts of $Sc_3N$@$I_h$-$C_{80}$ were got for the first time when $Sc_3N$@$I_h$-$C_{80}$ reacted with/without $Ph_2C$=O, PhC≡CPh or PhC≡N in the presence of tetrabutylammonium hydroxide (TBAOH) stored in $CH_3OH$ [82]. However, it was further observed that bis-methoxyl adducts of $Sc_3N$@$I_h$-$C_{80}$ form irrespective of the presence of other reagents, since TBAOH in $CH_3OH$ efficiently boosts the -$CH_3O$ addition [82]. Furthermore, single crystal X-ray diffraction analysis unambiguously determines the molecular structures of the products as 1,4- and 1,2-bis-methoxyl adducts, and the conformations of the two -$OCH_3$ groups in both bis-methoxyl adducts are obviously different bis-methoxyl, as shown in Figure 8a,b [82]. And, the planar $Sc_3N$ cluster in 1,4-bis-methoxyl adducts (**38**) is orthogonal with the plane crossing the addition sites, whereas the metal cluster in 1,2-bis-methoxyl adducts (**39**) is nearly parallel to the plane crossing the addition site, which indicates that exohedral modification is practical to control the orientation of the embedded cluster bis-methoxyl [82]. In addition, a possible mechanism related to an anion addition and a radical reaction was put forward (see Figure 8c), which opened up new horizons to the highly selective reactions between the methoxyl anion and EMFs [82]. Specifically, $CH_3O$- generated by deprotonates $CH_3OH$ via TBAOH prevailed over OH- (from TBAOH) in the o-DCB (o-dichlorobenzene) solution and attached to $Sc_3N$@$C_{80}$ to form the mono-anion $[Sc_3N@C_{80}(OCH_3)]^-$ firstly [82]. Then, the mono-anion was oxidized to [$Sc_3N$@$C_{80}$(OCH)•radical by $I_2$, and the dimethoxyfullerene anion $[Sc_3N@C_{80}(OCH_3)_2]^-$ formed by accepting another $CH_3O^-$ [82]. Finally, the ultimate 1,2- or 1,4-addition dimethoxyfullerene products were yielded through the oxidation of dimethoxyfullerene anion by $I_2$ [82].

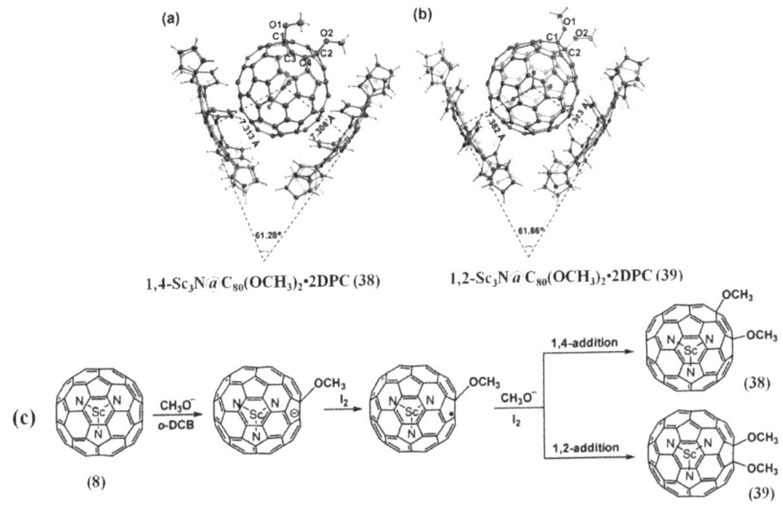


**Figure 8.** Single crystal X-ray structures of co-crystals 1,4-$Sc_3N@C_{80}(OCH_3)_2$•2DPC (**a**) and (**b**) 1,4-$Sc_3N@C_{80}(OCH_3)_2$•2DPC with thermal ellipsoids at the 10% probability level. (**c**) Plausible mechanism of the reaction between $Sc_3N@I_h$-$C_{80}$ and TBAOH/$CH_3OH$ in o-DCB. Reproduced from Ref. [82] with permission from the Royal Society of Chemistry.

## *2.9. Multicomponent Reactions*

Very recently, pyrazoline-fused metallofullerene derivative (**41**) as shown in Figure 9 was generated from a four-component cascade reaction of $Sc_3N@I_h$-$C_{80}$ with 3,6-di(pyridin-2-yl)-1,2,4,5-tetrazines (**40**), $H_2O$, and $O_2$ [83]. The [5,6] pyrazoline-fused structure is remarkably different from the reaction of a 1,2,4,5-tetrazine with $Sc_3C_2@I_h$-$C_{80}$ which provided a bis-fulleroid derivative [84]. Moreover, the reaction was extended to 3,6-bis(6-methylpyridin-2-yl)-1,2,4,5-tetrazine, and a similar [5,6] pyrazoline-fused derivative (**42**) was achieved [83]. Furthermore, a plausible formation mechanism of pyrazoline-fused metallofullerenes is proposed, which involved a complicated sequence of Diels–Alder reaction, retro Diels–Alder reaction with $N_2$ extrusion, hydration reaction, rearrangement reaction, and oxidative dehydrogenation reaction [83].

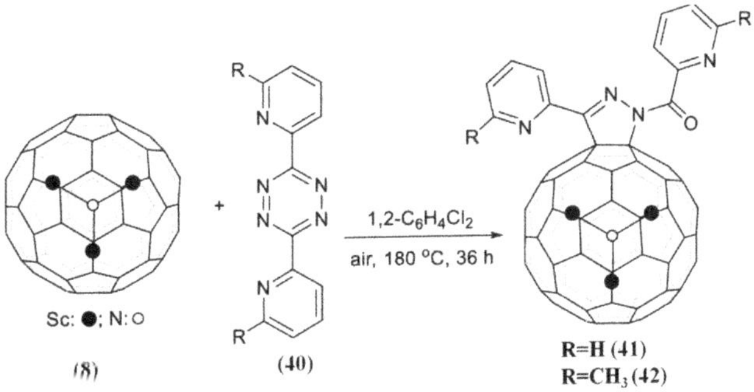


**Figure 9.** The reaction of $Sc_3N@I_h$-$C_{80}$ with tetrazines (**40**). Reproduced from Ref. [83] with permission. Copyright 2022 American Chemical Society.

**Table 1.** An exhaustive list of functionalizations of EMFs reported from 2019 to date.

| Reaction | Endohedral Metallofullerene | Reactants | Product | Addition Position | Number of Isomers | Ref. |
|---|---|---|---|---|---|---|
| Silylation and germylation | $Sc_3N@I_h$-$C_{80}$ | Disilirane | $Sc_3N@C_{80}Si_2Mes_4$ | [1,2], [1,4] | 2 | [40,43] |
| | | Silirane | $Sc_3N@C_{80}Si(Dep)_2CH_2CHtBp$ | [5,6]diastereomers,[6,6] | 3 | [43] |
| | | Digermirane | $Sc_3N@C_{80}Ge(Dep)_4$ | [1,4] | 1 | [43] |
| | $Sc_3N@D_{5h}$-$C_{80}$ | Disilirane, silirane, digermirane | No product | non-reactive | 0 | [43] |
| Prato reaction (1,3 dipolar cycloaddition) | $Y_2@C_{3v}(8)$-$C_{82}$ | Diphenylnitrilimine ylide | $Y_2@C_{82}N_2(C_6H_5)_2$ | [6,6] | 1 | [48] |
| | | N-benzylazomethine ylide | $Y_2@C_{82}NCH_2(C_6H_5)$ | [6,6] | 1 | [48] |
| | $Sc_3N@D_{3h}$-$C_{78}$ | Trans-phenyl-(3-phenyl-oxiranyl)-methanone | $Sc_3N@C_{78}(C_6H_5)CHOCHCO(C_6H_5)$ | cis-[6,6] | 2 | [49] |
| | $Gd_3N@I_h$-$C_{80}$ | N-ethylglycine, paraformaldehyde | $Gd_3N@C_{80}((CH_2)_2NEt)_2$ | [6,6] [6,6]-bis adduct | 2 | [54] |
| | | | $Gd_3N@C_{80}((CH_2)_2NEt)_3$ | [6,6] [6,6] [6,6] tris-adduct | 3 | [55] |
| | | | $Gd_3N@C_{80}((CH_2)_2NEt)_3$ | [6,6] [6,6] [5,6] tris-adduct | 1 | [55] |
| | | | $Gd_3N@C_{80}((CH_2)_2NEt)_4$ | all [6,6] tera-adduct | 1 | [55] |
| Bingel–Hirsch reaction | $Y@C_s(6)$-$C_{82}$ | Diethyl bromomalonate, DBU | $Y@C_{82}CBr(CCOOCH_2CH_3)_2$ | [5,6,6]-single bond | 3 | [58] |
| | $Sc_3N@I_h(7)$-$C_{80}$ | Diethyl bromomalonate, DBU | $Sc_3N@C_{80}$ $CBr(CCOOCH_2CH_3)_2$ | [5,6]-open, [6,6]-open | 2 | [62] |
| | $Lu_3N@I_h(7)$-$C_{80}$ | Diethyl bromomalonate, DBU | $Lu_3N@C_{80}$ $CBr(CCOOCH_2CH_3)_2$ | [6,6]-open | 1 | [62] |
| Carbene addition | $Th@C_{3v}(8)$-$C_{82}$ | 2-adamantane-2,30-[3H]-diazirine($AdN_2$) | $Th@C_{82}Ad$ | [6,6]-open, et al. | 3 | [69] |
| | $U@C_{2v}(9)$-$C_{82}$ | 2-adamantane-2,30-[3H]-diazirine($AdN_2$) | $U@C_{82}Ad$ | [6,6]-open, et al. | 4 | [69] |
| | $Dy_2TiC@I_h$-$C_{80}$ | 2-adamantane-2,3-[3H]-diazirine ($AdN_2$) | $Dy_2TiC@C_{80}$-Ad | [6,6]-open | 1 | [70] |

**Table 1.** *Cont.*

| Reaction | Endohedral Metallofullerene | Reactants | Product | Addition Position | Number of Isomers | Ref. |
|---|---|---|---|---|---|---|
| Radical addition reaction | $Ln_2@C_{80}$($Ln_2$ = $Y_2$, $Gd_2$, $Tb_2$, $Dy_2$, $Ho_2$, $Er_2$, TbY, TbGd) | $BrCH_2Ph$ | Ln2@C80(CH2Ph) | [5,6,6] | 1 | [74] |
| | $Dy@C_{2v}(5)$-$C_{80}$ | $BrCH_2Ph$ | Dy@C80(CH2Ph) | [5,6,6] | 1 | [75] |
| | $Er@C_{72}$ | 1,2,4-trichlorobenzene | $Er@C_{72}(C_6H_3Cl_2)$ | Unknown | 1 | [76] |
| | $Er@C_{76}$ | Iodobenzene | $Er@C_{76}(C_6H_5)$ | Unknown | 1 | [76] |
| | | $AgCOOCF_3$ | Er@C76(CF3)5 | Unknown | 1 | [76] |
| | $Er@C_{82}$ | $AgCOOCF_3$ | Er@C82(CF3)3 | Unknown | 1 | [76] |
| | | | Er@C82(CF3)5 | Unknown | 3 | [76] |
| | $Y@C_{2v}(9)$-$C_{82}$ | N-arylbezamidine | $Y@C_{82}$NPhCNPh | [5,6] | 1 | [77] |
| Electrophilic trifluoromethylation | $Tb_2@C_{80}^-$ | Umemoto reagent II | $Tb_2@C_{80}(CF_3)$ | [5,6,6] | 1 | [78] |
| | $Y_2@C_{80}^-$ | Umemoto reagent II | $Tb_2@C_{80}(CF_3)$ | [5,6,6] | 1 | [78] |
| Coordination reaction | $Y@C_{2v}(9)$-$C_{82}$ | $Re_2(CO)_{10}$ | $Re(CO)_5$-$\eta^1$-$Y@C_{2v}(9)$-$C_{82}$ | [5,6,6] | 1 | [80] |
| | $Eu@C_2(5)$-$C_{82}$ | $W(CO)_4(Ph_2PC_2H_4PPh_2)$ | $Eu@C_{82}$ $W(CO)_3(Ph_2PC_2H_4PPh_2)$ | $\eta^2$-[6,6]-bond | 1 | [81] |
| | $Eu@C_2(13)$-$C_{84}$ | $W(CO)_4(Ph_2PC_2H_4PPh_2)$ | $Eu@C_{84}$) $W(CO)_3(Ph_2PC_2H_4PPh_2)$ | $\eta^2$-[6,6]-bond | 1 | [81] |
| Methoxide reaction | $Sc_3N@I_h$-$C_{80}$ | TBAOH, $CH_3OH$ | $Sc_3N@C_{80}(OCH_3)_2$ | [1,2], [1,4] | 2 | [82] |

**Table 1.** *Cont.*

| Reaction | Endohedral Metallofullerene | Reactants | Product | Addition Position | Number of Isomers | Ref. |
|---|---|---|---|---|---|---|
| Multicomponent reactions | $Sc_3N@I_h$-$C_{80}$ | 3,6-di(pyridin-2-yl)-1,2,4,5-tetrazines, water, and oxygen | $Sc_3N@C_{80}CN{=}NCO(C_5NH_4)_2$ | [5,6]-closed | 1 | [83] |
| | | 3,6-bis(6-methylpyridin-2-yl)-1,2,4,5-tetrazine, water, and oxygen | $Sc_3N@C_{80}CN{=}NCO(C_5NCH_3)_2$ | [5,6]-closed | 1 | [83] |
| | $Lu_3N@I_h$-$C_{80}$ | tBuNC and EWG-bearing terminalalkynes | $Lu_3N@C_{80}CCR^1CONHtBuCOR^2$ | [6,6]-open | 1 | [85] |
| | | tBuNC and dimethyl acetylenedicarboxylate (DMAD) | $Lu_3N@C_{80}C(COOMe)C(COOMe)CNtBu$ | [6,6]-open | 1 | [85] |
| Difluoromethylenation reaction | $Sc_3N@I_h$-$C_{80}$ | $CF_2ClCOONa$ | $Sc_3N@C_{80}(CF_2)$ | [6,6]-open | 1 | [86] |
| | $Sc_3N@C_{78}$ | $CF_2ClCOONa$ | $Sc_3N@C_{78}(CF_2)$ | [6,6] | 1 | [87] |

Unexpected formation of metallofulleroids with [6,6]-open structure from the isocyanide-based multicomponent reactions of an isocyanide (e.g., tBuNC), various alkynes, and $Lu_3N@I_h$-$C_{80}$ have been reported recently [85]. A major product was obtained by using EWG (electron-withdrawing group)-bearing terminal alkynes, whereas no products were obtained by using disubstituted internal alkynes [85]; see Figure 10a. Notably, when $Lu_3N@Ih$-$C_{80}$ was replaced by $C_{60}$, there were no isolated products under any conditions [85]. Surprisingly, unlike internal alkynes, dimethyl acetylenedicarboxylate (DMAD) yielded a ketenimine metallofulleroid **45**, as shown in Figure 10b, while the addition pattern is greatly different from that of $C_{60}$ reacting with the same reactants [85]. Strikingly, a variable temperature single-crystal study of metallofulleroid **46** showed that the motion of the embedded cluster is significantly hindered by the close interaction with the exohedral organic appendant of a neighboring molecule [85]. Furthermore, DFT calculations suggested that the reaction mechanism involved three main steps including a barrierless anionic attack of the dipole, 3-exo-trig ring closure, and final hydration [85].

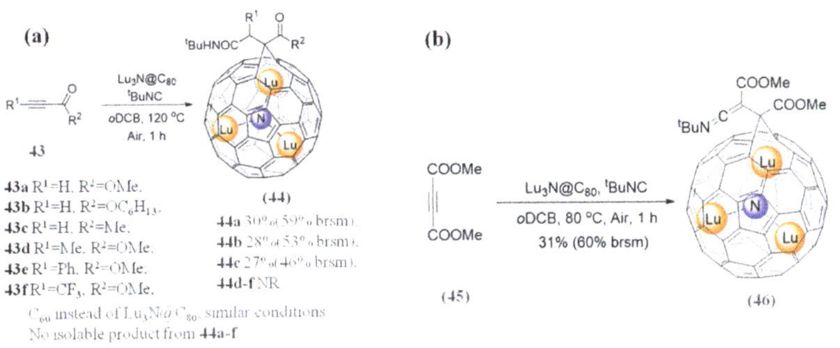


**Figure 10.** (**a**) Preparation of **44a–c** reacted from alkyl isocyanides and alkynes. Brsm = based on recovered starting material; NR = No reaction. (**b**) Synthesis of a ketenimine metallofulleroid **46**. Reproduced from Ref. [85] with permission. Copyright 2021 John Wiley and Sons.

## 2.10. *Difluoromethylenation Reaction*

Recently, difluoromethylenation of $Sc_3N@I_h$-$C_{80}$ reacting with $CF_2ClCOONa$ affords a sole $C_s$-symmetric $Sc_3N@C_{80}(CF_2)$ adduct [86]. Based on the $^{19}F$, $^{13}C$, and $^{45}Sc$ NMR spectroscopic, $Sc_3N@C_{80}(CF_2)$ was identified as the [6,6]-open structure with 2.2–2.3 Å between the bridgehead atoms, which is consistent with the DFT-optimized structure (see Figure 11) [86]. And the UV/vis spectra of $Sc_3N@C_{80}(CF_2)$ are similar to that of $Sc_3N@I_h$-$C_{80}$, thus indicating the [6,6]-open structure with weakly perturbed $\pi$-electron systems [86]. Furthermore, the electrochemical behavior of $Sc_3N@C_{80}(CF_2)$ resembles that of $Sc_3N@I_h$-$C_{80}$, in which both the HOMO and LUMO level of $Sc_3N@C_{80}(CF_2)$ downshift 0.1 eV compared to the pristine $Sc_3N@I_h$-$C_{80}$, resulting in essentially unchanged HOMO–LUMO gap [86]. It is worth noting that in the experiment the absence of the [5,6]-conformer, the more energetically favorable one based on the theoretical analysis, suggests that the adduct is kinetical rather than thermodynamical [86]. In addition, the nucleophilic route, which involves nucleophilic addition of $CF_2Cl^-$ anion followed by cyclopropanation via $Cl^-$ displacement, is considered to be better than the route of [2 + 1]-cycloaddition of $CF_2$ carbene [86].

Very recently, the difluoromethylenation reaction was extended to $Sc_3N@C_{78}$ and it showed higher reactivity than $Sc_3N@C_{80}$, in which the reaction temperature is lower [87]. Mono-adduct $Sc_3N@C_{78}(CF_2)$ was identified by the mass spectrum, but, unfortunately, the $Sc_3N@C_{78}(CF_2)$ degraded rapidly, which was different from $Sc_3N@C_{80}(CF_2)$ and the $CF_2$ adducts of empty fullerenes [87]. And, only the UV-Vis spectrum of the $Sc_3N@C_{78}(CF_2)$ was recorded and showed high similarity with the spectra of the Bingel–Hirsch mono-adduct and that of the major Prato mono-adduct, indicating that $Sc_3N@C_{78}(CF_2)$ has the same bond type, bond 66-6 [87]. Moreover, DFT results suggested that the open [6,6]-$Sc_3N@C_{78}(CF_2)$

is more stable and the regioselectivity is controlled kinetically as well as the $CF_2$ addition mechanism resembles that of $Sc_3N@C_{80}$ [87].

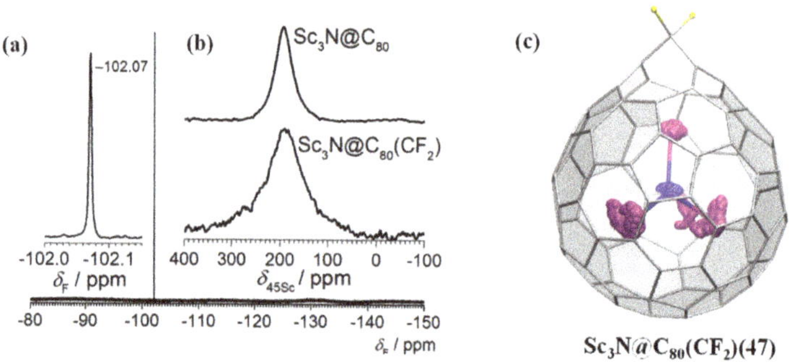


**Figure 11.** (**a**) $^{19}F$ NMR spectra of $Sc_3N@C_{80}(CF_2)$. (**b**) Room temperature $^{45}Sc$ NMR spectra of $Sc_3N@I_h{-}C_{80}$ (top) and $Sc_3N@C_{80}(CF_2)$ (bottom). (**c**) DFT molecular dynamics trajectories for $Sc_3N@C_{80}(CF_2)$ (100 ps), C atoms are magenta, and N is blue. Reproduced from Ref. [86] with permission from the Royal Society of Chemistry.

## 3. Conclusions and Outlook

In summary, EMFs exhibit active reactivities toward versatile functionalization reactions, which confer EMFs tunable properties. Prato reaction was extended to EMFs with low symmetry, such as $Sc_3N@C_{78}$ and $Y_2@C_{3v}(8)$-$C_{82}$; furthermore, the complicated bis-adducts and multi-adducts of Prato reaction were studied as well. Metal encapsulation induces a highly regioselective Bingel–Hirsch adducts of $Y@C_s(6)$-$C_{82}$ achieved and a new cationic metallofullerene route was put forward for Bingel–Hirsch reaction. Carbene additions of actinide EMFs and Ti-based EMFs were developed. A new non-cycloaddition of EMFs, methoxide reaction, was exploited. Furthermore, new complicated multi-component reactions instead of simple transplantation from empty fullerenes arise and acquire unexpected adducts, expanding new horizons in the functionalization of EMFs. Especially, besides radical reactions, new electrophilic trifluoromethylation was exploited to stabilize the missing EMFs.

High regioselectivity is frequently observed for EMFs due to strong metal–cage interactions. The endohedral cluster plays an important role in determining the regioselectivity and addition pattern. Furthermore, combined with calculated results, the regioselectivity of the reactions was plausibly interpreted by POAV, HOMOs/LUMOs, and charges. However, new credible, advanced, and reasonable theoretical calculation schemes are also expected. In turn, the addition groups affect the location, motion, orientation, and electronic state of the encapsulated clusters. As a result, the ultimate structures and properties of the derivatives of EMFs are together regulated by the intramolecular interplay of the cluster, carbon cage, and exohedral groups.

Recent functionalization reactions are mainly restricted to exploring the inherent chemical activities of EMFs or stabilizing EMFs. However, using reaction groups to modulate the magnetic, luminescence, or photovoltaic properties of EMFs is rarely reported. And it is not much clear whether and how the type or amount of reaction groups affects those properties of EMFs. Therefore, more efforts are needed to reveal the relationships between the reaction groups and those properties of EMFs and make metallofullerenes truly applicable materials. Functionalizations of EMFs are an attractive and popular area of fullerene research, which still holds great promise.

**Author Contributions:** Conceptualization, S.W.; Methodology, X.T., H.L., S.D. and S.W.; Writing—Original Draft, S.W.; Visualization, B.Y., X.L. and Y.H.; Writing—Review and Editing, X.Z., S.W. and F.J. All authors have read and agreed to the published version of the manuscript.

**Funding:** This work is supported by the National Natural Science Foundation of China (No. 22201028, No. 52002365); the China Postdoctoral Science Foundation (No. 2022M720587); the Postdoctoral Special Funding Program of the Chongqing Human Resources and Social Security Bureau (No. 2021XM1017); the Science and Technology Research Program of Chongqing Municipal Education Commission (NO. KJQN202100823); and the Natural Science Foundation of Chongqing Technology and Business University (2152014).

**Data Availability Statement:** No new data were created.

**Conflicts of Interest:** The authors declare no conflict of interest.

## References

1. Popov, A.A.; Yang, S.F.; Dunsch, L. Endohedral Fullerenes. *Chem. Rev.* **2013**, *113*, 5989–6113. [CrossRef] [PubMed]
2. Yang, S.F.; Wei, T.; Jin, F. When metal clusters meet carbon cages: Endohedral clusterfullerenes. *Chem. Soc. Rev.* **2017**, *46*, 5005–5058. [CrossRef] [PubMed]
3. Lu, X.; Feng, L.; Akasaka, T.; Nagase, S. Current status and future developments of endohedral metallofullerenes. *Chem. Soc. Rev.* **2012**, *41*, 7723–7760. [CrossRef] [PubMed]
4. Cai, W.T.; Chen, C.H.; Chen, N.; Echegoyen, L. Fullerenes as nanocontainers that stabilize unique actinide species inside: Structures, formation, and reactivity. *Acc. Chem. Res.* **2019**, *52*, 1824–1833. [CrossRef] [PubMed]
5. Meng, Q.Y.; Abella, L.; Yao, Y.R.; Sergentu, D.; Yang, W.; Liu, X.Y.; Zhuang, J.X.; Echegoyen, L.; Autschbach, J.; Chen, N. A charged diatomic triple-bonded U≡N species trapped in $C_{82}$ fullerene cages. *Nat. Commun.* **2022**, *13*, 7192. [CrossRef] [PubMed]
6. Zhuang, J.X.; Morales-Martínez, R.; Zhang, J.W.; Wang, Y.F.; Yao, Y.R.; Pei, C.Y.; Rodríguez-Fortea, A.; Wang, S.A.; Echegoyen, L.; Graaf, C.D.; et al. Characterization of a strong covalent $Th^{3+}$-$Th^{3+}$ bond inside an $I_h(7)$-$C_{80}$ fullerene cage. *Nat. Commun.* **2021**, *12*, 2372. [CrossRef] [PubMed]
7. Shinohara, H. Endohedral metallofullerenes. *Rep. Prog. Phys.* **2000**, *63*, 843–892. [CrossRef]
8. Xu, W.; Feng, L.; Calvaresi, M.; Liu, J.; Liu, Y.; Niu, B.; Shi, Z.J.; Lian, Y.F.; Zerbetto, F. An experimentally observed trimetallofullerene $Sm_3@I_h$-$C_0$. *J. Am. Chem. Soc.* **2013**, *135*, 4187–4190. [CrossRef]
9. Stevenson, S.; Rice, G.; Glass, T.; Harich, K.; Cromer, F.; Jordan, M.R.; Craft, J.; Hadju, E.; Bible, R.; Olmstead, M.M.; et al. Small-bandgap endohedral metallofullerenes in high yield and purity. *Nature* **1999**, *401*, 55–57. [CrossRef]
10. Wang, C.R.; Kai, T.; Tomiyama, T.; Yoshida, T.; Kobayashi, Y.; Nishibori, E.; Takata, M.; Sakata, M.; Shinohara, H. A scandium carbide endohedral metallofullerene: $(Sc_2C_2)@C_{84}$. *Angew. Chem. Int. Ed.* **2001**, *40*, 397–399. [CrossRef]
11. Svitova, A.L.; Ghiassi, K.B.; Schlesier, C.; Junghans, K.; Zhang, Y.; Olmstead, M.M.; Balch, A.L.; Dunsch, L.; Popov, A.A. Endohedral fullerene with μ3-carbido ligand and titanium-carbon double bond stabilized inside a carbon cage. *Nat. Commun.* **2014**, *5*, 3568. [CrossRef] [PubMed]
12. Krause, M.; Ziegs, F.; Popov, A.A.; Dunsch, L. Entrapped bonded hydrogen in a fullerene: The five-atom cluster $Sc_3CH$ in $C_{80}$. *Chem. Phys. Chem.* **2007**, *8*, 537–540. [CrossRef] [PubMed]
13. Wang, T.S.; Feng, L.; Wu, J.Y.; Xu, W.; Xiang, J.F.; Tan, K.; Ma, Y.H.; Zheng, J.P.; Jiang, L.; Lu, X.; et al. Planar quinary cluster inside a fullerene cage: Synthesis and structural characterizations of $Sc_3NC@C_{80}$-$I_h$. *J. Am. Chem. Soc.* **2010**, *132*, 16362–16364. [CrossRef]
14. Stevenson, S.; Mackey, M.A.; Stuart, M.A.; Phillips, J.P.; Easterling, M.L.; Chancellor, C.J.; Olmstead, M.M.; Balch, A.L. A distorted tetrahedral metal oxide cluster inside an icosahedral carbon cage. Synthesis, isolation, and structural characterization of $Sc_4(\mu_3\text{-}O)_2@I_h$-$C_{80}$. *J. Am. Chem. Soc.* **2008**, *130*, 11844–11845. [CrossRef] [PubMed]
15. Dunsch, L.; Yang, S.F.; Zhang, L.; Svitova, A.; Oswald, S.; Popov, A.A. Metal sulfide in a $C_{82}$ fullerene cage: A new form of endohedral clusterfullerenes. *J. Am. Chem. Soc.* **2010**, *132*, 5413–5421. [CrossRef] [PubMed]
16. Yang, S.F.; Chen, C.B.; Liu, F.P.; Xie, Y.P.; Li, F.Y.; Jiao, M.Z.; Suzuki, M.; Wei, T.; Wang, S.; Chen, Z.F.; et al. An improbable monometallic cluster entrapped in a popular fullerene cage: $YCN@C_s(6)$-$C_{82}$. *Sci. Rep.* **2013**, *3*, 1487. [CrossRef] [PubMed]
17. Chaur, M.N.; Valencia, R.; Rodríguez-Fortea, A.; Poblet, J.M.; Echegoyen, L. Trimetallic nitride endohedral fullerenes: Experimental and theoretical evidence for the $M_3N^{6+}@C_{2n}^{6-}$ model. *Angew. Chem. Int. Ed.* **2009**, *48*, 1425–1428. [CrossRef] [PubMed]
18. Rodríguez-Fortea, A.; Balch, A.L.; Poblet, J.M. Endohedral metallofullerenes: A unique host–guest association. *Chem. Soc. Rev.* **2011**, *40*, 3551–3563. [CrossRef]
19. Lu, X.; Bao, L.P.; Akasaka, T.; Nagase, S. Recent progress in the chemistry of endohedral metallofullerenes. *Chem. Commun.* **2014**, *50*, 14701–14715. [CrossRef]
20. Bao, L.P.; Peng, P.; Lu, X. Bonding inside and outside Fullerene Cages. *Acc. Chem. Res.* **2018**, *51*, 810–815. [CrossRef]
21. Yamada, M.; Tanabe, Y.; Dang, J.S.; Sato, S.; Mizorogi, N.; Hachiya, M.; Suzuki, M.; Abe, T.; Kurihara, H.; Maeda, Y.; et al. $D_{2d}(23)$-$C_{84}$ versus $Sc_2C_2@D_{2d}(23)$-$C_{84}$: Impact of Endohedral $Sc_2C_2$ Doping on Chemical Reactivity in the Photolysis of Diazirine. *J. Am. Chem. Soc.* **2016**, *138*, 16523–16532. [CrossRef] [PubMed]
22. Akasaka, T.; Kato, T.; Kobayashi, K.; Nagase, S.; Yamamoto, K.; Funasaka, H.; Takahashi, T. Exohedral adducts of $La@C_{82}$. *Nature* **1995**, *374*, 600–601. [CrossRef]

23. Yamada, M.; Nakahodo, T.; Wakahara, T.; Tsuchiya, T.; Maeda, Y.; Akasaka, T.; Kako, M.; Yoza, K.; Horn, E.; Mizorogi, N.; et al. Positional control of encapsulated atoms inside a fullerene cage by exohedral addition. *J. Am. Chem. Soc.* **2005**, *127*, 14570–14571. [CrossRef] [PubMed]
24. Maeda, Y.; Matsunaga, Y.; Wakahara, T.; Takahashi, S.; Tsuchiya, T.; Ishitsuka, M.O.; Hasegawa, T.; Akasaka, T.; Liu, M.T.H.; Kokura, K.; et al. Isolation and characterization of a carbene derivative of La@$C_{82}$. *J. Am. Chem. Soc.* **2004**, *126*, 6858–6859. [CrossRef] [PubMed]
25. Maeda, Y.; Miyashita, J.; Hasegawa, T.; Wakahara, T.; Tsuchiya, T.; Nakahodo, T.; Akasaka, T.; Mizorogi, N.; Kobayashi, K.; Nagase, S.; et al. Reversible and regioselective reaction of La@$C_{82}$ with cyclopentadiene. *J. Am. Chem. Soc.* **2005**, *127*, 12190–12191. [CrossRef] [PubMed]
26. Iezzi, E.B.; Duchamp, J.C.; Harich, K.; Glass, T.E.; Lee, H.M.; Olmstead, M.M.; Balch, A.L.; Dorn, H.C. A Symmetric Derivative of the Trimetallic Nitride Endohedral Metallofullerene, $Sc_3N@C_{80}$. *J. Am. Chem. Soc.* **2002**, *124*, 524–525. [CrossRef] [PubMed]
27. Yamada, M.; Wakahara, T.; Nakahodo, T.; Tsuchiya, T.; Maeda, Y.; Akasaka, T.; Yoza, K.; Horn, E.; Mizorogi, N.; Nagase, S. Synthesis and structural characterization of endohedral pyrrolidinodimetallofullerene: $La_2@C_{80}(CH_2)_2NTrt$. *J. Am. Chem. Soc.* **2006**, *128*, 1402–1403. [CrossRef] [PubMed]
28. Cardona, C.M.; Kitaygorodskiy, A.; Echegoyen, L. Trimetallic nitride endohedral metallofullerenes: Reactivity dictated by the encapsulated metal cluster. *J. Am. Chem. Soc.* **2005**, *127*, 10448–10453. [CrossRef] [PubMed]
29. Feng, L.; Nakahodo, T.; Wakahara, T.; Tsuchiya, T.; Maeda, Y.; Akasaka, T.; Kato, T.; Horn, E.; Yoza, K.; Mizorogi, N.; et al. A singly bonded derivative of endohedral metallofullerene: $La@C_{82}CBr(COOC_2H_5)_2$. *J. Am. Chem. Soc.* **2005**, *127*, 17136–17137. [CrossRef]
30. Lukoyanova, O.; Cardona, C.M.; Rivera, J.; Lugo-Morales, L.Z.; Chancellor, C.J.; Olmstead, M.M.; Rodríguez-Fortea, A.; Poblet, J.M.; Balch, A.L.; Echegoyen, L. "Open Rather than Closed" malonate methano-fullerene derivatives. The formation of methanofulleroid adducts of $Y_3N@C_{80}$. *J. Am. Chem. Soc.* **2007**, *129*, 10423–10430. [CrossRef]
31. Bao, L.P.; Chen, M.Q.; Pan, C.W.; Yamaguchi, T.; Kato, T.; Olmstead, M.M.; Balch, A.L.; Akasaka, T.; Lu, X. Crystallographic Evidence for Direct Metal–Metal Bonding in a Stable Open-Shell $La_2@I_h$-$C_{80}$ Derivative. *Angew. Chem. Int. Ed.* **2016**, *128*, 4314–4318. [CrossRef]
32. Jin, F.; Tamm, N.B.; Troyanov, S.I.; Yang, S.F. Steering the Geometry of Butterfly-Shaped Dimetal Carbide Cluster within a Carbon Cage via Trifluoromethylation of $Y_2C_2@C_{82}$(6). *J. Am. Chem. Soc.* **2018**, *140*, 3496–3499. [CrossRef] [PubMed]
33. Iiduka, Y.; Wakahara, T.; Nakahodo, T.; Tsuchiya, T.; Sakuraba, A.; Maeda, Y.; Akasaka, T.; Yoza, K.; Horn, E.; Kato, T.; et al. Structural determination of metallofullerene $Sc_3C_{82}$ revisited: a surprising finding. *J. Am. Chem. Soc.* **2005**, *127*, 12500–12501. [CrossRef]
34. Kurihara, H.; Lu, X.; Iiduka, Y.; Mizorogi, N.; Slanina, Z.; Tsuchiya, T.; Akasaka, T.; Nagase, S. Sc2C2@C80 Rather than Sc2@C82: Templated Formation of Unexpected C2v(5)-C80 and Temperature-Dependent Dynamic Motion of Internal Sc2C2 Cluster. *J. Am. Chem. Soc.* **2011**, *133*, 2382–2385.
35. Jin, P.; Li, Y.; Magagula, S.; Chen, Z.F. Exohedral functionalization of endohedral metallofullerenes: Interplay between inside and outside. *Coordin. Chem. Rev.* **2019**, *388*, 406–439. [CrossRef]
36. Akasaka, T.; Kato, T.; Nagase, S.; Kobayashi, K.; Yamamoto, K.; Funasaka, H.; Takahashi, T. Chemical derivatization of endohedral metallofullerene La@$C_{82}$ with digermirane. *Tetrahedron* **1996**, *52*, 5015–5020. [CrossRef]
37. Yamada, M.; Feng, L.; Wakahara, T.; Tsuchiya, T.; Maeda, Y.; Lian, Y.; Kako, M.; Akasaka, T.; Kato, T.; Kobayashi, K.; et al. Synthesis and characterization of exohedrally silylated M@C82 (M = Y and La). *J. Phys. Chem. B* **2005**, *109*, 6049–6051.
38. Wakahara, T.; Yamada, M.; Takahashi, S.; Nakahodo, T.; Tsuchiya, T.; Maeda, Y.; Akasaka, T.; Kako, M.; Yoza, K.; Horn, E. Two-dimensional hopping motion of encapsulated La atoms in silylated $La_2@C_{80}$. *Chem. Commun.* **2007**, 2680–2682. [CrossRef]
39. Yamada, M.; Wakahara, T.; Tsuchiya, T.; Maeda, Y.; Kako, M.; Akasaka, T.; Yoza, K.; Horn, E.; Mizorogi, N.; Nagase, S. Location of the metal atoms in $Ce_2@C_{78}$ and its bis-silylated derivative. *Chem. Commun.* **2008**, 558–560. [CrossRef]
40. Wakahara, T.; Iiduka, Y.; Ikenaga, O.; Nakahodo, T.; Sakuraba, A.; Tsuchiya, T.; Maeda, Y.; Kako, M.; Akasaka, T.; Yoza, K. Characterization of the bis-silylated endofullerene $Sc_3N@C_{80}$. *J. Am. Chem. Soc.* **2006**, *128*, 9919–9925. [CrossRef]
41. Kako, M.; Miyabe, M.; Sato, K.; Suzuki, M.; Mizorogi, N.; Wang, W.-W.; Yamada, M.; Maeda, Y.; Olmstead, M.M.; Balch, A.L.; et al. Preparation, structural determination, and characterization of electronic properties of bis-silylated and bis-germylated $Lu_3N@I_h$-$C_{80}$. *Chem. Eur. J.* **2015**, *21*, 16411–16420. [CrossRef] [PubMed]
42. Kako, M.; Nagase, S.; Akasaka, T. Functionalization of endohedral metallofullerenes with reactive silicon and germanium compounds. *Molecules* **2017**, *22*, 1179. [CrossRef] [PubMed]
43. Kako, M.; Miyabe, K.; Fukazawa, S.; Kanzawa, S.; Yasui, M.; Yamada, M.; Maeda, Y.; Slanina, Z.; Uhlík, F.; Adamowicz, L.; et al. Photoreactions of $Sc_3N@C_{80}$ with disilirane, silirane, and digermirane: A photochemical method to separate $I_h$ and $D_{5h}$ Isomers. *Photochem* **2022**, *2*, 122–137. [CrossRef]
44. Cao, B.P.; Wakahara, T.; Maeda, Y.; Han, A.H.; Akasaka, T.; Kato, T.; Kobayashi, K.; Nagase, S. Lanthanum Endohedral Metallofulleropyrrolidines: Synthesis, Isolation, and EPR Characterization. *Chem. Eur. J.* **2004**, *10*, 716–720. [CrossRef] [PubMed]
45. Yamada, M.; Okamura, M.; Sato, S.; Someya, C.I.; Mizorogi, N.; Tsuchiya, T.; Akasaka, T.; Kato, T.; Nagase, S. Two Regioisomers of Endohedral Pyrrolidinodimetallofullerenes $M_2@I_h$-$C_{80}(CH_2)_2NTrt$ (M=La, Ce; Trt=trityl): Control of Metal Atom Positions by Addition Positions. *Chem.- Eur. J.* **2009**, *15*, 10533–10542. [CrossRef] [PubMed]
46. Cai, T.; Ge, Z.X.; Iezzi, E.B.; Glass, T.E.; Harich, K.; Gibson, H.W.; Dorn, H.C. Synthesis and characterization of the first trimetallic nitride templated pyrrolidino endohedral metallofullerenes. *Chem. Commun.* **2005**, 3594–3596. [CrossRef] [PubMed]

47. Wang, T.; Wu, J.; Xu, W.; Xiang, J.; Lu, X.; Li, B.; Jiang, L.; Shu, C.; Wang, C. Spin Divergence Induced by Exohedral Modification: ESR Study of $Sc_3C_2@C_{80}$ Fulleropyrrolidine. *Angew. Chem. Int. Ed.* **2010**, *49*, 1830–1833. [CrossRef]
48. Yu, B.; Shen, W.Q.; Yang, L.; Liu, Y.C.; Pan, C.W.; Cong, H.L.; Jin, P.; Lu, X. Regioselective synthesis, crystallographic characterization, and electrochemical properties of pyrazole- and pyrrole-ring-fused derivatives of $Y_2@C_{3v}(8)$-$C_{82}$. *Chem. Eur. J.* **2020**, *26*, 2464–2469. [CrossRef]
49. Chen, M.Q.; Guan, R.N.; Li, B.; Yang, L.; Niu, C.; Jin, P.; Wang, G.W.; Yang, S.F. Anomalous *cis*-conformation regioselectivity of heterocycle-fused $Sc_3N@D_{3h}$-$C_{78}$ derivatives. *Angew. Chem. Int. Ed.* **2021**, *60*, 7880–7886. [CrossRef]
50. Miao, C.B.; Tian, Z.Y.; Ruan, X.J.; Sun, X.Q.; Yang, H.T. Reaction of [60]fullerene with poxides under photo-irradiation: Synthesis of $C_{60}$-fused tetrahydrofuran derivatives. *Heterocycles* **2011**, *83*, 1615–1620. [CrossRef]
51. Wang, G.W.; Wu, P.; Yang, H.T.; Miao, C.B.; Xu, Y. Novel cycloaddition reaction of [60] fullerene with carbonyl ylides generated from epoxides. *J. Org. Chem.* **2006**, *71*, 4346–4348. [CrossRef] [PubMed]
52. Aroua, S.; Garcia-Borràs, M.; Bölter, M.F.; Osuna, S.; Yamakoshi, Y. Endohedral metal-induced regioselective formation of bis-Prato adduct of $Y_3N@I_h$-$C_{80}$ and $Gd_3N@I_h$-$C_{80}$. *J. Am. Chem. Soc.* **2015**, *137*, 58–61. [CrossRef] [PubMed]
53. Cerón, M.R.; Izquierdo, M.; Garcia-Borràs, M.; Lee, S.S.; Stevenson, S.; Osuna, S.; Echegoyen, L. Bis-1,3-dipolar cycloadditions on endohedral fullerenes $M_3N@I_h$-$C_{80}$ (M = Sc, Lu): Remarkable endohedral-cluster regiochemical control. *J. Am. Chem. Soc.* **2015**, *137*, 11775–11782. [CrossRef] [PubMed]
54. Semivrazhskaya, O.; Romero-Rivera, A.; Aroua, S.; Troyanov, S.I.; Garcia-Borràs, M.; Stevenson, S.; Osuna, S.; Yamakoshi, Y. Structures of $Gd_3N@C_{80}$ Prato bis-adducts: Crystal structure, thermal isomerization, and computational study. *J. Am. Chem. Soc.* **2019**, *141*, 10988–10993. [CrossRef] [PubMed]
55. Semivrazhskaya, O.; Aroua, S.; Yulikov, M.; Romero-Rivera, A.; Stevenson, S.; Garcia-Borràs, M.; Osuna, S.; Yamakoshi, Y. Regioselective synthesis and characterization of tris- and tetra-Prato adducts of $M_3N@C_{80}$ (M = Y, Gd). *J. Am. Chem. Soc.* **2020**, *142*, 12954–12965. [CrossRef] [PubMed]
56. Bolskar, R.D.; Benedetto, A.F.; Husebo, L.O.; Price, R.E.; Jackson, E.F.; Wallace, S.; Wilson, L.J.; Alford, J.M. First soluble $M@C_{60}$ derivatives provide enhanced access to metallofullerenes and permit in vivo evaluation of $Gd@C_{60}[C(COOH)_2]_{10}$ as a MRI contrast agent. *J. Am. Chem. Soc.* **2003**, *125*, 5471–5478. [CrossRef] [PubMed]
57. Feng, L.; Wakahara, T.; Nakahodo, T.; Tsuchiya, T.; Piao, Q.; Maeda, Y.; Lian, Y.F.; Akasaka, T.; Horn, E.; Yoza, K.; et al. The bingel monoadducts of $La@C_{82}$: Synthesis, characterization, and electrochemistry. *Chem. Eur. J.* **2006**, *12*, 5578–5586. [CrossRef]
58. Shen, W.Q.; Yang, L.; Li, B.; Jin, P.; Yu, B.; Cong, H.L.; Akasaka, T.; Lu, X. Metal-encapsulation induces a highly regioselective Bingel–Hirsch reaction of the labile $Y@C_s(6)$-$C_{82}$. *Chem. Commun.* **2020**, *56*, 14357–14360. [CrossRef]
59. Li, F.F.; Rodriguez-Fortea, A.; Poblet, J.M.; Echegoyen, L. Reactivity of Metallic Nitride Endohedral Metallofullerene Anions: Electrochemical Synthesis of a $Lu_3N@I_h$-$C_{80}$ Derivative. *J. Am. Chem. Soc.* **2011**, *133*, 2760–2765. [CrossRef]
60. Chaur, M.N.; Melin, F.; Elliott, B.; Athans, A.J.; Walker, K.; Holloway, B.C.; Echegoyen, L. The influence of cage size on the reactivity of trimetallic nitride metallofullerenes: A mono- and bis-methanoadduct of $Gd_3N@C_{80}$ and a monoadduct of $Gd_3N@C_{84}$. *Chem. Commun.* **2008**, 2665–2667. [CrossRef]
61. Wang, S.; Huang, J.; Gao, C.L.; Jin, F.; Li, Q.X.; Xie, S.Y.; Yang, S.F. Singly bonded monoadduct rather than methanofullerene: Manipulating the addition pattern of trimetallic nitride clusterfullerene through one endohedral metal atom substitution. *Chem. Eur. J.* **2016**, *22*, 8309–8315. [CrossRef] [PubMed]
62. Hu, Y.J.; Solé-Daura, A.; Yao, Y.R.; Liu, X.C.; Liu, S.J.; Yu, A.; Peng, P.; Poblet, J.M.; Rodríguez-Fortea, A.; Echegoyen, L.; et al. Chemical reactions of cationic metallofullerenes: An alternative route for exohedral functionalization. *Chem. Eur. J.* **2020**, *26*, 1748–1753. [CrossRef] [PubMed]
63. Yamada, M.; Someya, C.; Wakahara, T.; Tsuchiya, T.; Maeda, Y.; Akasaka, T.; Yoza, K.; Horn, E.; Liu, M.T.H.; Mizorogi, N.; et al. Metal Atoms Collinear with the Spiro Carbon of 6,6-Open Adducts, $M_2@C_{80}$(Ad) (M = La and Ce, Ad = Adamantylidene). *J. Am. Chem. Soc.* **2008**, *130*, 1171–1176. [CrossRef] [PubMed]
64. Iiduka, Y.; Wakahara, T.; Nakajima, K.; Nakahodo, T.; Tsuchiya, T.; Maeda, Y.; Akasaka, T.; Yoza, K.; Liu, M.T.H.; Mizorogi, N.; et al. Experimental and Theoretical Studies of the Scandium Carbide Endohedral Metallofullerene $Sc_2C_2@C_{82}$ and Its Carbene Derivative. *Angew. Chem. Int. Ed.* **2007**, *46*, 5562–5564. [CrossRef] [PubMed]
65. Yamada, M.; Abe, T.; Saito, C.; Yamazaki, T.; Sato, S.; Mizorogi, N.; Slanina, Z.; Uhlík, F.; Suzuki, M.; Maeda, Y.; et al. Adamantylidene Addition to $M_3N@I_h$-$C_{80}$ (M=Sc, Lu) and $Sc_3N@D_{5h}$-$C_{80}$: Synthesis and Crystallographic Characterization of the [5,6]-Open and [6,6]-Open Adducts. *Chem. Eur. J.* **2017**, *23*, 6552–6561. [CrossRef] [PubMed]
66. Chen, M.Q.; Bao, L.P.; Ai, M.; Shen, W.Q.; Lu, X. $Sc_3N@I_h$-$C_{80}$ as a novel lewis acid to trap abnormal N-heterocyclic carbene unprecedented formation of a singly bonded [6, 6, 6]-adduct. *Chem. Sci.* **2016**, *7*, 2331–2334. [CrossRef]
67. Chen, M.Q.; Shen, W.Q.; Peng, P.; Bao, L.P.; Zhao, S.S.; Xie, Y.P.; Jin, P.; Fang, H.Y.; Li, F.F.; Lu, X. Evidence of oxygen activation in the reaction between an N-heterocyclic carbene and $M_3N@I_h(7)$-$C_{80}$: An unexpected method of steric hindrance release. *J. Org. Chem.* **2017**, *82*, 3500–3505. [CrossRef] [PubMed]
68. Shen, W.Q.; Yang, L.; Wu, Y.; Bao, L.P.; Li, Y.; Jin, P.; Fang, H.Y.; Xie, Y.P.; Lu, X. Reactions between N-heterocyclic carbene and lutetium-metallofullerenes: High regioselectivity directed by electronic effect in addition to steric hindrance. *J. Org. Chem.* **2019**, *84*, 606–612. [CrossRef]

69. Liu, X.Y.; Li, B.; Yang, W.; Yao, Y.R.; Yang, L.; Zhuang, J.X.; Li, X.M.; Jin, P.; Chen, N. Synthesis and characterization of carbene derivatives of Th@$C_{3v}$(8)-$C_{82}$ and U@$C_{2v}$(9)-$C_{82}$: Exceptional chemical properties induced by strong actinide-carbon cage interaction. *Chem. Sci.* **2021**, *12*, 2488–2497. [CrossRef]
70. Chen, M.Q.; Zhao, Y.X.; Jin, F.; Li, M.Y.; Guan, R.N.; Xin, J.P.; Yao, Y.R.; Zhao, X.; Wang, G.W.; Zhang, Q.Y.; et al. Decisive role of non-rare earth metals in high-regioselectivity addition of $\mu_3$-carbido clusterfullerene. *Inorg. Chem. Front.* **2022**, *9*, 5688–5696. [CrossRef]
71. Nikawa, H.; Kikuchi, T.; Wakahara, T.; Nakahodo, T.; Tsuchiya, T.; Rahman, G.M.A.; Akasaka, T.; Maeda, Y.; Yoza, K.; Horn, E.; et al. Missing Metallofullerene La@$C_{74}$. *J. Am. Chem. Soc.* **2005**, *127*, 9684–9685. [CrossRef] [PubMed]
72. Shu, C.Y.; Cai, T.; Xu, L.S.; Zuo, T.M.; Reid, J.; Harich, K.; Dorn, H.C.; Gibson, H.W. Manganese(III)-Catalyzed Free Radical Reactions on Trimetallic Nitride Endohedral Metallofullerenes. *J. Am. Chem. Soc.* **2007**, *129*, 15710–15717. [CrossRef] [PubMed]
73. Jin, F.; Wang, S.; Tamm, N.B.; Yang, S.F.; Troyanov, S.I. Synthesis, Isolation and Trifluoromethylation of Two Isomers of $C_{84}$-based Monometallic Cyanide Clusterfullerenes: Interplay between the Endohedral Cluster with the Exohedral Addend. *Angew. Chem. Int. Ed.* **2017**, *56*, 11990–11994. [CrossRef] [PubMed]
74. Liu, F.P.; Velkos, G.; Krylov, D.S.; Spree, L.; Zalibera, M.; Ray, R.; Samoylova1, N.A.; Chen, C.H.; Rosenkranz, M.; Schiemenz1, S.; et al. Air-stable redox-active nanomagnets with lanthanide spins radical-bridged by a metal–metal bond. *Nat. Commun.* **2019**, *10*, 571. [CrossRef] [PubMed]
75. Han, X.Y.; Xin, J.P.; Yao, Y.R.; Liang, Z.H.; Qiu, Y.F.; Chen, M.Q.; Yang, S.F. Capturing the long-sought Dy@$C_{2v}$(5)-$C_{80}$ via benzyl radical stabilization. *Nanomaterials* **2022**, *12*, 3291. [CrossRef] [PubMed]
76. Xu, D.; Jiang, Y.H.; Wang, Y.Y.; Zhou, T.H.; Shi, Z.J.; Omachi, H.; Shinohara, H.; Sun, B.Y.; Wang, Z.Y. Turning on the near-infrared photoluminescence of erbium metallofullerenes by covalent modification. *Inorg. Chem.* **2019**, *58*, 14325–14330. [CrossRef] [PubMed]
77. Li, J.; Yu, P.Y.; Lai, P.; Zou, J.J.; Liu, Z.; Yi, X.G.; Wang, W.; Pan, C.W. Regioselective Radical Reaction of Monometallofullerene Y@C2v(9)-C82 With N-arylbenzamidine Mediated by Silver Carbonate. *Front. Chem.* **2020**, *8*, 593602. [CrossRef] [PubMed]
78. Wang, Y.F.; Velkos, G.; Israel, N.J.; Rosenkranz, M.; Büchner, B.; Liu, F.P.; Popov, A.A. Electrophilic trifluoromethylation of dimetallofullerene anions en route to air-stable single-molecule magnets with high blocking temperature of magnetization. *J. Am. Chem. Soc.* **2021**, *143*, 18139–18149. [CrossRef]
79. Chen, C.-H.; Yeh, W.Y.; Liu, Y.H.; Lee, G.H. [($\mu$-H)$_3$Re$_3$(CO)$_9$($\eta^2$, $\eta^2$, $\eta^2$-Sc$_2$C$_2$@$C_{3v}$(8)-$C_{82}$)]: Face-capping cluster complex of an endohedral fullerene. *Angew. Chem. Int. Ed.* **2012**, *51*, 3046–13049. [CrossRef]
80. Xie, Y.P.; Pan, C.W.; Bao, L.P.; Slanina, Z.; Akasaka, T.; Lu, X. Regioselective coordination of $Re_2(CO)_{10}$ to Y@$C_{2v}$(9)-$C_{82}$: An unprecedented η1 complex stabilized by intramolecular electron transfer. *Organometallics* **2019**, *38*, 2259–2263. [CrossRef]
81. Bao, L.P.; Yu, P.Y.; Li, Y.; Pan, C.W.; Shen, W.Q.; Jin, P.; Liang, S.Q.; Lu, X. Highly regioselective complexation of tungsten with Eu@$C_{82}$/Eu@$C_{84}$: Interplay between endohedral and exohedral metallic units induced by electron transfer. *Chem. Sci.* **2019**, *10*, 4945–4950. [CrossRef] [PubMed]
82. Hu, Y.J.; Yao, Y.R.; Liu, X.C.; Yu, A.; Xie, X.M.; Abella, L.; Rodríguez-Fortea, A.; Poblet, J.M.; Akasaka, T.; Peng, P.; et al. Unexpected formation of 1,2- and 1,4-bismethoxyl $Sc_3N$@$I_h$-$C_{80}$ derivatives via regioselective anion addition: An unambiguous structural identification and mechanism study. *Chem. Sci.* **2021**, *12*, 8123–8130. [CrossRef] [PubMed]
83. Niu, C.; Liu, Z.; Chen, M.Q.; Yang, S.F.; Wang, G.W. Unexpected formation of pyrazoline-fused metallofullerenes from the multicomponent cascade reaction of $Sc_3N$@$I_h$-$C_{80}$ with tetrazines, water, and oxygen. *Org. Lett.* **2022**, *24*, 3493–3498. [CrossRef] [PubMed]
84. Kurihara, H.; Iiduka, Y.; Rubin, Y.; Waelchli, M.; Mizorogi, N.; Slanina, Z.; Tsuchiya, T.; Nagase, S.; Akasaka, T. Unexpected formation of a $Sc_3C_2$@$C_{80}$ bisfulleroid derivative. *J. Am. Chem. Soc.* **2012**, *134*, 4092–4095. [CrossRef] [PubMed]
85. Li, Y.B.; Emge, T.J.; Moreno-Vicente, A.; Kopcha, W.P.; Sun, Y.; Mansoor, I.F.; Lipke, M.C.; Hall, G.S.; Poblet, J.M.; Rodríguez-Fortea, A.; et al. Unexpected formation of metallofulleroids from multicomponent reactions, with crystallographic and computational studies of the cluster motion. *Angew. Chem. Int. Ed.* **2021**, *60*, 25269–25273. [CrossRef] [PubMed]
86. Pykhova, A.D.; Semivrazhskaya, O.O.; Samoylova, N.A.; Rybalchenko, A.V.; Rosenkranz, M.; Ioffe, I.N.; Popov, A.A.; Goryunkov, A.A. Addition of $CF_2$ group to endohedral fullerene $Sc_3N$@$I_h$-$C_{80}$. *Dalton Trans.* **2020**, *49*, 9137–9147. [CrossRef] [PubMed]
87. Pykhova, A.D.; Olesya, O. Semivrazhskaya, O.O.; Samoylova, N.A.; Popov, A.A.; Ioffe, I.N.; Goryunkov, A.A. Regioselective $CF_2$ functionalization of $Sc_3N$@$D_{3h}$(5)-$C_{78}$. *Dalton Trans.* **2022**, *51*, 1182–1190. [CrossRef]

MDPI AG
Grosspeteranlage 5
4052 Basel
Switzerland
Tel.: +41 61 683 77 34

*Inorganics* Editorial Office
E-mail: inorganics@mdpi.com
www.mdpi.com/journal/inorganics

www.ingramcontent.com/pod-product-compliance
Lightning Source LLC
Chambersburg PA
CBHW061439210926
51850CB00008B/1551
* 9 7 8 3 7 2 5 8 6 7 7 8 3 *